W9-CEP-769

Introductory Linear Algebra with Applications

Under the Editorship
of the Late Carl B. Allendoerfer

Bernard Kolman
Drexel University

Introductory Linear Algebra with Applications

Macmillan Publishing Co., Inc.
New York

Collier Macmillan Publishers
London

Printed in the United States of America

Macmillan Publishing Co., Inc.
866 Third Avenue, New York, New York 10022

Collier Macmillan Canada, Ltd.

Library of Congress Cataloging in Publication Data

Kolman, Bernard, (date)
Introductory linear algebra with applications.

Includes index.
1. Algebras, Linear. I. Title.
QA184.K67 512'.5 75-4928
ISBN 0-02-365950-5

Printing: 3 4 5 6 7 8 Year: 7 8 9 0 1 2

To my children, Lisa and Stephen

Preface

Linear algebra is fast becoming a standard part of the undergraduate mathematical training of a diverse number of students for at least two reasons. First, few subjects can claim to have such widespread applications in other areas of mathematics—multivariable calculus, differential equations, and probability theory, for example—as well as in physics, biology, chemistry, economics, psychology, sociology, and all fields of engineering. Second, the subject provides the student with his first introduction to postulational or axiomatic mathematics.

In a semester or quarter course in linear algebra the instructor and student are frequently faced with several problems. The important subject of eigenvalues is often studied in too much haste to do the material justice. The student seldom sees any meaningful applications of linear algebra, and he rarely gets any feel for the widespread use, in conjunction with computers, of the new area of numerical linear algebra. In this book we have tried to deal with these problems.

This book presents a brief introduction to linear algebra and to some of its significant applications and is designed for a semester or quarter course at the freshman or sophomore level. If most of the applications are covered, there is enough material for a year course. Calculus is not a prerequisite.

The emphasis is on the computational and geometrical aspects of the subject, keeping the abstraction down to a minimum. Thus we sometimes omit proofs of difficult or less rewarding theorems, while amply illustrating them with examples. The proofs that are included are presented at a level appropriate for the student. We have also concentrated our attention on the essential areas of linear algebra; the book does not attempt to cover the subject exhaustively.

Chapter 1 deals with matrices and their properties; Chapter 2 covers determinants in a brief manner. In Chapter 3, on vectors and vector spaces, and Chapter 4, on linear transformations and matrices, we explore some of the geometric ideas of the subject. Although the general definition of a vector space is given and the results are stated in terms of vector spaces, all examples considered are n-space and its subspaces. Thus R^n is viewed as a generalization of R^2 and R^3, the familiar 2- and 3-spaces. Chapter 5, on eigenvalues and eigenvectors, settles the diagonalization problem for symmetric matrices. Chapter 6 contains an introduction to linear programming, an extremely important application of linear algebra. Chapter 7 covers several other diverse applications of linear algebra: lines and planes, quadratic forms, graph theory, the theory of games, least squares, and linear economic models. This material includes applications to the social sciences and to economics. In Chapter 8, on numerical linear algebra, we deal rather briefly with numerical methods commonly used in linear algebra to solve linear systems of equations and to find eigenvalues and eigenvectors of matrices. These methods are widely used in conjunction with computers. However, the examples and exercises presented here can all be worked with desk or pocket-size calculators.

It would be difficult, at this time, to ignore the use of the computer in linear algebra, since most computations involving linear algebra in real problems are carried out on computers. This relationship is briefly explored in the appendix. First, we provide a list of 53 *computer projects* grouped by sections corresponding to the text. These projects are of widely differing degrees of difficulty, and the student is expected to have adequate programming skills to handle these tasks. Next, we discuss the availability of *canned software* to implement linear algebra techniques. The use of many of these programs requires some programming skill, although some of them only require that the user know how to use a keypunch machine to punch up the data. Finally, we outline some of the features of APL, a programming language that is especially suitable for matrix manipulations and is easy to learn; we also make several remarks about BASIC, a general-purpose computer language that is being widely used with time-sharing systems.

The exercises in this book are grouped into two classes. The first class, "Exercises," contains routine exercises. The second class, "Theoretical Exercises," contains exercises that fill in gaps in some of the proofs and amplify material in the text. These exercises can be used to raise the level of the course and to challenge the more capable and interested student.

For maximum flexibility the applications have all been grouped together as Chapters 6, 7, and 8. These three chapters are almost entirely independent, the one exception being Section 7.4, "The Theory of Games," which requires Chapter 6 as preliminary study. The first five chapters, which constitute the basic linear algebra material, can be comfortably covered in a semester or quarter (see a suggested pace below), and there should be time left over to do some of the applications. These can either be covered after the first five chapters have been studied or they can be taken up after the required material has been developed. The accompanying chart gives the prerequisites for each of the applications. Moreover, some of the applications, in particular those in Chapters 7 and 8, can also be used as independent student projects. The author's classroom experience with the latter approach has met with favorable student reaction. Thus the instructor can be quite selective both in the choice of material and in the method of study of the applications.

Application	**Prerequisite**
Chapter 6	Section 1.3
Section 7.1	Chapter 3
Section 7.2	Chapter 5
Section 7.3	Section 1.2
Section 7.4	Chapter 6
Section 7.5	Section 1.3
Section 7.6	Chapter 1
Section 8.1	None
Section 8.2	Section 1.3
Section 8.3	Chapter 5

Suggested Pace for Basic Material (Chapters 1–5)

Chapter 1	6 lectures
Chapter 2	4 lectures
Chapter 3	10 lectures (omit Section 3.3)
Chapter 4	5 lectures
Chapter 5	5 lectures
	30 lectures

This schedule, which can be readily modified by spending more time on the theoretical exercises and proofs, emphasizes the essential material in each section.

I should like to express my thanks to Professors William Arendt, University of Missouri, and David A. Shedler, Virginia Commonwealth University, for thoroughly reviewing the entire manuscript. Their numerous suggestions and constructive criticisms resulted in many improvements.

Professors Robert E. Beck, Villanova University, and Charles S. Duris and John H. Staib, both at Drexel University, provided much valuable help with sections of Chapters 7 and 8. I gratefully acknowledge their help.

My thanks also go to Alan Wlasuk and Robert Fini for providing solutions to many of the exercises.

A special expression of appreciation goes to my typist: Miss Susan R. Gershuni, who, as on other occasions, cheerfully and skillfully typed the entire manuscript and its several revisions.

Finally, I should like to thank Everett W. Smethurst, Senior Editor, Mrs. Elaine W. Wetterau, Production Supervisor, and the entire staff of Macmillan Publishing Co., Inc., for their interest and unfailing cooperation in all phases of this project.

B. K.

Contents

Introductory Linear Algebra with Applications

1 Linear Equations and Matrices

1.1 Linear Systems

A good many problems in the natural and social sciences as well as in engineering and science deal with equations relating two sets of variables. An equation of the type

$$y = ax,$$

expressing the variable y in terms of the variable x and the constant a, is called a linear equation. The word "linear" is used here because the graph of the above equation is a straight line. Similarly, the equation

$$y = a_1x_1 + a_2x_2 + \cdots + a_nx_n, \tag{1}$$

expressing y in terms of the variables $x_1, x_2, \ldots, x_n$ and the constants $a_1, a_2, \ldots, a_n$, is called a **linear equation**. In many applications we are given y and must find numbers $x_1, x_2, \ldots, x_n$ satisfying (1).

A **solution** to a linear equation (1) is an ordered collection of n numbers $s_1, s_2, \ldots, s_n$ which have the property that (1) is satisfied when $x_1 = s_1$, $x_2 = s_2, \ldots, x_n = s_n$ are substituted in (1).

Thus $x_1 = 2$, $x_2 = 3$, and $x_3 = -4$ is a solution to the linear equation

$$6x_1 - 3x_2 + 4x_3 = -13,$$

because

$$6(2) - 3(3) + 4(-4) = -13.$$

More generally, a **system of *m* linear equations** in ***n* unknowns**, or simply a **linear system**, is a set of m linear equations each in n unknowns. A linear system can be conveniently denoted by

$$\begin{aligned} a_{11}x_1 + a_{12}x_2 + \cdots + a_{1n}x_n &= y_1 \\ a_{21}x_1 + a_{22}x_2 + \cdots + a_{2n}x_n &= y_2 \\ &\vdots \\ a_{m1}x_1 + a_{m2}x_2 + \cdots + a_{mn}x_n &= y_m. \end{aligned} \tag{2}$$

Thus the ith equation is

$$a_{i1}x_1 + a_{i2}x_2 + \cdots + a_{in}x_n = y_i.$$

In (2) the a_{ij} are known constants. Given values of $y_1, y_2, \ldots, y_n$, we want to find values of $x_1, x_2, \ldots, x_n$ that will satisfy each equation in (2).

A **solution** to a linear system (2) is an ordered collection of n numbers $s_1, s_2, \ldots, s_n$, which have the property that each equation in (2) is satisfied when $x_1 = s_1$, $x_2 = s_2, \ldots, x_n = s_n$ are substituted in (2).

The reader has already had experience in solving linear systems by the method of *elimination*. That is, we eliminate some of the unknowns by adding a multiple of one equation to another equation. Most likely, the reader has confined his earlier work in this area to linear systems in which $m = n$, that is, linear systems having as many equations as unknowns. In this course we shall broaden our outlook by dealing with systems in which we have $m = n$, $m < n$, and $m > n$. Indeed, there are numerous applications in which $m \neq n$. If we deal with two, three, or four unknowns, we shall often write them as x, y, z, and w.

Example 1. Consider the linear system

$$\begin{aligned} x - 3y &= -3 \\ 2x + y &= 8. \end{aligned} \tag{3}$$

To eliminate x, we subtract twice the first equation from the second, obtaining

$$7y = 14,$$

an equation having no x term. We have eliminated the unknown x. Then solving for y, we have

$$y = 2,$$

and substituting into the first equation of (3), we obtain

$$x = 3.$$

Then $x = 3, y = 2$ is the only solution to the given linear system.

Example 2. Consider the linear system

$$\begin{aligned} x - 3y &= -7 \\ 2x - 6y &= 7. \end{aligned} \tag{4}$$

Again, we decide to eliminate x. We subtract twice the first equation from the second one, obtaining

$$0 = 21,$$

which makes no sense. This means that (4) has no solution. We might have come to the same conclusion from observing that in (4) the left side of the second equation is twice the left side of the first equation, but the right side of the second equation is not twice the right side of the first equation.

Example 3. Consider the linear system

$$\begin{aligned} x + 2y + 3z &= 6 \\ 2x - 3y + 2z &= 14 \\ 3x + y - z &= -2. \end{aligned} \tag{5}$$

To eliminate x, we subtract twice the first equation from the second one and three times the first equation from the third one, obtaining

$$\begin{aligned} -7y - 4z &= 2 \\ -5y - 10z &= -20. \end{aligned} \tag{6}$$

This is a system of two equations in the unknowns y and z. We divide the second equation of (6) by -5, obtaining

$$\begin{aligned} -7y - 4z &= 2 \\ y + 2z &= 4, \end{aligned}$$

which we write, by interchanging equations, as

$$\begin{aligned} y + 2z &= 4 \\ -7y - 4z &= 2. \end{aligned} \tag{7}$$

We now eliminate y in (7) by adding 7 times the first equation to the second one, to obtain

$$10z = 30,$$

or

$$z = 3. \tag{8}$$

Substituting this value of z into the first equation of (7), we find $y = -2$. Substituting these values of y and z into the first equation of (5), we find $x = 1$. We might further observe that our elimination procedure has effectively produced the following linear system:

$$\begin{aligned} x + 2y + 3z &= 6 \\ y + 2z &= 4 \\ z &= 3, \end{aligned} \tag{9}$$

obtained by using the first equations of (5) and (7) as well as (8). The importance of the procedure lies in the fact that the linear systems (5) and (9) have exactly the same solutions. Of course, (9) can be solved quite easily.

Example 4. Consider the linear system

$$\begin{aligned} x + 2y - 3z &= -4 \\ 2x + y - 3z &= 4. \end{aligned} \tag{10}$$

Eliminating x, we subtract twice the first equation from the second equation, to obtain

$$-3y + 3z = 12. \tag{11}$$

We must now solve (11). A solution is

$$y = z - 4,$$

where z can be any real number. Then from the first equation of (10),

$$\begin{aligned} x &= -4 - 2y + 3z \\ &= -4 - 2(z - 4) + 3z \\ &= \quad z + 4. \end{aligned}$$

Thus a solution to the linear system (10) is

$$\begin{aligned} x &= z + 4 \\ y &= z - 4 \\ z &= \text{any real number.} \end{aligned}$$

This means that the linear system (10) has infinitely many solutions. Every time we assign a value to z we obtain another solution to (10). Thus, if $z = 1$, then

$$x = 5, \quad y = -3, \quad \text{and} \quad z = 1$$

is a solution, while if $z = -2$, then

$$x = 2, \quad y = -6, \quad \text{and} \quad z = -2$$

is another solution.

Example 5. Consider the linear system

$$\begin{aligned} x + 2y &= 10 \\ 2x - 2y &= -4 \\ 3x + 5y &= 26. \end{aligned} \tag{12}$$

Eliminating x, we subtract twice the first equation from the second one to obtain $-6y = -24$, or

$$y = 4.$$

Subtracting three times the first equation from the third one, we obtain $-y = -4$, or

$$y = 4.$$

Thus we have been led to the system

$$\begin{aligned} x + 2y &= 10 \\ y &= 4 \\ y &= 4, \end{aligned} \tag{13}$$

which has the same solutions as (12). Substituting $y = 4$ in the first equation of (13), we obtain $x = 2$. Hence $x = 2, y = 4$ is a solution to (12).

Example 6. Consider the linear system

$$\begin{aligned} x + 2y &= 10 \\ 2x - 2y &= -4 \\ 3x + 5y &= 20. \end{aligned} \tag{14}$$

To eliminate x, we subtract twice the first equation from the second one to obtain $-6y = -24$, or

$$y = 4.$$

Subtracting three times the first equation from the third one, we obtain $-y = -10$, or

$$y = 10.$$

Thus we have been led to the system

$$\begin{aligned} x + 2y &= 10 \\ y &= 4 \\ y &= 10, \end{aligned} \tag{15}$$

which has the same solutions as (14). Since (15) has no solutions, we conclude that (14) has no solutions.

We have seen that the method of elimination consists of repeatedly performing the following operations:

1. Interchange two equations.
2. Multiply an equation by a nonzero constant.
3. Add a multiple of one equation to another equation.

It is not difficult to show (Exercises T.1 through T.3) that the method of elimination yields another linear system having exactly the same solutions as the given system. The new linear system can then be solved quite readily.

1.1 Exercises

In Exercises 1 through 13 solve the given linear system by the method of elimination.

1. $$\begin{aligned} x + 2y &= 8 \\ 3x - 4y &= 4. \end{aligned}$$

2. $$\begin{aligned} 2x - 3y + 4z &= -12 \\ x - 2y + z &= -5 \\ 3x + y + 2z &= 1. \end{aligned}$$

3. $$\begin{aligned} 3x + 2y + z &= 2 \\ 4x + 2y + 2z &= 8 \\ x - y + z &= 4. \end{aligned}$$

4. $$\begin{aligned} x + y &= 5 \\ 3x + 3y &= 10. \end{aligned}$$

5. $$\begin{aligned} x + 2y + 3z &= -6 \\ 2x - 3y - 4z &= 15 \\ 3x + 4y + 5z &= -8. \end{aligned}$$

6. $$\begin{aligned} x + y - 2z &= 5 \\ 2x + 3y + 4z &= 2. \end{aligned}$$

7. $$\begin{aligned} x + 4y - z &= 12 \\ 3x + 8y - 2z &= 4. \end{aligned}$$

8. $$\begin{aligned} 3x + 4y - z &= 8 \\ 6x + 8y - 2z &= 3. \end{aligned}$$

9. $$\begin{aligned} x + y + 3z &= 12 \\ 2x + 2y + 6z &= 6. \end{aligned}$$

10. $$\begin{aligned} x + y &= 1 \\ 2x - y &= 5 \\ 3x + 4y &= 2. \end{aligned}$$

11. $$\begin{aligned} 2x + 3y &= 13 \\ x - 2y &= 3 \\ 5x + 2y &= 27. \end{aligned}$$

12. $$\begin{aligned} x - 5y &= 6 \\ 3x + 2y &= 1 \\ 5x + 2y &= 1. \end{aligned}$$

13. $$\begin{aligned} x + 3y &= -4 \\ 2x + 5y &= -8 \\ x + 3y &= -5. \end{aligned}$$

Theoretical Exercises

T.1. Show that the linear system obtained by interchanging two equations in (2) has exactly the same solutions as (2).

T.2. Show that the linear system obtained by multiplying an equation in (2) by a nonzero constant has exactly the same solutions as (2).

T.3. Show that the linear system obtained by adding a multiple of an equation in (2) to another equation in (2) has exactly the same solutions as (2).

1.2 Matrices

In this section we define matrices and study some of their properties. Matrices enable us to write linear systems in a compact form and make it easy to automate the elimination method, on an electronic computer, to obtain a fast and efficient procedure for solution. Their use is not, however, merely that of a convenient notation. We now develop operations on matrices and will work with matrices according to the rules they obey; this will enable us to solve systems of linear equations and do other computational problems in a fast and efficient manner. Of course, as any good definition should do, the notion of a matrix provides not only a new way of looking at old problems but also gives rise to a great many new questions, some of which we study in this book.

DEFINITION An $m \times n$ **matrix A** is a rectangular array of mn real (or complex) numbers arranged in m horizontal **rows** and n vertical **columns**:

$$\mathbf{A} = \begin{bmatrix} a_{11} & a_{12} & \cdots & a_{1n} \\ a_{21} & a_{22} & \cdots & a_{2n} \\ \vdots & \vdots & & \vdots \\ a_{m1} & a_{m2} & \cdots & a_{mn} \end{bmatrix}. \tag{1}$$

The ***i*th row** of **A** is

$$\begin{bmatrix} a_{i1} \; a_{i2} \cdots a_{in} \end{bmatrix} \qquad (1 \leqslant i \leqslant m);$$

the ***j*th column** of **A** is

$$\begin{bmatrix} a_{1j} \\ a_{2j} \\ \vdots \\ a_{mj} \end{bmatrix} \qquad (1 \leqslant j \leqslant n).$$

We shall say that **A is *m* by *n***. If $m = n$, we say that **A** is a **square matrix of order *n***, and that the numbers $a_{11}, a_{22}, \ldots, a_{nn}$ form the **main diagonal** of **A**. We refer to the number a_{ij}, which is in the ith row and jth column of **A**, as **the *i*, *j*th element of A**, or the **(*i*, *j*) entry of A**, and we often write (1) as

$$\mathbf{A} = \left[a_{ij} \right].$$

In this book we restrict our attention to matrices all of whose elements are real numbers. However, matrices with complex entries are studied and are important in applications.

Example 1. Let

$$\mathbf{A} = \begin{bmatrix} 1 & 2 & 3 \\ -1 & 0 & 1 \end{bmatrix}, \quad \mathbf{B} = \begin{bmatrix} 1 & 4 \\ 2 & -3 \end{bmatrix}, \quad \mathbf{C} = \begin{bmatrix} 1 \\ -1 \\ 2 \end{bmatrix},$$

$$\mathbf{D} = \begin{bmatrix} 1 & 1 & 0 \\ 2 & 0 & 1 \\ 3 & -1 & 2 \end{bmatrix}, \quad \mathbf{E} = [3], \quad \text{and} \quad \mathbf{F} = [-1 \quad 0 \quad 2].$$

Then $\mathbf{A}$ is a 2×3 matrix with $a_{12} = 2$, $a_{13} = 3$, $a_{22} = 0$, and $a_{23} = 1$; $\mathbf{B}$ is a 2×2 matrix with $b_{11} = 1$, $b_{12} = 4$, $b_{21} = 2$, and $b_{22} = -3$; $\mathbf{C}$ is a 3×1 matrix with $c_{11} = 1$, $c_{21} = -1$, and $c_{31} = 2$; $\mathbf{D}$ is a 3×3 matrix; $\mathbf{E}$ is a 1×1 matrix; and $\mathbf{F}$ is a 1×3 matrix.

DEFINITION A square matrix $\mathbf{A} = [a_{ij}]$ for which every term off the main diagonal is zero, that is, $a_{ij} = 0$ for $i \neq j$, is called a **diagonal matrix**.

Example 2

$$\mathbf{G} = \begin{bmatrix} 4 & 0 \\ 0 & -2 \end{bmatrix} \quad \text{and} \quad \mathbf{H} = \begin{bmatrix} -3 & 0 & 0 \\ 0 & -2 & 0 \\ -0 & 0 & 4 \end{bmatrix}$$

are diagonal matrices.

DEFINITION A diagonal matrix $\mathbf{A} = [a_{ij}]$, for which all terms on the main diagonal are equal, that is, $a_{ij} = c$ for $i = j$ and $a_{ij} = 0$ for $i \neq j$, is called a **scalar matrix**.

Example 3. The following are scalar matrices:

$$\mathbf{I}_3 = \begin{bmatrix} 1 & 0 & 0 \\ 0 & 1 & 0 \\ 0 & 0 & 1 \end{bmatrix}, \quad \mathbf{J} = \begin{bmatrix} -2 & 0 \\ 0 & -2 \end{bmatrix}.$$

Whenever a new object is introduced in mathematics, one must determine when two such objects are equal. For example, in the set of all rational numbers, the numbers $\frac{2}{3}$ and $\frac{4}{6}$ are called equal, although they are not represented in the same manner. What we have in mind is the definition that a/b equals c/d when $ad = bc$. Accordingly, we now have the following definition.

DEFINITION Two $m \times n$ matrices $\mathbf{A} = [a_{ij}]$ and $\mathbf{B} = [b_{ij}]$ are said to be **equal** if $a_{ij} = b_{ij}$, $1 \leqslant i \leqslant m$, $1 \leqslant j \leqslant n$, that is, if corresponding elements agree.

Example 4. The matrices

$$\mathbf{A} = \begin{bmatrix} 1 & 2 & -1 \\ 2 & -3 & 4 \\ 0 & -4 & 5 \end{bmatrix} \quad \text{and} \quad \mathbf{B} = \begin{bmatrix} 1 & 2 & w \\ 2 & x & 4 \\ y & -4 & z \end{bmatrix}$$

are equal if and only if $w = -1$, $x = -3$, $y = 0$, and $z = 5$.

We shall now define a number of operations that will produce new matrices out of given matrices. These operations are useful in the applications of matrices.

Matrix Addition

DEFINITION If $\mathbf{A} = [a_{ij}]$ and $\mathbf{B} = [b_{ij}]$ are $m \times n$ matrices, then the **sum** of $\mathbf{A}$ and $\mathbf{B}$ is the matrix $\mathbf{C} = [c_{ij}]$, defined by

$$c_{ij} = a_{ij} + b_{ij} \qquad (1 \leqslant i \leqslant m,\ 1 \leqslant j \leqslant n).$$

That is, $\mathbf{C}$ is obtained by adding corresponding elements of $\mathbf{A}$ and $\mathbf{B}$.

Example 5. Let

$$\mathbf{A} = \begin{bmatrix} 1 & -2 & 4 \\ 2 & -1 & 3 \end{bmatrix} \quad \text{and} \quad \mathbf{B} = \begin{bmatrix} 0 & 2 & -4 \\ 1 & 3 & 1 \end{bmatrix}.$$

Then

$$\mathbf{A} + \mathbf{B} = \begin{bmatrix} 1+0 & -2+2 & 4-4 \\ 2+1 & -1+3 & 3+1 \end{bmatrix} = \begin{bmatrix} 1 & 0 & 0 \\ 3 & 2 & 4 \end{bmatrix}.$$

It should be noted that the sum of the matrices $\mathbf{A}$ and $\mathbf{B}$ is defined only when $\mathbf{A}$ and $\mathbf{B}$ have the same number of rows and the same number of columns, that is, only when $\mathbf{A}$ and $\mathbf{B}$ are of the same size. We shall now establish the convention that when $\mathbf{A} + \mathbf{B}$ is formed, both $\mathbf{A}$ and $\mathbf{B}$ are of the same size. The basic properties of matrix addition are contained in the following theorem.

THEOREM 1.1 (Properties of Matrix Addition)

(a) $\mathbf{A} + \mathbf{B} = \mathbf{B} + \mathbf{A}$.

(b) $\mathbf{A} + (\mathbf{B} + \mathbf{C}) = (\mathbf{A} + \mathbf{B}) + \mathbf{C}$.

(c) *There is a unique* $m \times n$ *matrix* $\mathbf{0}$ *such that*

$$\mathbf{A} + \mathbf{0} = \mathbf{A} \tag{2}$$

for any $m \times n$ *matrix* $\mathbf{A}$. *The matrix* $\mathbf{0}$ *is called the* $m \times n$ **zero matrix**.

(d) *For each* $m \times n$ *matrix* $\mathbf{A}$, *there is a unique* $m \times n$ *matrix* $\mathbf{D}$ *such that*

$$\mathbf{A} + \mathbf{D} = \mathbf{0}. \tag{3}$$

We shall write $\mathbf{D}$ *as* $-\mathbf{A}$, *so that* (3) *can be written as*

$$\mathbf{A} + (-\mathbf{A}) = \mathbf{0}.$$

The matrix $-\mathbf{A}$ *is called the* **negative** *of* $\mathbf{A}$.

proof

(a) To establish (a) we must prove that the i, jth element of $\mathbf{A} + \mathbf{B}$ equals the i, jth element of $\mathbf{B} + \mathbf{A}$. The i, jth element of $\mathbf{A} + \mathbf{B}$ is $a_{ij} + b_{ij}$; the i, jth element of $\mathbf{B} + \mathbf{A}$ is $b_{ij} + a_{ij}$. Since

$$a_{ij} + b_{ij} = b_{ij} + a_{ij} \qquad (1 \leqslant i \leqslant m, 1 \leqslant j \leqslant n),$$

the result follows.

(b) Exercise T.1.

(c) Let $\mathbf{U} = [u_{ij}]$. Then

$$\mathbf{A} + \mathbf{U} = \mathbf{A}$$

if and only if

$$a_{ij} + u_{ij} = a_{ij},$$

which holds if and only if $u_{ij} = 0$. Thus $\mathbf{U}$ is the $m \times n$ matrix all of whose elements are zero; $\mathbf{U}$ is denoted by $\mathbf{0}$.

(d) Exercise T.1.

Example 6. To illustrate (c) of Theorem 1.1, we note that the 2×2 zero matrix is

$$\begin{bmatrix} 0 & 0 \\ 0 & 0 \end{bmatrix}.$$

If

$$\mathbf{A} = \begin{bmatrix} 4 & -1 \\ 2 & 3 \end{bmatrix},$$

we have

$$\begin{bmatrix} 4 & -1 \\ 2 & 3 \end{bmatrix} + \begin{bmatrix} 0 & 0 \\ 0 & 0 \end{bmatrix} = \begin{bmatrix} 4+0 & -1+0 \\ 2+0 & 3+0 \end{bmatrix} = \begin{bmatrix} 4 & -1 \\ 2 & 3 \end{bmatrix}.$$

The 2×3 zero matrix is

$$\begin{bmatrix} 0 & 0 & 0 \\ 0 & 0 & 0 \end{bmatrix}.$$

Example 7. To illustrate (d) of Theorem 1.1, let

$$\mathbf{A} = \begin{bmatrix} 2 & 3 & 4 \\ -4 & 5 & -2 \end{bmatrix}.$$

Then

$$-\mathbf{A} = \begin{bmatrix} -2 & -3 & -4 \\ 4 & -5 & 2 \end{bmatrix}.$$

We shall write $\mathbf{A} + (-\mathbf{B})$ as $\mathbf{A} - \mathbf{B}$, and refer to this as the **difference** between $\mathbf{A}$ and $\mathbf{B}$.

Example 8. Let

$$\mathbf{A} = \begin{bmatrix} 3 & -2 & 5 \\ -1 & 2 & 3 \end{bmatrix} \quad \text{and} \quad \mathbf{B} = \begin{bmatrix} 2 & 3 & 2 \\ -3 & 4 & 6 \end{bmatrix}.$$

Then

$$\mathbf{A} - \mathbf{B} = \begin{bmatrix} 3-2 & -2-3 & 5-2 \\ -1+3 & 2-4 & 3-6 \end{bmatrix} = \begin{bmatrix} 1 & -5 & 3 \\ 2 & -2 & -3 \end{bmatrix}.$$

Matrix Multiplication

DEFINITION If $\mathbf{A} = [a_{ij}]$ is an $m \times p$ matrix and $\mathbf{B} = [b_{ij}]$ is a $p \times n$ matrix, then the **product** of **A** and **B** is the $m \times n$ matrix $\mathbf{C} = [c_{ij}]$, defined by

$$c_{ij} = a_{i1}b_{1j} + a_{i2}b_{2j} + \cdots + a_{ip}b_{pj} \qquad (1 \leqslant i \leqslant m,\ 1 \leqslant j \leqslant n). \quad (4)$$

Equation (4) says that the i, jth element in the product matrix is computed by adding all products obtained by multiplying each element in the ith row of **A** by the corresponding element in the jth column of **B**; this is shown in the following matrix. Thus the product of **A** and **B** can be defined only when the number of rows of **B** is exactly the same as the number of columns of **A**.

$$\begin{bmatrix} a_{11} & a_{12} & \cdots & a_{1p} \\ a_{21} & a_{22} & \cdots & a_{2p} \\ \vdots & \vdots & & \vdots \\ a_{i1} & a_{i2} & & a_{ip} \\ \vdots & \vdots & & \vdots \\ a_{m1} & a_{m2} & \cdots & a_{mp} \end{bmatrix} \begin{bmatrix} b_{11} & b_{12} & \cdots & b_{1j} & \cdots & b_{1n} \\ b_{21} & b_{22} & \cdots & b_{2j} & \cdots & b_{2n} \\ \vdots & \vdots & & & & \vdots \\ b_{p1} & b_{p2} & \cdots & b_{pj} & \cdots & b_{pn} \end{bmatrix}$$

$$= \begin{bmatrix} c_{11} & c_{12} & \cdots & c_{1n} \\ c_{21} & c_{22} & \cdots & c_{2n} \\ \vdots & \vdots & c_{ij} & \vdots \\ c_{m1} & c_{m2} & \cdots & c_{mn} \end{bmatrix}.$$

Example 9. Let

$$\mathbf{A} = \begin{bmatrix} 1 & 2 & -1 \\ 3 & 1 & 4 \end{bmatrix} \quad \text{and} \quad \mathbf{B} = \begin{bmatrix} -2 & 5 \\ 4 & -3 \\ 2 & 1 \end{bmatrix}.$$

Then

$$\mathbf{AB} = \begin{bmatrix} (1)(-2) + (2)(4) + (-1)(2) & (1)(5) + (2)(-3) + (-1)(1) \\ (3)(-2) + (1)(4) + (4)(2) & (3)(5) + (1)(-3) + (4)(1) \end{bmatrix}$$

$$= \begin{bmatrix} 4 & -2 \\ 6 & 16 \end{bmatrix}.$$

The algebraic properties of matrix addition, as described in Theorem 1.1, are similar to those satisfied by the real numbers. Multiplication of matrices requires much more care than their addition, since the algebraic properties of matrix multiplication differ from those satisfied by the real numbers.

If $\mathbf{A}$ is an $m \times p$ matrix and $\mathbf{B}$ is a $p \times n$ matrix, then $\mathbf{AB}$ is defined and is an $m \times n$ matrix. However, in considering $\mathbf{BA}$, several situations may occur:

1. $\mathbf{BA}$ may not be defined; this will take place if $n \neq m$.
2. If $\mathbf{BA}$ is defined, which means that $m = n$, then $\mathbf{BA}$ is $p \times p$ while $\mathbf{AB}$ is $m \times m$; thus, if $m \neq p$, $\mathbf{AB}$ and $\mathbf{BA}$ are of different sizes.
3. If $\mathbf{BA}$ and $\mathbf{AB}$ are both of the same size, they may still be unequal.

Example 10. If $\mathbf{A}$ is a 2×3 matrix and $\mathbf{B}$ is a 3×4 matrix, then $\mathbf{AB}$ is a 2×4 matrix while $\mathbf{BA}$ is undefined.

Example 11. Let $\mathbf{A}$ be 2×3 and let $\mathbf{B}$ be 3×2. Then $\mathbf{AB}$ is 2×2 while $\mathbf{BA}$ is 3×3.

Example 12. Let

$$\mathbf{A} = \begin{bmatrix} 1 & 2 \\ -1 & 3 \end{bmatrix} \quad \text{and} \quad \mathbf{B} = \begin{bmatrix} 2 & 1 \\ 0 & 1 \end{bmatrix}.$$

Then

$$\mathbf{AB} = \begin{bmatrix} 2 & 3 \\ -2 & 2 \end{bmatrix} \quad \text{while} \quad \mathbf{BA} = \begin{bmatrix} 1 & 7 \\ -1 & 3 \end{bmatrix}.$$

Thus $\mathbf{AB} \neq \mathbf{BA}$.

We also note two other peculiarities of matrix multiplication. If a and b are real numbers, then $ab = 0$ can hold only if a or b is zero. However, this is not true for matrices.

Example 13. If

$$\mathbf{A} = \begin{bmatrix} 1 & 2 \\ 2 & 4 \end{bmatrix} \quad \text{and} \quad \mathbf{B} = \begin{bmatrix} 4 & -6 \\ -2 & 3 \end{bmatrix},$$

then neither **A** nor **B** is the zero matrix, but

$$\mathbf{AB} = \begin{bmatrix} 0 & 0 \\ 0 & 0 \end{bmatrix}.$$

If a, b, and c are real numbers for which $ab = ac$ and $a \neq 0$, it then follows that $b = c$. That is, we can cancel a out. However, the cancellation law does not hold for matrices, as the following example shows.

Example 14. If

$$\mathbf{A} = \begin{bmatrix} 1 & 2 \\ 2 & 4 \end{bmatrix}, \quad \mathbf{B} = \begin{bmatrix} 2 & 1 \\ 3 & 2 \end{bmatrix}, \quad \text{and} \quad \mathbf{C} = \begin{bmatrix} -2 & 7 \\ 5 & -1 \end{bmatrix},$$

then

$$\mathbf{AB} = \mathbf{AC} = \begin{bmatrix} 8 & 5 \\ 16 & 10 \end{bmatrix},$$

but $\mathbf{B} \neq \mathbf{C}$.

However, matrix multiplication does satisfy many familiar properties of multiplication.

THEOREM 1.2 (Properties of Matrix Multiplication)

(a) *If* **A**, **B**, *and* **C** *are* $m \times p$, $p \times q$, *and* $q \times n$ *matrices, respectively, then*

$$\mathbf{A}(\mathbf{BC}) = (\mathbf{AB})\mathbf{C}.$$

(b) *If* **A** *is an* $m \times p$ *matrix and* **B** *and* **C** *are* $p \times n$ *matrices, then*

$$\mathbf{A}(\mathbf{B} + \mathbf{C}) = \mathbf{AB} + \mathbf{AC}.$$

(c) *If* **A** *and* **B** *are* $m \times p$ *matrices and* **C** *is a* $p \times n$ *matrix, then*

$$(\mathbf{A} + \mathbf{B})\mathbf{C} = \mathbf{AC} + \mathbf{BC}.$$

proof

(a) We omit a general proof. Exercise T.2 asks the reader to prove the result if $m = 2$, $p = 3$, $q = 4$, and $n = 3$.

(b) Exercise T.3.

(c) Exercise T.3.

Example 15. Let

$$\mathbf{A} = \begin{bmatrix} 5 & 2 & 3 \\ 2 & -3 & 4 \end{bmatrix}, \quad \mathbf{B} = \begin{bmatrix} 2 & -1 & 1 & 0 \\ 0 & 2 & 2 & 2 \\ 3 & 0 & -1 & 3 \end{bmatrix},$$

and

$$\mathbf{C} = \begin{bmatrix} 1 & 0 & 2 \\ 2 & -3 & 0 \\ 0 & 0 & 3 \\ 2 & 1 & 0 \end{bmatrix}.$$

Then

$$\mathbf{A}(\mathbf{BC}) = \begin{bmatrix} 5 & 2 & 3 \\ 2 & -3 & 4 \end{bmatrix} \begin{bmatrix} 0 & 3 & 7 \\ 8 & -4 & 6 \\ 9 & 3 & 3 \end{bmatrix} = \begin{bmatrix} 43 & 16 & 56 \\ 12 & 30 & 8 \end{bmatrix}$$

and

$$(\mathbf{AB})\mathbf{C} = \begin{bmatrix} 19 & -1 & 6 & 13 \\ 16 & -8 & -8 & 6 \end{bmatrix} \begin{bmatrix} 1 & 0 & 2 \\ 2 & -3 & 0 \\ 0 & 0 & 3 \\ 2 & 1 & 0 \end{bmatrix} = \begin{bmatrix} 43 & 16 & 56 \\ 12 & 30 & 8 \end{bmatrix}.$$

Example 16. Let

$$\mathbf{A} = \begin{bmatrix} 2 & 2 & 3 \\ 3 & -1 & 2 \end{bmatrix}, \quad \mathbf{B} = \begin{bmatrix} 1 & 0 \\ 2 & 2 \\ 3 & -1 \end{bmatrix}, \quad \text{and} \quad \mathbf{C} = \begin{bmatrix} -1 & 2 \\ 1 & 0 \\ 2 & -2 \end{bmatrix}.$$

Then

$$\mathbf{A}(\mathbf{B} + \mathbf{C}) = \begin{bmatrix} 2 & 2 & 3 \\ 3 & -1 & 2 \end{bmatrix} \begin{bmatrix} 0 & 2 \\ 3 & 2 \\ 5 & -3 \end{bmatrix} = \begin{bmatrix} 21 & -1 \\ 7 & -2 \end{bmatrix}$$

and

$$\mathbf{AB} + \mathbf{AC} = \begin{bmatrix} 15 & 1 \\ 7 & -4 \end{bmatrix} + \begin{bmatrix} 6 & -2 \\ 0 & 2 \end{bmatrix} = \begin{bmatrix} 21 & -1 \\ 7 & -2 \end{bmatrix}.$$

The $n \times n$ diagonal matrix

$$\mathbf{I}_n = \begin{bmatrix} 1 & 0 & \cdots & 0 \\ 0 & 1 & \cdots & 0 \\ \vdots & & & \vdots \\ 0 & & \cdots & 1 \end{bmatrix},$$

all of whose diagonal elements are 1, is called the **identity matrix of order** n. If $\mathbf{A}$ is an $m \times n$ matrix, then it is easy to verify (Exercise T.4) that

$$\mathbf{I}_m\mathbf{A} = \mathbf{A}\mathbf{I}_n = \mathbf{A}.$$

Example 17. The identity matrix $\mathbf{I}_2$ of order 2 is

$$\mathbf{I}_2 = \begin{bmatrix} 1 & 0 \\ 0 & 1 \end{bmatrix}.$$

If

$$\mathbf{A} = \begin{bmatrix} 4 & -2 & 3 \\ 5 & 0 & 2 \end{bmatrix},$$

then

$$\mathbf{I}_2\mathbf{A} = \mathbf{A}.$$

The identity matrix $\mathbf{I}_3$ of order 3 is

$$\mathbf{I}_3 = \begin{bmatrix} 1 & 0 & 0 \\ 0 & 1 & 0 \\ 0 & 0 & 1 \end{bmatrix}.$$

Then

$$\mathbf{A}\mathbf{I}_3 = \mathbf{A}.$$

Linear Systems

Consider the linear system of m equations in n unknowns

$$\begin{aligned} a_{11}x_1 + a_{12}x_2 + \cdots + a_{1n}x_n &= b_1 \\ a_{21}x_1 + a_{22}x_2 + \cdots + a_{2n}x_n &= b_2 \\ \vdots \qquad\qquad\qquad & \quad \vdots \\ a_{m1}x_1 + a_{m2}x_2 + \cdots + a_{mn}x_n &= b_m. \end{aligned} \tag{5}$$

Now define the following matrices:

$$\mathbf{A} = \begin{bmatrix} a_{11} & a_{12} & \cdots & a_{1n} \\ a_{21} & a_{22} & \cdots & a_{2n} \\ \vdots & \vdots & & \vdots \\ a_{m1} & a_{m2} & \cdots & a_{mn} \end{bmatrix}, \qquad \mathbf{X} = \begin{bmatrix} x_1 \\ x_2 \\ \vdots \\ x_n \end{bmatrix}, \qquad \mathbf{B} = \begin{bmatrix} b_1 \\ b_2 \\ \vdots \\ b_m \end{bmatrix}.$$

Then the linear system (5) can be written in matrix form as

$$\mathbf{AX} = \mathbf{B}.$$

The matrix **A** is called the **coefficient matrix** of the linear system (5) and the matrix

$$\left[\begin{array}{cccc|c} a_{11} & a_{12} & \cdots & a_{1n} & b_1 \\ a_{21} & a_{22} & & a_{2n} & b_2 \\ \vdots & \vdots & & \vdots & \vdots \\ a_{m1} & a_{m2} & \cdots & a_{mn} & b_m \end{array}\right],$$

obtained by adjoining **B** to **A**, is called the **augmented matrix** of the linear system (5). The augmented matrix of (5) will be written as $[A \,\vdots\, B]$. Conversely, any matrix with more than one column is the augmented matrix of a linear system. The coefficient and augmented matrices will play key roles in our method for solving linear systems.

Example 18. Consider the linear system

$$\begin{aligned} 2x + 3y - 4z &= 5 \\ -2x \qquad + z &= 7 \\ 3x + 2y + 2z &= 3. \end{aligned}$$

Letting

$$\mathbf{A} = \begin{bmatrix} 2 & 3 & -4 \\ -2 & 0 & 1 \\ 3 & 2 & 2 \end{bmatrix}, \quad \mathbf{X} = \begin{bmatrix} x \\ y \\ z \end{bmatrix}, \quad \text{and} \quad \mathbf{B} = \begin{bmatrix} 5 \\ 7 \\ 3 \end{bmatrix},$$

we can write the given linear system in matrix form as

$$\mathbf{AX} = \mathbf{B}.$$

The coefficient matrix is **A** and the augmented matrix is

$$\left[\begin{array}{rrr|r} 2 & 3 & -4 & 5 \\ -2 & 0 & 1 & 7 \\ 3 & 2 & 2 & 3 \end{array}\right].$$

Example 19. The matrix

$$\left[\begin{array}{rrr|r} 2 & -1 & 3 & 4 \\ 3 & 0 & 2 & 5 \\ -2 & 1 & 4 & 6 \end{array}\right]$$

is the augmented matrix of the linear system

$$\begin{aligned} 2x - y + 3z &= 4 \\ 3x + 2z &= 5 \\ -2x + y + 4z &= 6. \end{aligned}$$

Scalar Multiplication

DEFINITION If $\mathbf{A} = [a_{ij}]$ is an $m \times n$ matrix and r is a real number, then the **scalar multiple** of **A** by r, $r\mathbf{A}$ is the $m \times n$ matrix $\mathbf{B} = [b_{ij}]$, where

$$b_{ij} = ra_{ij} \qquad (1 \leqslant i \leqslant m, 1 \leqslant j \leqslant n).$$

That is, **B** is obtained by multiplying each element of **A** by r.

Example 20. If $r = -3$ and

$$\mathbf{A} = \begin{bmatrix} 4 & -2 & 3 \\ 2 & -5 & 0 \\ 3 & 6 & -2 \end{bmatrix},$$

then

$$r\mathbf{A} = \begin{bmatrix} -12 & 6 & -9 \\ -6 & 15 & 0 \\ -9 & -18 & 6 \end{bmatrix}.$$

THEOREM 1.3 (Properties of Scalar Multiplication). *Let r and s be real numbers.*

(a) *If $\mathbf{A}$ is an $m \times n$ matrix, then $r(s\mathbf{A}) = (rs)\mathbf{A}$.*
(b) *If $\mathbf{A}$ is an $m \times n$ matrix, then $(r + s)\mathbf{A} = r\mathbf{A} + s\mathbf{A}$.*
(c) *If $\mathbf{A}$ and $\mathbf{B}$ are $m \times n$ matrices, then $r(\mathbf{A} + \mathbf{B}) = r\mathbf{A} + r\mathbf{B}$.*
(d) *If $\mathbf{A}$ is an $m \times p$ matrix and $\mathbf{B}$ is a $p \times n$ matrix, then $\mathbf{A}(r\mathbf{B}) = r(\mathbf{AB})$.*

proof. Exercise T.5.

Example 21. Let $r = -2$,

$$\mathbf{A} = \begin{bmatrix} 1 & 2 & 3 \\ -2 & 0 & 1 \end{bmatrix} \quad \text{and} \quad \mathbf{B} = \begin{bmatrix} 2 & -1 \\ 1 & 4 \\ 0 & -2 \end{bmatrix}.$$

Then

$$\mathbf{A}(r\mathbf{B}) = \begin{bmatrix} 1 & 2 & 3 \\ -2 & 0 & 1 \end{bmatrix} \begin{bmatrix} -4 & 2 \\ -2 & -8 \\ 0 & 4 \end{bmatrix} = \begin{bmatrix} -8 & -2 \\ 8 & 0 \end{bmatrix}$$

and

$$r(\mathbf{AB}) = (-2)\begin{bmatrix} 4 & 1 \\ -4 & 0 \end{bmatrix} = \begin{bmatrix} -8 & -2 \\ 8 & 0 \end{bmatrix},$$

which illustrates (d) of Theorem 1.3.

It is easy to show that $(-1)\mathbf{A} = -\mathbf{A}$ (Exercise T.6).

The Transpose of a Matrix

DEFINITION If $\mathbf{A} = [a_{ij}]$ is an $m \times n$ matrix, then the $n \times m$ matrix $\mathbf{A}^T = [a_{ij}^T]$, where

$$a_{ij}^T = a_{ji} \qquad (1 \leqslant i \leqslant m, 1 \leqslant j \leqslant n)$$

is called the **transpose of A**. Thus the transpose of **A** is obtained by interchanging the rows and columns of **A**.

Example 22. Let

$$\mathbf{A} = \begin{bmatrix} 4 & -2 & 3 \\ 0 & 5 & -2 \end{bmatrix}, \quad \mathbf{B} = \begin{bmatrix} 6 & 2 & -4 \\ 3 & -1 & 2 \\ 0 & 4 & 3 \end{bmatrix}, \quad \mathbf{C} = \begin{bmatrix} 5 & 4 \\ -3 & 2 \\ 2 & -3 \end{bmatrix},$$

$$\mathbf{D} = \begin{bmatrix} 3 & -5 & 1 \end{bmatrix}, \quad \text{and} \quad \mathbf{E} = \begin{bmatrix} 2 \\ -1 \\ 3 \end{bmatrix}.$$

Then

$$\mathbf{A}^T = \begin{bmatrix} 4 & 0 \\ -2 & 5 \\ 3 & -2 \end{bmatrix}, \quad \mathbf{B}^T = \begin{bmatrix} 6 & 3 & 0 \\ 2 & -1 & 4 \\ -4 & 2 & 3 \end{bmatrix},$$

$$\mathbf{C}^T = \begin{bmatrix} 5 & -3 & 2 \\ 4 & 2 & -3 \end{bmatrix}, \quad \mathbf{D}^T = \begin{bmatrix} 3 \\ -5 \\ 1 \end{bmatrix}, \quad \text{and} \quad \mathbf{E}^T = \begin{bmatrix} 2 & -1 & 3 \end{bmatrix}.$$

THEOREM 1.4. *If r is a scalar and* **A** *and* **B** *are matrices, then*

(a) $(\mathbf{A}^T)^T = \mathbf{A}$.
(b) $(\mathbf{A} + \mathbf{B})^T = \mathbf{A}^T + \mathbf{B}^T$.
(c) $(\mathbf{AB})^T = \mathbf{B}^T\mathbf{A}^T$.
(d) $(r\mathbf{A})^T = r\mathbf{A}^T$.

proof. We leave the proofs of (a), (b), and (d) as an exercise (Exercise T.7) and only prove (c) here. Thus let $\mathbf{A} = [a_{ij}]$ be $m \times p$ and let $\mathbf{B} = [b_{ij}]$

be $p \times n$. The i, jth element of $(\mathbf{AB})^T$ is c_{ij}^T. Now

$$\begin{aligned} c_{ij}^T = c_{ji} &= a_{j1}b_{1i} + a_{j2}b_{2i} + \cdots + a_{jp}b_{pi} \\ &= a_{1j}^T b_{i1}^T + a_{2j}^T b_{i2}^T + \cdots + a_{pj}^T b_{ip}^T \\ &= b_{i1}^T a_{1j}^T + b_{i2}^T a_{2j}^T + \cdots + b_{ip}^T a_{pj}^T, \end{aligned}$$

which is the i, jth element of $\mathbf{B}^T\mathbf{A}^T$.

Example 23. Let

$$\mathbf{A} = \begin{bmatrix} 1 & 3 & 2 \\ 2 & -1 & 3 \end{bmatrix} \quad \text{and} \quad \mathbf{B} = \begin{bmatrix} 0 & 1 \\ 2 & 2 \\ 3 & -1 \end{bmatrix}.$$

Then

$$(\mathbf{AB})^T = \begin{bmatrix} 12 & 7 \\ 5 & -3 \end{bmatrix}$$

and

$$\mathbf{B}^T\mathbf{A}^T = \begin{bmatrix} 0 & 2 & 3 \\ 1 & 2 & -1 \end{bmatrix} \begin{bmatrix} 1 & 2 \\ 3 & -1 \\ 2 & 3 \end{bmatrix} = \begin{bmatrix} 12 & 7 \\ 5 & -3 \end{bmatrix}.$$

DEFINITION A matrix $\mathbf{A} = [a_{ij}]$ is called **symmetric** if

$$\mathbf{A}^T = \mathbf{A}.$$

That is, $\mathbf{A}$ is symmetric if it is a square matrix for which

$$a_{ij} = a_{ji} \qquad \text{(Exercise T.8).}$$

If $\mathbf{A}$ is symmetric, then the elements of $\mathbf{A}$ are symmetric with respect to the main diagonal of $\mathbf{A}$.

Example 24. The matrices

$$\mathbf{A} = \begin{bmatrix} 1 & 2 & 3 \\ 2 & 4 & 5 \\ 3 & 5 & 6 \end{bmatrix} \quad \text{and} \quad \mathbf{I}_3 = \begin{bmatrix} 1 & 0 & 0 \\ 0 & 1 & 0 \\ 0 & 0 & 1 \end{bmatrix}$$

are symmetric.

Summation Notation

We shall sometimes use the **summation notation** and we now review this useful and compact notation.

By $\sum_{i=1}^{n} a_i$ we mean the expression

$$a_1 + a_2 + \cdots + a_n.$$

The letter i is called the **index of summation**; it is a dummy variable that can be replaced by another letter and we can then write

$$\sum_{i=1}^{n} a_i = \sum_{j=1}^{n} a_j = \sum_{k=1}^{n} a_k.$$

Example 25. If

$$a_1 = 3, \quad a_2 = 4, \quad a_3 = 5, \quad \text{and} \quad a_4 = 8,$$

then

$$\sum_{i=1}^{4} a_i = 3 + 4 + 5 + 8 = 20.$$

Example 26. By $\sum_{i=1}^{n} r_i a_i$ we mean

$$r_1 a_1 + r_2 a_2 + \cdots + r_n a_n.$$

It is not difficult to show (Exercise T.23) that the summation notation satisfies the following properties:

(i) $\sum_{i=1}^{n} (r_i + s_i) a_i = \sum_{i=1}^{n} r_i a_i + \sum_{i=1}^{n} s_i a_i.$

(ii) $\sum_{i=1}^{n} c(r_i a_i) = c\left(\sum_{i=1}^{n} r_i a_i\right).$

Example 27. We can write Equation (4), for the i, jth element in the product of the matrices **A** and **B**, in terms of the summation notation as

$$c_{ij} = \sum_{k=1}^{p} a_{ik}b_{kj} \qquad (1 \leqslant i \leqslant m, 1 \leqslant j \leqslant n).$$

It is also possible to form double sums. Thus by $\sum_{j=1}^{m} \sum_{i=1}^{n} a_{ij}$ we mean that we first sum on i and then sum the resulting expression on j.

Example 28. If $n = 2$ and $m = 3$, we have

$$\begin{aligned}\sum_{j=1}^{3} \sum_{i=1}^{2} a_{ij} &= \sum_{j=1}^{3} (a_{1j} + a_{2j}) \\ &= \sum_{j=1}^{3} a_{1j} + \sum_{j=1}^{3} a_{2j} \\ &= (a_{11} + a_{12} + a_{13}) + (a_{21} + a_{22} + a_{23}).\end{aligned}$$

It is not difficult to show (Exercise T.24) that

$$\sum_{i=1}^{n} \sum_{j=1}^{m} a_{ij} = \sum_{j=1}^{m} \sum_{i=1}^{n} a_{ij}.$$

1.2 Exercises

1. Let

$$\mathbf{A} = \begin{bmatrix} 2 & -3 & 5 \\ 6 & -5 & 4 \end{bmatrix}, \quad \mathbf{B} = \begin{bmatrix} 4 \\ -3 \\ 5 \end{bmatrix}, \quad \text{and} \quad \mathbf{C} = \begin{bmatrix} 7 & 3 & 2 \\ -4 & 3 & 5 \\ 6 & 2 & -1 \end{bmatrix}.$$

(a) What is a_{12}, a_{22}, a_{23}?
(b) What is b_{11}, b_{31}?
(c) What is c_{13}, c_{31}, c_{33}?

In Exercises 2 through 6 let

$$\mathbf{A} = \begin{bmatrix} 1 & 2 & 3 \\ 2 & 1 & 4 \end{bmatrix}, \quad \mathbf{B} = \begin{bmatrix} 1 & 0 \\ 2 & 1 \\ 3 & 2 \end{bmatrix}, \quad \mathbf{C} = \begin{bmatrix} 3 & -1 & 3 \\ 4 & 1 & 5 \\ 2 & 1 & 3 \end{bmatrix},$$

$$\mathbf{D} = \begin{bmatrix} 3 & -2 \\ 2 & 4 \end{bmatrix}, \quad \mathbf{E} = \begin{bmatrix} 2 & -4 & 5 \\ 0 & 1 & 4 \\ 3 & 2 & 1 \end{bmatrix}, \quad \text{and} \quad \mathbf{F} = \begin{bmatrix} -4 & 5 \\ 2 & 3 \end{bmatrix}.$$

2. If possible, compute
(a) $\mathbf{C} + \mathbf{E}$.
(b) $\mathbf{AB}$ and $\mathbf{BA}$.
(c) $2\mathbf{D} - 3\mathbf{F}$.
(d) $\mathbf{CB} + \mathbf{D}$.
(e) $\mathbf{AB} + \mathbf{DF}$.
(f) $(3)(2\mathbf{A})$ and $6\mathbf{A}$.

3. If possible, compute
(a) $\mathbf{A}(\mathbf{BD})$ and $(\mathbf{AB})\mathbf{D}$.
(b) $\mathbf{A}(\mathbf{C} + \mathbf{E})$ and $\mathbf{AC} + \mathbf{AE}$.
(c) $3\mathbf{A} + 2\mathbf{A}$ and $5\mathbf{A}$.
(d) $\mathbf{DF} + \mathbf{AB}$.
(e) $\mathbf{EF} + 2\mathbf{A}$.
(f) $(-4)(3\mathbf{C})$ and $(-12)\mathbf{C}$.

4. (a) $2\mathbf{B} - 3\mathbf{F}$.
(b) $\mathbf{EB} + \mathbf{FA}$.
(c) $\mathbf{A}(\mathbf{B} + \mathbf{D})$.
(d) $(\mathbf{D} + \mathbf{F})\mathbf{A}$.
(e) $3\mathbf{F} + 4\mathbf{F}$ and $7\mathbf{F}$.
(f) $\mathbf{AC} + \mathbf{DE}$.

5. If possible, compute
(a) $\mathbf{A}^T$ and $(\mathbf{A}^T)^T$.
(b) $(\mathbf{C} + \mathbf{E})^T$ and $\mathbf{C}^T + \mathbf{E}^T$.
(c) $(\mathbf{AB})^T$ and $\mathbf{B}^T\mathbf{A}^T$.
(d) $(2\mathbf{D} + 3\mathbf{F})^T$.
(e) $(2\mathbf{BC} + \mathbf{F})^T$.
(f) $\mathbf{B}^T\mathbf{C} + \mathbf{A}$.

6. If possible, compute
(a) $(3\mathbf{C} - 2\mathbf{E})^T\mathbf{B}$.
(b) $\mathbf{A}^T(\mathbf{D} + \mathbf{F})$.
(c) $(2\mathbf{E})\mathbf{A}^T$.
(d) $(\mathbf{BC})^T$ and $\mathbf{C}^T\mathbf{B}^T$.
(e) $(\mathbf{B}^T + \mathbf{A})\mathbf{C}$.
(f) $(\mathbf{D}^T + \mathbf{E})\mathbf{F}$.

7. Verify Theorem 1.1 for

$$\mathbf{A} = \begin{bmatrix} 1 & 2 & -2 \\ 3 & 4 & 5 \end{bmatrix}, \qquad \mathbf{B} = \begin{bmatrix} 2 & 0 & 1 \\ 3 & -2 & 5 \end{bmatrix},$$

and

$$\mathbf{C} = \begin{bmatrix} -4 & -6 & 1 \\ 2 & 3 & 0 \end{bmatrix}.$$

8. Verify (a) of Theorem 1.2 for

$$\mathbf{A} = \begin{bmatrix} 1 & 3 \\ 2 & -1 \end{bmatrix}, \quad \mathbf{B} = \begin{bmatrix} -1 & 3 & 2 \\ 1 & -3 & 4 \end{bmatrix},$$

and

$$\mathbf{C} = \begin{bmatrix} 1 & 0 \\ 3 & -1 \\ 1 & 2 \end{bmatrix}.$$

9. Verify (b) of Theorem 1.2 for

$$\mathbf{A} = \begin{bmatrix} 1 & -3 \\ -3 & 4 \end{bmatrix}, \quad \mathbf{B} = \begin{bmatrix} 2 & -3 & 2 \\ 3 & -1 & -2 \end{bmatrix},$$

and

$$\mathbf{C} = \begin{bmatrix} 0 & 1 & 2 \\ 1 & 3 & -2 \end{bmatrix}.$$

10. Verify (a), (b), and (c) of Theorem 1.3 for $r = 6$, $s = -2$, and

$$\mathbf{A} = \begin{bmatrix} 4 & 2 \\ 1 & -3 \end{bmatrix}, \quad \mathbf{B} = \begin{bmatrix} 0 & 2 \\ -4 & 3 \end{bmatrix}.$$

11. Verify (d) of Theorem 1.3 for $r = -3$ and

$$\mathbf{A} = \begin{bmatrix} 1 & 3 \\ 2 & -1 \end{bmatrix}, \quad \mathbf{B} = \begin{bmatrix} -1 & 3 & 2 \\ 1 & -3 & 4 \end{bmatrix}.$$

12. Verify (b) and (d) of Theorem 1.4 for $r = -4$ and

$$\mathbf{A} = \begin{bmatrix} 1 & 3 & 2 \\ 2 & 1 & -3 \end{bmatrix}, \quad \mathbf{B} = \begin{bmatrix} 4 & 2 & -1 \\ -2 & 1 & 5 \end{bmatrix}.$$

13. Verify (c) of Theorem 1.4 for

$$\mathbf{A} = \begin{bmatrix} 1 & 3 & 2 \\ 2 & 1 & -3 \end{bmatrix}, \quad \mathbf{B} = \begin{bmatrix} 3 & -1 \\ 2 & 4 \\ 1 & 2 \end{bmatrix}.$$

14. If $\begin{bmatrix} a+b & c+d \\ c-d & a-b \end{bmatrix} = \begin{bmatrix} 4 & 6 \\ 10 & 2 \end{bmatrix}$, find a, b, c, and d.

15. If $\begin{bmatrix} a+2b & 2a-b \\ 2c+d & c-2d \end{bmatrix} = \begin{bmatrix} 4 & -2 \\ 4 & -3 \end{bmatrix}$, find a, b, c, and d.

16. Let $\mathbf{A} = \begin{bmatrix} 1 & 2 \\ 3 & 2 \end{bmatrix}$ and $\mathbf{B} = \begin{bmatrix} 2 & -1 \\ -3 & 4 \end{bmatrix}$. Show that $\mathbf{AB} \neq \mathbf{BA}$.

17. If $\mathbf{A} = \begin{bmatrix} -2 & 3 \\ 2 & -3 \end{bmatrix}$ and $\mathbf{B} = \begin{bmatrix} 3 & 6 \\ 2 & 4 \end{bmatrix}$, show that $\mathbf{AB} = \mathbf{0}$.

18. If $\mathbf{A} = \begin{bmatrix} -2 & 3 \\ 2 & -3 \end{bmatrix}$, $\mathbf{B} = \begin{bmatrix} -1 & 3 \\ 2 & 0 \end{bmatrix}$, and $\mathbf{C} = \begin{bmatrix} -4 & -3 \\ 0 & -4 \end{bmatrix}$, show that $\mathbf{AB} = \mathbf{AC}$.

19. If $\mathbf{A} = \begin{bmatrix} 0 & 1 \\ 1 & 0 \end{bmatrix}$, show that $\mathbf{A}^2 = \mathbf{I}_2$.

20. Consider the following linear system:

$$\begin{aligned} 2x \qquad\qquad\quad + w &= 7 \\ 3x + 2y + 3z \qquad &= -2 \\ 2x + 3y - 4z \qquad &= 3 \\ x + \qquad 3z \qquad &= 5. \end{aligned}$$

(a) Find the coefficient matrix.
(b) Write the linear system in matrix form.
(c) Find the augmented matrix.

21. Write the linear system with augmented matrix

$$\left[\begin{array}{rrrr|r} -2 & -1 & 0 & 4 & 5 \\ -3 & 2 & 7 & 8 & 3 \\ 1 & 0 & 0 & 2 & 4 \\ 3 & 0 & 1 & 3 & 6 \end{array}\right].$$

22. Write the linear system with augmented matrix

$$\left[\begin{array}{rrr|r} 2 & 0 & -4 & 3 \\ 0 & 1 & 2 & 5 \\ 1 & 3 & 4 & -1 \end{array}\right].$$

23. Consider the following linear system:

$$\begin{aligned} 3x - y + 2z &= 4 \\ 2x + y \qquad &= 2 \\ y + 3z &= 7 \\ 4x \qquad - z &= 4. \end{aligned}$$

(a) Find the coefficient matrix.
(b) Write the linear system in matrix form.
(c) Find the augmented matrix.

Theoretical Exercises

T.1. Prove properties (b) and (d) of Theorem 1.1.

T.2. If $\mathbf{A} = [a_{ij}]$ is a 2×3 matrix, $\mathbf{B} = [b_{ij}]$ is a 3×4 matrix, and $\mathbf{C} = [c_{ij}]$ is a 4×3 matrix, show that

$$\mathbf{A}(\mathbf{BC}) = (\mathbf{AB})\mathbf{C}.$$

T.3. Prove properties (b) and (c) of Theorem 1.2.

T.4. If $\mathbf{A}$ is an $m \times n$ matrix, show that

$$\mathbf{AI}_n = \mathbf{I}_m\mathbf{A} = \mathbf{A}.$$

T.5. Prove Theorem 1.3.

T.6. Show that $(-1)\mathbf{A} = -\mathbf{A}$.

T.7. Complete the proof of Theorem 1.4.

T.8. Show that $\mathbf{A}$ is symmetric if and only if $a_{ij} = a_{ji}$.

T.9. Prove that if $\mathbf{A}$ is symmetric, then $\mathbf{A}^T$ is symmetric.

T.10. Let $\mathbf{A}$ and $\mathbf{B}$ be symmetric matrices.
(a) Show that $\mathbf{A} + \mathbf{B}$ is symmetric.
(b) Show that $\mathbf{AB}$ is symmetric if and only if $\mathbf{AB} = \mathbf{BA}$.

T.11. A matrix $\mathbf{A} = [a_{ij}]$ is called **skew symmetric** if $\mathbf{A}^T = -\mathbf{A}$. Show that $\mathbf{A}$ is skew symmetric if and only if $a_{ij} = -a_{ji}$.

T.12. If $\mathbf{A}$ is an $n \times n$ matrix, prove that
(a) $\mathbf{AA}^T$ and $\mathbf{A}^T\mathbf{A}$ are symmetric.
(b) $\mathbf{A} + \mathbf{A}^T$ is symmetric.
(c) $\mathbf{A} - \mathbf{A}^T$ is skew symmetric.

T.13. Prove that if $\mathbf{A}$ is an $n \times n$ matrix, then $\mathbf{A}$ can be written uniquely as $\mathbf{A} = \mathbf{S} + \mathbf{K}$, where $\mathbf{S}$ is symmetric and $\mathbf{K}$ is skew symmetric.

T.14. Prove that if $\mathbf{A}$ is an $n \times n$ scalar matrix, then $\mathbf{A} = r\mathbf{I}_n$ for some real number r.

T.15. (a) Prove that if $\mathbf{A}$ has a row of zeros, then $\mathbf{AB}$ has a row of zeros.
(b) Prove that if $\mathbf{B}$ has a column of zeros, then $\mathbf{AB}$ has a column of zeros.

T.16. Prove that the sum and product of diagonal matrices is diagonal.

T.17. Prove that the sum and product of scalar matrices is scalar.

T.18. A matrix $\mathbf{A} = [a_{ij}]$ is called **upper triangular** if $a_{ij} = 0$ for $i > j$. It is called **lower triangular** if $a_{ij} = 0$ for $i < j$.
(a) Prove that the sum and product of upper triangular matrices is upper triangular.
(b) Prove that the sum and product of lower triangular matrices is lower triangular.

T.19. Show that

$$\mathbf{I}_n^T = \mathbf{I}_n.$$

T.20. Show that if $\mathbf{AX} = \mathbf{B}$ is a linear system that has more than one solution, then it has infinitely many solutions.

(*Hint:* If $\mathbf{X}_1$ and $\mathbf{X}_2$ are solutions, consider $\mathbf{X}_3 = r\mathbf{X}_1 + s\mathbf{X}_2$, where $r + s = 1$.)

T.21. Show that the jth column of the matrix product $\mathbf{AB}$ is equal to the matrix product $\mathbf{AB}_j$, where $\mathbf{B}_j$ is the jth column of $\mathbf{B}$.

T.22. Let $\mathbf{A}$ be an $m \times n$ matrix. Show that if $r\mathbf{A} = \mathbf{0}$, then $r = 0$ or $\mathbf{A} = \mathbf{0}$.

T.23. Show that the summation notation satisfies the following properties:

(i) $\sum_{i=1}^{n} (r_i + s_i)a_i = \sum_{i=1}^{n} r_i a_i + \sum_{i=1}^{n} s_i a_i.$

(ii) $\sum_{i=1}^{n} c(r_i a_i) = c\left(\sum_{i=1}^{n} r_i a_i\right).$

T.24. Show that

$$\sum_{i=1}^{n} \sum_{j=1}^{m} a_{ij} = \sum_{j=1}^{m} \sum_{i=1}^{n} a_{ij}.$$

1.3 Solutions of Equations

In this section we shall systematize the familiar method of elimination of variables for the solution of linear systems and thus obtain a useful method for solving such systems. This method starts with the augmented matrix of the given linear system and obtains a matrix of a certain special form. This new matrix represents a linear system that has exactly the same solutions as the given system. However, this new linear system can be solved very easily. Our discussion will apply to arbitrary matrices that are not necessarily being viewed as augmented matrices of linear systems.

DEFINITION An $m \times n$ matrix $\mathbf{A}$ is said to be in **reduced row echelon form** if:

(a) Each of the first k rows $(1 \leqslant k \leqslant m)$ has at least one nonzero element, and if $k < m$, then rows $k + 1, k + 2, \ldots, m$ consist entirely of zeros (that is, the rows consisting entirely of zeros, if any, are at the bottom of the matrix).

(b) Counting from left to right, the first nonzero element in each of the first k rows is a 1.

(c) If the leading 1 in row i, $1 \leqslant i \leqslant k$, occurs in column j_i, then $j_1 < j_2 < \cdots < j_k$.
(d) All the elements in column j_i are zero except for the 1 in row i.

If $k = m$, then there are no rows all of whose elements are zero.

Example 1. The matrices

$$\mathbf{A} = \begin{bmatrix} 1 & 0 & 0 & 4 \\ 0 & 1 & 0 & 5 \\ 0 & 0 & 1 & 2 \end{bmatrix}, \qquad \mathbf{B} = \begin{bmatrix} 1 & 0 & 0 & 0 & 2 \\ 0 & 0 & 1 & 0 & 1 \\ 0 & 0 & 0 & 1 & 0 \end{bmatrix},$$

and

$$\mathbf{C} = \begin{bmatrix} 1 & 0 & 0 & 0 & 0 \\ 0 & 0 & 1 & 0 & 0 \\ 0 & 0 & 0 & 0 & 1 \\ 0 & 0 & 0 & 0 & 0 \\ 0 & 0 & 0 & 0 & 0 \end{bmatrix}$$

are in reduced row echelon form.

Example 2. The matrices

$$\mathbf{A} = \begin{bmatrix} 1 & 2 & 0 & 4 \\ 0 & 0 & 0 & 0 \\ 0 & 0 & 1 & -3 \end{bmatrix}, \qquad \mathbf{B} = \begin{bmatrix} 1 & 0 & 3 & 4 \\ 0 & 2 & -2 & 5 \\ 0 & 0 & 1 & 2 \end{bmatrix},$$

$$\mathbf{C} = \begin{bmatrix} 1 & 0 & 3 & 4 \\ 0 & 1 & -2 & 5 \\ 0 & 1 & 2 & 2 \\ 0 & 0 & 0 & 0 \end{bmatrix}, \quad \text{and} \quad \mathbf{D} = \begin{bmatrix} 1 & 2 & 3 & 4 \\ 0 & 1 & -2 & 5 \\ 0 & 0 & 1 & 2 \\ 0 & 0 & 0 & 0 \end{bmatrix}$$

are not in reduced row echelon form, since they fail to satisfy (a), (b), (c), and (d), respectively.

We shall now turn to the discussion of how to transform a given matrix to a matrix in reduced row echelon form.

DEFINITION An **elementary row operation** on an $m \times n$ matrix $\mathbf{A} = [a_{ij}]$ is any one of the following operations:
(a) Interchange rows r and s of $\mathbf{A}$. That is, replace $a_{r1}, a_{r2}, \ldots, a_{rn}$ by $a_{s1}, a_{s2}, \ldots, a_{sn}$ and $a_{s1}, a_{s2}, \ldots, a_{sn}$ by $a_{r1}, a_{r2}, \ldots, a_{rn}$.

(b) Multiply row r of $\mathbf{A}$ by $c \neq 0$. That is, replace $a_{r1}, a_{r2}, \ldots, a_{rn}$ by $ca_{r1}, ca_{r2}, \ldots, ca_{rn}$.

(c) Add d times row r of $\mathbf{A}$ to row s of $\mathbf{A}$, $r \neq s$. That is, replace $a_{s1}, a_{s2}, \ldots, a_{sn}$ by $a_{s1} + da_{r1}, a_{s2} + da_{r2}, \ldots, a_{sn} + da_{rn}$.

Example 3. Let

$$\mathbf{A} = \begin{bmatrix} 0 & 0 & 1 & 2 \\ 2 & 3 & 0 & -2 \\ 3 & 3 & 6 & -9 \end{bmatrix}.$$

Interchanging rows 1 and 3 of $\mathbf{A}$, we obtain

$$\mathbf{B} = \begin{bmatrix} 3 & 3 & 6 & -9 \\ 2 & 3 & 0 & -2 \\ 0 & 0 & 1 & 2 \end{bmatrix}.$$

Multiplying the third row of $\mathbf{A}$ by $\frac{1}{3}$, we obtain

$$\mathbf{C} = \begin{bmatrix} 0 & 0 & 1 & 2 \\ 2 & 3 & 0 & -2 \\ 1 & 1 & 2 & -3 \end{bmatrix}.$$

Adding -2 times row 2 of $\mathbf{A}$ to row 3 of $\mathbf{A}$, we obtain

$$\mathbf{D} = \begin{bmatrix} 0 & 0 & 1 & 2 \\ 2 & 3 & 0 & -2 \\ -1 & -3 & 6 & -5 \end{bmatrix}.$$

DEFINITION An $m \times n$ matrix $\mathbf{A}$ is said to be **row equivalent** to an $m \times n$ matrix $\mathbf{B}$ if $\mathbf{B}$ results from $\mathbf{A}$ by a finite sequence of elementary row operations.

Example 4. The matrix

$$\mathbf{A} = \begin{bmatrix} 1 & 2 & 4 & 3 \\ 2 & 1 & 3 & 2 \\ 1 & -1 & 2 & 3 \end{bmatrix} \quad \text{is row equivalent to} \quad \mathbf{D} = \begin{bmatrix} 2 & 4 & 8 & 6 \\ 1 & -1 & 2 & 3 \\ 4 & -1 & 7 & 8 \end{bmatrix},$$

because if we add twice row 3 of $\mathbf{A}$ to its second row, we obtain

$$\mathbf{B} = \begin{bmatrix} 1 & 2 & 4 & 3 \\ 4 & -1 & 7 & 8 \\ 1 & -1 & 2 & 3 \end{bmatrix}.$$

Interchanging rows 2 and 3 of **B**, we obtain

$$\mathbf{C} = \begin{bmatrix} 1 & 2 & 4 & 3 \\ 1 & -1 & 2 & 3 \\ 4 & -1 & 7 & 8 \end{bmatrix}.$$

Multiplying row 1 of **C** by 2, we obtain **D**.

It is easy to show (Exercise T.2) that (1) every matrix is row equivalent to itself; (2) if **A** is row equivalent to **B**, then **B** is row equivalent to **A**; and (3) if **A** is row equivalent to **B** and **B** is row equivalent to **C**, then **A** is row equivalent to **C**. In view of (2), both statements, "**A** is row equivalent to **B**" and "**B** is row equivalent to **A**," can be replaced by "**A** and **B** are row equivalent."

THEOREM 1.5. *Every nonzero $m \times n$ matrix is row equivalent to a unique matrix in reduced row echelon form.*

We shall illustrate the proof of the theorem by giving the steps that must be carried out on a specific matrix **A** to obtain a matrix in reduced row echelon form that is row equivalent to **A**. We omit the proof that the matrix thus obtained is unique. We use the following example to illustrate the steps involved.

Example 5

$$\mathbf{A} = \begin{bmatrix} 0 & 2 & 3 & -4 & 1 \\ 0 & 0 & 2 & 3 & 4 \\ 2 & 2 & -5 & 2 & 4 \\ 2 & 0 & -6 & 9 & 7 \end{bmatrix}.$$

STEP 1. Find the first (counting from left to right) column in **A** not all of whose entries are zero. This column is called the **pivotal column**.

$$\mathbf{A} = \begin{bmatrix} 0 & 2 & 3 & -4 & 1 \\ 0 & 0 & 2 & 3 & 4 \\ 2 & 2 & -5 & 2 & 4 \\ 2 & 0 & -6 & 9 & 7 \end{bmatrix}.$$

Pivotal column of **A** (arrow pointing to the first column)

STEP 2. Identify the first (counting from top to bottom) nonzero entry in the pivotal column. This element is called the **pivot**, which we circle in **A**.

$$\mathbf{A} = \begin{bmatrix} 0 & 2 & 3 & -4 & 1 \\ 0 & 0 & 2 & 3 & 4 \\ \textcircled{2} & 2 & -5 & 2 & 4 \\ 2 & 0 & -6 & 9 & 7 \end{bmatrix}.$$

Pivot → (third row, first column)

STEP 3. Interchange, if necessary, the first row with the row where the pivot occurs so that the pivot is now in the first row. Call the new matrix $\mathbf{A}_1$.

$$\mathbf{A}_1 = \begin{bmatrix} \textcircled{2} & 2 & -5 & 2 & 4 \\ 0 & 0 & 2 & 3 & 4 \\ 0 & 2 & 3 & -4 & 1 \\ 2 & 0 & -6 & 9 & 7 \end{bmatrix}$$

The first and third rows of **A** were interchanged.

STEP 4. Divide the first row of $\mathbf{A}_1$ by the pivot. Thus the entry in the first row and pivotal column (where the pivot was located) is now a 1. Call the new matrix $\mathbf{A}_2$.

$$\mathbf{A}_2 = \begin{bmatrix} 1 & 1 & -\frac{5}{2} & 1 & 2 \\ 0 & 0 & 2 & 3 & 4 \\ 0 & 2 & 3 & -4 & 1 \\ 2 & 0 & -6 & 9 & 7 \end{bmatrix}$$

The first row of $\mathbf{A}_1$ was divided by 2.

STEP 5. Add multiples of the first row of $\mathbf{A}_2$ to all other rows to make all entries in the pivotal column, except the entry where the pivot was located, equal to zero. Thus all entries in the pivotal column and rows 2, 3, . . . , m, are zero. Call the new matrix $\mathbf{A}_3$.

$$\mathbf{A}_3 = \begin{bmatrix} 1 & 1 & -\frac{5}{2} & 1 & 2 \\ 0 & 0 & 2 & 3 & 4 \\ 0 & 2 & 3 & -4 & 1 \\ 0 & -2 & -1 & 7 & 3 \end{bmatrix}$$

-2 times the first row of $\mathbf{A}_2$ was added to its fourth row.

STEP 6. Identify $\mathbf{B}$ as the $(m - 1) \times n$ submatrix of $\mathbf{A}_3$ obtained by deleting the first row of $\mathbf{A}_3$; do not erase the first row of $\mathbf{A}_3$. Repeat steps 1–5 on $\mathbf{B}$.

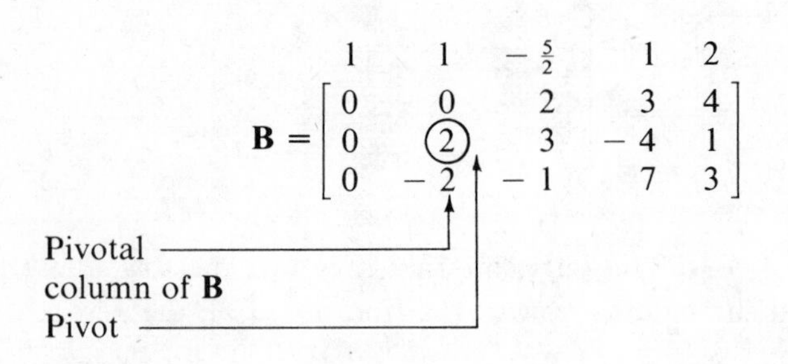

$$\begin{array}{ccccc} 1 & 1 & -\frac{5}{2} & 1 & 2 \end{array}$$
$$\mathbf{B}_1 = \begin{bmatrix} 0 & \textcircled{2} & 3 & -4 & 1 \\ 0 & 0 & 2 & 3 & 4 \\ 0 & -2 & -1 & 7 & 3 \end{bmatrix}$$

The first and second rows of $\mathbf{B}$ were interchanged.

$$\begin{array}{ccccc} 1 & 1 & -\frac{5}{2} & 1 & 2 \end{array}$$
$$\mathbf{B}_2 = \begin{bmatrix} 0 & 1 & \frac{3}{2} & -2 & \frac{1}{2} \\ 0 & 0 & 2 & 3 & 4 \\ 0 & -2 & -1 & 7 & 3 \end{bmatrix}$$

The first row of $\mathbf{B}_1$ was divided by 2.

$$\begin{array}{ccccc} 1 & 1 & -\frac{5}{2} & 1 & 2 \end{array}$$
$$\mathbf{B}_3 = \begin{bmatrix} 0 & 1 & \frac{3}{2} & -2 & \frac{1}{2} \\ 0 & 0 & 2 & 3 & 4 \\ 0 & 0 & 2 & 3 & 4 \end{bmatrix}$$

2 times the first row of $\mathbf{B}_2$ was added to its third row.

STEP 7. Add multiples of the first row of $\mathbf{B}_3$ to all the rows of $\mathbf{A}_3$ above the first row of $\mathbf{B}_3$ so that all entries in the pivotal column, except for the pivot, become zero.

$$\begin{array}{ccccc} 1 & 0 & -4 & 3 & \frac{3}{2} \end{array}$$
$$\mathbf{B}_3 = \begin{bmatrix} 0 & 1 & \frac{3}{2} & -2 & \frac{1}{2} \\ 0 & 0 & 2 & 3 & 4 \\ 0 & 0 & 2 & 3 & 4 \end{bmatrix}$$

(-1) times the first row of $\mathbf{B}_3$ was added to the shaded row.

STEP 8. Identify $\mathbf{C}$ as the $(m-2) \times n$ submatrix of $\mathbf{B}_3$ obtained by deleting the first row of $\mathbf{B}_3$; do not erase the first row of $\mathbf{B}_3$. Repeat steps 1–7 on $\mathbf{C}$.

$$\begin{matrix} 1 & 0 & -4 & 3 & \frac{3}{2} \\ 0 & 1 & \frac{3}{2} & -2 & \frac{1}{2} \end{matrix}$$
$$\mathbf{C} = \begin{bmatrix} 0 & 0 & \textcircled{2} & 3 & 4 \\ 0 & 0 & 2 & 3 & 4 \end{bmatrix}.$$

Pivotal column of $\mathbf{C}$ (points to the third column)

Pivot (points to the circled entry 2)

$$\begin{matrix} 1 & 0 & -4 & 3 & \frac{3}{2} \\ 0 & 1 & \frac{3}{2} & -2 & \frac{1}{2} \end{matrix}$$
$$\mathbf{C}_1 = \mathbf{C}_2 = \begin{bmatrix} 0 & 0 & 1 & \frac{3}{2} & 2 \\ 0 & 0 & 2 & 3 & 4 \end{bmatrix}$$

No rows of $\mathbf{C}$ had to be interchanged. The first row of $\mathbf{C}$ was divided by 2.

$$\begin{matrix} 1 & 0 & -4 & 3 & \frac{3}{2} \\ 0 & 1 & \frac{3}{2} & -2 & \frac{1}{2} \end{matrix}$$
$$\mathbf{C}_3 = \begin{bmatrix} 0 & 0 & 1 & \frac{3}{2} & 2 \\ 0 & 0 & 0 & 0 & 0 \end{bmatrix}$$

-2 times the first row of $\mathbf{C}_2$ was added to its second row.

$$\begin{matrix} 1 & 0 & 0 & 9 & \frac{19}{2} \\ 0 & 1 & 0 & -\frac{17}{4} & -\frac{5}{2} \end{matrix}$$
$$\mathbf{C}_3 = \begin{bmatrix} 0 & 0 & 1 & \frac{3}{2} & 2 \\ 0 & 0 & 0 & 0 & 0 \end{bmatrix}$$

4 times the first row of $\mathbf{C}_3$ was added to the first shaded row; $-\frac{3}{2}$ times the first row of $\mathbf{C}_3$ was added to the second shaded row.

The final matrix,

$$\begin{bmatrix} 1 & 0 & 0 & 9 & \frac{19}{2} \\ 0 & 1 & 0 & -\frac{17}{4} & -\frac{5}{2} \\ 0 & 0 & 1 & \frac{3}{2} & 2 \\ 0 & 0 & 0 & 0 & 0 \end{bmatrix},$$

is in reduced row echelon form.

We now apply these results to the solution of linear systems.

THEOREM 1.6. *Let* **AX** = **B** *and* **CX** = **D** *be two linear systems each of m equations in n unknowns. If the augmented matrices* $[A \mid B]$ *and* $[C \mid D]$ *of these systems are row equivalent, then both linear systems have exactly the same solutions.*

proof. This follows from the definition of row equivalence and from the fact that the three elementary row operations on the augmented matrix turn out to be the three manipulations on a linear system, discussed in Section 1.1, yielding a linear system having the same solutions as the given system. We also note that if one system has no solution, then the other system has no solution.

COROLLARY 1.1. *If* **A** *and* **B** *are row equivalent $m \times n$ matrices, then the linear systems* **AX** = **0** *and* **BX** = **0** *have exactly the same solutions.*

proof. Exercise T.3.

The following examples illustrate the method of solving linear systems, which is called **Gauss–Jordan reduction**.

Example 6. Consider the linear system

$$\begin{aligned} x + 2y + 3z &= 9 \\ 2x - y + z &= 8 \\ 3x \phantom{{}+2y} - z &= 3. \end{aligned} \tag{1}$$

The augmented matrix of this linear system is

$$\left[\begin{array}{rrr|r} 1 & 2 & 3 & 9 \\ 2 & -1 & 1 & 8 \\ 3 & 0 & -1 & 3 \end{array}\right],$$

which is row equivalent to the matrix (verify),

$$\left[\begin{array}{rrr|r} 1 & 0 & 0 & 2 \\ 0 & 1 & 0 & -1 \\ 0 & 0 & 1 & 3 \end{array}\right] \tag{2}$$

in reduced row echelon form. The linear system represented by (2) is

$$\begin{aligned} x \phantom{{}+y+z} &= 2 \\ y \phantom{{}+z} &= -1 \\ z &= 3 \end{aligned}$$

so that the unique solution to the given linear system (1) is

$$x = 2$$
$$y = -1$$
$$z = 3.$$

Example 7. Consider the linear system

$$\begin{aligned} x + y + 2z - 5w &= 3 \\ 2x + 5y - z - 9w &= -3 \\ 2x + y - z + 3w &= -11 \\ x - 3y + 2z + 7w &= -5. \end{aligned} \tag{3}$$

The augmented matrix of this linear system is

$$\left[\begin{array}{cccc|c} 1 & 1 & 2 & -5 & 3 \\ 2 & 5 & 1 & -9 & -3 \\ 2 & 1 & -1 & 3 & -11 \\ 1 & -3 & 2 & 7 & -5 \end{array}\right],$$

which is row equivalent to

$$\left[\begin{array}{cccc|c} 1 & 0 & 0 & 2 & -5 \\ 0 & 1 & 0 & -3 & 2 \\ 0 & 0 & 1 & -2 & 3 \\ 0 & 0 & 0 & 0 & 0 \end{array}\right]. \tag{4}$$

The linear system represented by (4) is

$$\begin{aligned} x \qquad + 2w &= -5 \\ y \quad - 3w &= 2 \\ z - 2w &= 3. \end{aligned}$$

This system can be solved, obtaining

$$\begin{aligned} x &= -5 - 2r \\ y &= 2 + 3r \\ z &= 3 + 2r \\ w &= r, \end{aligned} \tag{5}$$

where r is any real number. Thus (5) is the solution to the given linear system. Since r can be assigned any real number, the given linear system (3) has infinitely many solutions.

Example 8. Consider the linear system

$$\begin{aligned} x + 2y + 3z + 4v + 5w &= 6 \\ x + 3y + 5z + 7v + 4w &= 13 \\ x + 4y + 7z + 10v + 3w &= 20 \\ x + 2y + 4z + 6v + 8w &= 13. \end{aligned} \tag{6}$$

The augmented matrix of this linear system is

$$\left[\begin{array}{ccccc|c} 1 & 2 & 3 & 4 & 5 & 6 \\ 1 & 3 & 5 & 7 & 4 & 13 \\ 1 & 4 & 7 & 10 & 3 & 20 \\ 1 & 2 & 4 & 6 & 8 & 13 \end{array}\right],$$

which is row equivalent to

$$\left[\begin{array}{ccccc|c} 1 & 0 & 0 & 0 & 10 & -1 \\ 0 & 1 & 0 & -1 & -7 & -7 \\ 0 & 0 & 1 & 2 & 3 & 7 \\ 0 & 0 & 0 & 0 & 0 & 0 \end{array}\right]. \tag{7}$$

The linear system represented by (7) is

$$\begin{aligned} x \qquad\qquad + 10w &= -1 \\ y \qquad - v - 7w &= -7 \\ z + 2v + 3w &= 7, \end{aligned}$$

which can be solved, obtaining

$$\begin{aligned} x &= -1 - 10s \\ y &= -7 + r + 7s \\ z &= 7 - 2r - 3s \\ v &= r \\ w &= s, \end{aligned} \tag{8}$$

where r and s are any real numbers. Thus (8) is the solution to the given linear system (6).

Example 9. Consider the linear system

$$\begin{aligned} x + 2y + 3z + 4w &= 5 \\ x + 3y + 5z + 7w &= 11 \\ x \qquad - z - 2w &= -6. \end{aligned} \tag{9}$$

The augmented matrix of this linear system is

$$\left[\begin{array}{cccc|c} 1 & 2 & 3 & 4 & 5 \\ 1 & 3 & 5 & 7 & 11 \\ 1 & 0 & -1 & -2 & -6 \end{array}\right],$$

which is row equivalent to

$$\left[\begin{array}{cccc|c} 1 & 0 & -1 & -2 & 0 \\ 0 & 1 & 2 & 3 & 0 \\ 0 & 0 & 0 & 0 & 1 \end{array}\right]. \tag{10}$$

The last equation of the linear system represented by (10) is

$$0x + 0y + 0z + 0w = 1,$$

which has no solution for any x, y, z, and w. Consequently, the given linear system (9) has no solution.

The last example is characteristic of the way in which a linear system has no solution. That is, a linear system $\mathbf{AX} = \mathbf{B}$ has no solution if and only if its augmented matrix is row equivalent to a matrix in reduced row echelon form, which has a row whose first n elements are zero and whose $(n + 1)$th element is 1 (Exercise T.4). The Gauss–Jordan reduction method is frequently used in actual practice, and computer codes implementing it are widely available.

Homogeneous Systems

A linear system of the form

$$\begin{aligned} a_{11}x_1 + a_{12}x_2 + \cdots + a_{1n}x_n &= 0 \\ a_{21}x_1 + a_{22}x_2 + \cdots + a_{2m}x_n &= 0 \\ \vdots \qquad\qquad\qquad & \quad \vdots \\ a_{m1}x_1 + a_{m2}x_2 + \cdots + a_{mn}x_n &= 0 \end{aligned} \tag{11}$$

is called a **homogeneous system.** We can also write (11) in matrix form as

$$\mathbf{AX} = \mathbf{0}. \tag{12}$$

The solution,

$$x_1 = x_2 = \cdots = x_n = 0,$$

for the homogeneous system (12) is called the **trivial solution**. A solution $x_1, x_2, \ldots, x_n$ to a homogeneous system in which not all the x_i are zero is called a **nontrivial solution.**

Example 10. Consider the homogeneous system

$$\begin{aligned} x + 2y + 3z &= 0 \\ -x + 3y + 2z &= 0 \\ 2x + y - 2z &= 0. \end{aligned} \tag{13}$$

The augmented matrix of this system,

$$\left[\begin{array}{rrr|r} 1 & 2 & 3 & 0 \\ -1 & 3 & 2 & 0 \\ 2 & 1 & -2 & 0 \end{array}\right],$$

is row equivalent to

$$\left[\begin{array}{rrr|r} 1 & 0 & 0 & 0 \\ 0 & 1 & 0 & 0 \\ 0 & 0 & 1 & 0 \end{array}\right],$$

which is in reduced row echelon form. Hence the solution to (13) is

$$x = y = z = 0,$$

which means that the given homogeneous system (13) has only the trivial solution.

Example 11. Consider the homogeneous system

$$\begin{aligned} x + y + z + w &= 0 \\ x \qquad\quad + w &= 0 \\ x + 2y + z \qquad &= 0. \end{aligned} \tag{14}$$

The augmented matrix of this system,

$$\left[\begin{array}{cccc|c} 1 & 1 & 1 & 1 & 0 \\ 1 & 0 & 0 & 1 & 0 \\ 1 & 2 & 1 & 0 & 0 \end{array}\right],$$

is row equivalent to

$$\left[\begin{array}{cccc|c} 1 & 0 & 0 & 1 & 0 \\ 0 & 1 & 0 & -1 & 0 \\ 0 & 0 & 1 & 1 & 0 \end{array}\right],$$

which is in reduced row echelon form. Hence the solution to (14) is

$$\begin{aligned} x &= -r \\ y &= r \\ z &= -r \\ w &= r, \end{aligned}$$

where r is any real number.

Example 11 shows that a homogeneous system may have a nontrivial solution. The following theorem tells of one case when this occurs.

THEOREM 1.7. *A homogeneous system of m equations in n unknowns always has a nontrivial solution if $m < n$, that is, if the number of unknowns exceeds the number of equations.*

proof. Let **C** be a matrix in reduced row echelon form that is row equivalent to **A**. Then the homogeneous systems $\mathbf{AX} = \mathbf{0}$ and $\mathbf{CX} = \mathbf{0}$ have the same solutions. If we let r be the number of nonzero rows of **C**, then $r \leqslant m$. If $m < n$, then $r < n$. We are now effectively solving a homogeneous system of r equations in n unknowns and can solve for $x_1, \ldots, x_r$ in terms of the remaining $n - r$ unknowns $x_{r+1}, \ldots, x_n$, the latter being free to take on any values that we please. Thus, by letting one of the unknowns $x_{r+1}, \ldots, x_n$ be nonzero, we obtain a nontrivial solution of $\mathbf{CX} = \mathbf{0}$ and thus of $\mathbf{AX} = \mathbf{0}$.

In Section 8.2 we shall return to the problem of solving linear systems from a computer point of view.

1.3 Exercises

1. Which of the following matrices are in reduced row echelon form?

$$\mathbf{A} = \begin{bmatrix} 1 & 0 & 0 & 0 & -3 \\ 0 & 0 & 1 & 0 & 4 \\ 0 & 0 & 0 & 1 & 2 \end{bmatrix}, \quad \mathbf{B} = \begin{bmatrix} 0 & 1 & 0 & 0 & 5 \\ 0 & 0 & 1 & 0 & -4 \\ 0 & 0 & 0 & -1 & 3 \end{bmatrix},$$

$$\mathbf{C} = \begin{bmatrix} 0 & 1 & 0 & 0 & 5 \\ 0 & 0 & 1 & 0 & 4 \\ 0 & 1 & 0 & -2 & 3 \end{bmatrix}, \quad \mathbf{D} = \begin{bmatrix} 0 & 1 & 0 & 0 & 2 \\ 0 & 0 & 0 & 0 & -1 \\ 0 & 0 & 0 & 1 & 4 \\ 0 & 0 & 0 & 0 & 0 \\ 0 & 0 & 0 & 0 & 1 \end{bmatrix},$$

$$\mathbf{E} = \begin{bmatrix} 1 & 0 & 0 & 0 & 2 \\ 0 & 0 & 0 & 1 & 0 \\ 0 & 0 & 1 & 2 & 3 \\ 0 & 0 & 0 & 0 & 0 \end{bmatrix}, \quad \mathbf{F} = \begin{bmatrix} 0 & 0 & 0 & 0 & 0 \\ 0 & 0 & 1 & 2 & -3 \\ 0 & 0 & 0 & 1 & 0 \\ 0 & 0 & 0 & 0 & 0 \end{bmatrix},$$

$$\mathbf{G} = \begin{bmatrix} 0 & 0 & 0 & 0 & 1 \\ 0 & 1 & 0 & 0 & 2 \\ 0 & 0 & 0 & 1 & -1 \\ 0 & 0 & 0 & 0 & 0 \end{bmatrix}, \quad \mathbf{H} = \begin{bmatrix} 1 & 0 & 0 & 1 \\ 0 & 1 & 0 & 2 \\ 0 & 0 & 0 & -1 \\ 0 & 0 & 0 & 0 \end{bmatrix}.$$

2. Let

$$\mathbf{A} = \begin{bmatrix} 1 & 0 & 3 \\ -3 & 1 & 4 \\ 4 & 2 & 2 \\ 5 & -1 & 5 \end{bmatrix}.$$

Find the matrices obtained by performing the following elementary row operations on **A**.

(a) Interchanging the second and fourth rows.

(b) Multiplying the third row by 3.

(c) Adding -3 times the first row to the fourth row.

3. Let

$$\mathbf{A} = \begin{bmatrix} 2 & 0 & 4 & 2 \\ 3 & -2 & 5 & 6 \\ -1 & 3 & 1 & 1 \end{bmatrix}.$$

Find the matrices obtained by performing the following elementary row operations on **A**.

(a) Interchanging the second and third rows.

(b) Multiplying the second row by -4.

(c) Adding 2 times the third row to the first row.

4. Find three matrices that are row equivalent to

$$\mathbf{A} = \begin{bmatrix} 2 & -1 & 3 & 4 \\ 0 & 1 & 2 & -1 \\ 5 & 2 & -3 & 4 \end{bmatrix}.$$

5. Find three matrices that are row equivalent to

$$\begin{bmatrix} 4 & 3 & 7 & 5 \\ -1 & 2 & -1 & 3 \\ 2 & 0 & 1 & 4 \end{bmatrix}.$$

6. If

$$\mathbf{A} = \begin{bmatrix} 0 & 0 & -1 & 2 & 3 \\ 0 & 2 & 3 & 4 & 5 \\ 0 & 1 & 3 & -1 & 2 \\ 0 & 3 & 2 & 4 & 1 \end{bmatrix},$$

find a matrix **C** in reduced row echelon form that is row equivalent to **A**.

7. If

$$\mathbf{A} = \begin{bmatrix} 1 & -2 & 0 & 2 \\ 2 & -3 & -1 & 5 \\ 1 & 3 & 2 & 5 \\ 1 & 1 & 0 & 2 \end{bmatrix},$$

find a matrix **C** in reduced row echelon form that is row equivalent to **A**.

In Exercises 8 through 10 find all solutions to the given linear systems.

8. (a)
$$\begin{aligned} x + y + 2z &= -1 \\ x - 2y + z &= -5 \\ 3x + y + z &= 3. \end{aligned}$$

(b)
$$\begin{aligned} x + y + 3z + 2w &= 7 \\ 2x - y + 4w &= 8 \\ 3y + 6z &= 8. \end{aligned}$$

(c)
$$\begin{aligned} x + 2y - 4z &= 3 \\ x - 2y + 3z &= -1 \\ 2x + 3y - z &= 5 \\ 4x + 3y - 2z &= 7 \\ 5x + 2y - 6z &= 7. \end{aligned}$$

9. (a)
$$\begin{aligned} x + y + 2z + 3w &= 13 \\ x - 2y + z + w &= 8 \\ 3x + y + z - w &= 1 \end{aligned}$$

(b)
$$\begin{aligned} x + y + z &= 1 \\ x + y - 2z &= 3 \\ 2x + y + z &= 2. \end{aligned}$$

(c)
$$\begin{aligned} 2x + y + z - 2w &= 1 \\ 3x - 2y + z - 6w &= -2 \\ x + y - z - w &= -1 \\ 6x + z - 9w &= -2 \\ 5x - y + 2z - 8w &= 3. \end{aligned}$$

10. (a)
$$\begin{aligned} 2x - y + z &= 3 \\ x - 3y + z &= 4 \\ -5x - 2z &= -5. \end{aligned}$$

(b)
$$\begin{aligned} x + y + z + w &= 6 \\ 2x + y - z &= 3 \\ 3x + y + 2w &= 6. \end{aligned}$$

(c)
$$\begin{aligned} 2x - y + z &= 3 \\ 3x + y - 2z &= -2 \\ x - y + z &= 7 \\ x + 5y + 7z &= 13 \\ x - 7y - 5z &= 12. \end{aligned}$$

In Exercises 11 through 14 find all values of a for which the resulting linear system has (1) no solution, (2) a unique solution, and (3) infinitely many solutions.

11.
$$\begin{aligned} x + y - z &= 2 \\ x + 2y + z &= 3 \\ x + y + (a^2 - 5)z &= a. \end{aligned}$$

12.
$$\begin{aligned} x + y + z &= 2 \\ 2x + 3y + 2z &= 5 \\ 2x + 3y + (a^2 - 1)z &= a + 1. \end{aligned}$$

13.
$$\begin{aligned} x + y + z &= 2 \\ x + 2y + z &= 3 \\ x + y + (a^2 - 5)z &= a. \end{aligned}$$

14.
$$\begin{aligned} x + y &= 3 \\ x + (a^2 - 8)y &= a. \end{aligned}$$

In Exercises 15 through 18 solve the linear system with given augmented matrix.

15. (a) $\left[\begin{array}{ccc|c} 1 & 1 & 1 & 0 \\ 1 & 1 & 0 & 3 \\ 0 & 1 & 1 & 1 \end{array}\right]$. (b) $\left[\begin{array}{ccc|c} 1 & 2 & 3 & 0 \\ 1 & 1 & 1 & 0 \\ 1 & 1 & 2 & 0 \\ 1 & 3 & 3 & 0 \end{array}\right]$.

16. (a) $\left[\begin{array}{ccc|c} 1 & 2 & 3 & 0 \\ 1 & 1 & 1 & 0 \\ 5 & 7 & 9 & 0 \end{array}\right]$. (b) $\left[\begin{array}{ccc|c} 1 & 2 & 1 & 7 \\ 2 & 0 & 1 & 4 \\ 1 & 0 & 2 & 5 \\ 1 & 2 & 3 & 11 \\ 2 & 1 & 4 & 12 \end{array}\right]$.

17. (a) $\left[\begin{array}{cccc|c} 1 & 2 & 3 & 1 & 8 \\ 1 & 3 & 0 & 1 & 7 \\ 1 & 0 & 2 & 1 & 3 \end{array}\right]$. (b) $\left[\begin{array}{ccc|c} 1 & -2 & 3 & 4 \\ 2 & -1 & -3 & 5 \\ 3 & 0 & 1 & 2 \\ 3 & -3 & 0 & 7 \end{array}\right]$.

18. (a) $\left[\begin{array}{ccc|c} 4 & 2 & -1 & 5 \\ 3 & 3 & 6 & 1 \\ 5 & 1 & -8 & 8 \end{array}\right]$. (b) $\left[\begin{array}{cccc|c} 1 & 1 & 3 & -3 & 0 \\ 0 & 2 & 1 & -3 & 3 \\ 1 & 0 & 2 & -1 & -1 \end{array}\right]$.

Theoretical Exercises

T.1. Show that properties (a), (b), and (c) of the definition of the reduced row echelon form of a matrix **A** imply that all the elements in column j_i and rows $i + 1, i + 2, \ldots, m$ of **A** are zero.

T.2. Prove:
(a) Every matrix is row equivalent to itself.
(b) If **A** is row equivalent to **B**, then **B** is row equivalent to **A**.
(c) If **A** is row equivalent to **B** and **B** is row equivalent to **C**, then **A** is row equivalent to **C**.

T.3. Prove Corollary 1.1.

T.4. Prove that the linear system $\mathbf{AX} = \mathbf{B}$ has no solution if and only if its augmented matrix is row equivalent to a matrix in reduced row echelon form that has a row whose first n elements are zero and whose $(n + 1)$th element is 1.

T.5. Let

$$\mathbf{A} = \begin{bmatrix} a & b \\ c & d \end{bmatrix}.$$

Show that **A** is row equivalent to $\mathbf{I}_2$ if and only if $ad - bc \neq 0$.

T.6. Let

$$\mathbf{A} = \begin{bmatrix} a & b \\ c & d \end{bmatrix}.$$

Show that the homogeneous system $\mathbf{AX} = \mathbf{0}$ has only the trivial solution if and only if $ad - bc \neq 0$.

T.7. Let **A** be an $n \times n$ matrix in reduced row echelon form. Show that if $\mathbf{A} \neq \mathbf{I}_n$, then **A** has a row consisting entirely of zeros.

T.8. Show that the values of λ for which the homogeneous system

$$\begin{aligned} (a - \lambda)x + \qquad\quad dy &= 0 \\ cx + (b - \lambda)y &= 0 \end{aligned}$$

has a nontrivial solution satisfy the *equation* $(a - \lambda)(b - \lambda) - cd = 0$.

(*Hint:* See Exercise T.6.)

1.4 The Inverse of a Matrix

In this section we restrict our attention to square matrices and formulate the notion corresponding to the reciprocal of a nonzero real number.

DEFINITION An $n \times n$ matrix **A** is called **nonsingular** (or **invertible**) if there exists an $n \times n$ matrix **B** such that

$$\mathbf{AB} = \mathbf{BA} = \mathbf{I}_n.$$

The matrix **B** is called an **inverse** of **A**. If there exists no such matrix **B**, then **A** is called **singular** (or **noninvertible**).

Example 1. Let

$$\mathbf{A} = \begin{bmatrix} 2 & 3 \\ 2 & 2 \end{bmatrix} \quad \text{and} \quad \mathbf{B} = \begin{bmatrix} -1 & \frac{3}{2} \\ 1 & -1 \end{bmatrix}.$$

Since

$$\mathbf{AB} = \mathbf{BA} = \mathbf{I}_2,$$

we conclude that **B** is an inverse of **A** and that **A** is nonsingular.

Not every matrix has an inverse. For instance, consider the following example.

Example 2. Let

$$\mathbf{A} = \begin{bmatrix} 1 & 2 \\ 2 & 4 \end{bmatrix}.$$

Suppose that

$$\mathbf{B} = \begin{bmatrix} a & b \\ c & d \end{bmatrix}$$

is an inverse of **A**. Then

$$\mathbf{AB} = \begin{bmatrix} 1 & 2 \\ 2 & 4 \end{bmatrix}\begin{bmatrix} a & b \\ c & d \end{bmatrix} = \mathbf{I}_2 = \begin{bmatrix} 1 & 0 \\ 0 & 1 \end{bmatrix},$$

and so

$$\begin{bmatrix} a + 2c & b + 2d \\ 2a + 4c & 2b + 4d \end{bmatrix} = \begin{bmatrix} 1 & 0 \\ 0 & 1 \end{bmatrix}.$$

Equating corresponding elements of these two matrices, we obtain the linear system

$$\begin{aligned} a + 2c &= 1 \\ 2a + 4c &= 0, \end{aligned}$$

which must be solved for a and c. However, this linear system has no solution. Hence there is no such matrix $\mathbf{B}$ and $\mathbf{A}$ is a singular matrix.

THEOREM 1.8. *If a matrix has an inverse, then the inverse is unique.*

proof. Let $\mathbf{B}$ and $\mathbf{C}$ be inverses of $\mathbf{A}$. Then

$$\mathbf{BA} = \mathbf{AC} = \mathbf{I}_n.$$

Now

$$\mathbf{B} = \mathbf{BI}_n = \mathbf{B(AC)} = \mathbf{(BA)C} = \mathbf{I}_n\mathbf{C} = \mathbf{C},$$

which completes the proof.

We shall now write the inverse of $\mathbf{A}$, if it exists, as $\mathbf{A}^{-1}$. Thus

$$\mathbf{AA}^{-1} = \mathbf{A}^{-1}\mathbf{A} = \mathbf{I}_n.$$

Example 3. Let

$$\mathbf{A} = \begin{bmatrix} 1 & 2 \\ 3 & 4 \end{bmatrix}.$$

To find $\mathbf{A}^{-1}$, we let

$$\mathbf{A}^{-1} = \begin{bmatrix} a & b \\ c & d \end{bmatrix}.$$

We then have

$$\mathbf{AA}^{-1} = \begin{bmatrix} 1 & 2 \\ 3 & 4 \end{bmatrix}\begin{bmatrix} a & b \\ c & d \end{bmatrix} = \mathbf{I}_2 = \begin{bmatrix} 1 & 0 \\ 0 & 1 \end{bmatrix},$$

or

$$\begin{bmatrix} a + 2c & b + 2d \\ 3a + 4c & 3b + 4d \end{bmatrix} = \begin{bmatrix} 1 & 0 \\ 0 & 1 \end{bmatrix}.$$

Hence we solve the linear systems

$$\begin{aligned} a + 2c &= 1 \\ 3a + 4c &= 0 \end{aligned} \qquad \text{and} \qquad \begin{aligned} b + 2d &= 0 \\ 3b + 4d &= 1. \end{aligned}$$

The solutions are $a = -2$, $c = \frac{3}{2}$, $b = 1$, and $d = -\frac{1}{2}$. Thus $\mathbf{A}$ is nonsingular and

$$\mathbf{A}^{-1} = \begin{bmatrix} -2 & 1 \\ \frac{3}{2} & -\frac{1}{2} \end{bmatrix}.$$

The method used in Example 3 to find the inverse of a matrix is not a very efficient one. We shall soon modify it and thereby obtain a much faster method.

The following theorem deals with some useful properties of the inverse.

THEOREM 1.9

(a) *If* $\mathbf{A}$ *is a nonsingular matrix, then* $\mathbf{A}^{-1}$ *is nonsingular and*

$$(\mathbf{A}^{-1})^{-1} = \mathbf{A}.$$

(b) *If* $\mathbf{A}$ *and* $\mathbf{B}$ *are nonsingular matrices, then* $\mathbf{AB}$ *is nonsingular and*

$$(\mathbf{AB})^{-1} = \mathbf{B}^{-1}\mathbf{A}^{-1}.$$

(c) *If* $\mathbf{A}$ *is a nonsingular matrix, then*

$$(\mathbf{A}^T)^{-1} = (\mathbf{A}^{-1})^T.$$

proof

(a) $\mathbf{A}^{-1}$ is nonsingular if we can find a matrix $\mathbf{B}$ such that

$$\mathbf{A}^{-1}\mathbf{B} = \mathbf{B}\mathbf{A}^{-1} = \mathbf{I}_n.$$

Now

$$\mathbf{A}^{-1}\mathbf{A} = \mathbf{A}\mathbf{A}^{-1} = \mathbf{I}_n.$$

Thus $\mathbf{B} = \mathbf{A}$ is the inverse of $\mathbf{A}^{-1}$, and hence

$$(\mathbf{A}^{-1})^{-1} = \mathbf{A}.$$

(b) We have

$$(\mathbf{AB})(\mathbf{B}^{-1}\mathbf{A}^{-1}) = \mathbf{A}(\mathbf{B}\mathbf{B}^{-1})\mathbf{A}^{-1} = \mathbf{A}\mathbf{I}_n\mathbf{A}^{-1} = \mathbf{A}\mathbf{A}^{-1} = \mathbf{I}_n$$

and

$$(\mathbf{B}^{-1}\mathbf{A}^{-1})(\mathbf{AB}) = \mathbf{B}^{-1}(\mathbf{A}^{-1}\mathbf{A})\mathbf{B} = \mathbf{B}^{-1}\mathbf{I}_n\mathbf{B} = \mathbf{B}^{-1}\mathbf{B} = \mathbf{I}_n.$$

Therefore, **AB** is nonsingular. Since the inverse of a matrix is unique, we conclude that

$$(\mathbf{AB})^{-1} = \mathbf{B}^{-1}\mathbf{A}^{-1}.$$

(c) We have

$$\mathbf{AA}^{-1} = \mathbf{A}^{-1}\mathbf{A} = \mathbf{I}_n.$$

Taking transposes, we obtain

$$(\mathbf{AA}^{-1})^T = (\mathbf{A}^{-1}\mathbf{A})^T = \mathbf{I}_n^T = \mathbf{I}_n.$$

Then

$$(\mathbf{A}^{-1})^T\mathbf{A}^T = \mathbf{A}^T(\mathbf{A}^{-1})^T = \mathbf{I}_n.$$

This equation implies that

$$(\mathbf{A}^T)^{-1} = (\mathbf{A}^{-1})^T.$$

COROLLARY 1.2. *If* $\mathbf{A}_1, \mathbf{A}_2, \ldots, \mathbf{A}_r$ *are* $n \times n$ *nonsingular matrices, then* $\mathbf{A}_1\mathbf{A}_2 \cdots \mathbf{A}_r$ *is nonsingular and*

$$(\mathbf{A}_1\mathbf{A}_2 \cdots \mathbf{A}_r)^{-1} = \mathbf{A}_r^{-1}\mathbf{A}_{r-1}^{-1} \cdots \mathbf{A}_1^{-1}.$$

proof. Exercise T.1.

A Practical Method for Finding $\mathbf{A}^{-1}$

We shall now develop a practical method for finding $\mathbf{A}^{-1}$. If **A** is a given $n \times n$ matrix, we are looking for an $n \times n$ matrix $\mathbf{B} = [b_{ij}]$ such that

$$\mathbf{AB} = \mathbf{BA} = \mathbf{I}_n.$$

Let the columns of $\mathbf{B}$ be denoted by the $n \times 1$ matrices $\mathbf{X}_1, \mathbf{X}_2, \ldots, \mathbf{X}_n$, where

$$\mathbf{X}_j = \begin{bmatrix} b_{1j} \\ b_{2j} \\ \vdots \\ b_{ij} \\ \vdots \\ b_{nj} \end{bmatrix} \qquad (1 \leqslant j \leqslant n).$$

Let the columns of $\mathbf{I}_n$ be denoted by the $n \times 1$ matrices $\mathbf{E}_1, \mathbf{E}_2, \ldots, \mathbf{E}_n$. Thus

$$\mathbf{E}_j = \begin{bmatrix} 0 \\ 0 \\ \vdots \\ 1 \\ 0 \\ \vdots \\ 0 \end{bmatrix} \leftarrow j\text{th row}.$$

By Exercise T.21 of Section 1.2 the jth column of $\mathbf{AB}$ is the $n \times 1$ matrix $\mathbf{AX}_j$. Since equal matrices must agree column by column, it follows that the problem of finding an $n \times n$ matrix $\mathbf{B} = \mathbf{A}^{-1}$ such that

$$\mathbf{AB} = \mathbf{I}_n \tag{1}$$

is equivalent to the problem of finding n matrices (each $n \times 1$) $\mathbf{X}_1, \mathbf{X}_2, \ldots, \mathbf{X}_n$ such that

$$\mathbf{AX}_j = \mathbf{E}_j \qquad (1 \leqslant j \leqslant n). \tag{2}$$

Thus finding $\mathbf{B}$ is equivalent to solving n linear systems (each is n equations in n unknowns). This is precisely what we did in Example 3. Each of these systems can be solved by the Gauss–Jordan method. To solve the first linear system, we form the augmented matrix $[\mathbf{A} \mid \mathbf{E}_1]$ and put it into reduced row echelon form. We do the same with

$$[\mathbf{A} \mid \mathbf{E}_2], \ldots, [\mathbf{A} \mid \mathbf{E}_n].$$

However, if we observe that the coefficient matrix of each of these n linear systems is always $\mathbf{A}$, we can solve all these systems simultaneously. We form the $n \times 2n$ matrix

$$\left[\mathbf{A} \mid \mathbf{E}_1 \ \mathbf{E}_2 \ \ldots \ \mathbf{E}_n\right] = \left[\mathbf{A} \mid \mathbf{I}_n\right]$$

and transform it into reduced row echelon form $[\mathbf{C} \mid \mathbf{D}]$. The $n \times n$ matrix $\mathbf{C}$ is the reduced-row-echelon-form matrix that is row equivalent to $\mathbf{A}$. Let $\mathbf{D}_1, \mathbf{D}_2, \ldots, \mathbf{D}_n$ be the n columns of $\mathbf{D}$. Then the matrix $[\mathbf{C} \mid \mathbf{D}]$ gives rise to the n linear systems

$$\mathbf{C}\mathbf{X}_j = \mathbf{D}_j \qquad (1 \leqslant j \leqslant n) \tag{3}$$

or to the matrix equation

$$\mathbf{C}\mathbf{B} = \mathbf{D}. \tag{4}$$

There are now two possible cases.

CASE 1. $\mathbf{C} = \mathbf{I}_n$. Then Equation (3) becomes

$$\mathbf{I}_n\mathbf{X}_j = \mathbf{X}_j = \mathbf{D}_j,$$

and $\mathbf{B} = \mathbf{D}$, so we have obtained $\mathbf{A}^{-1}$.

CASE 2. $\mathbf{C} \neq \mathbf{I}_n$. It then follows from Exercise T.7 in Section 1.3 that $\mathbf{C}$ has a row consisting entirely of zeros. From Exercise T.15 in Section 1.2, we observe that the product $\mathbf{CB}$ in Equation (4) has a row of zeros. The matrix $\mathbf{D}$ in (4) arose from $\mathbf{I}_n$ by a sequence of elementary row operations, and it is intuitively clear that $\mathbf{D}$ cannot have a row of zeros. The statement that $\mathbf{D}$ cannot have a row of zeros can be rigorously established at this point, but we shall ask the reader to accept the argument now. In Section 2.2, an argument using determinants will show the validity of the result. This contradiction shows that $\mathbf{A}$ is a singular matrix.

The practical method for finding $\mathbf{A}^{-1}$ consists in forming the $n \times 2n$ matrix $[\mathbf{A} \mid \mathbf{I}_n]$ and performing elementary row operations that transform it to $[\mathbf{I}_n \mid \mathbf{A}^{-1}]$. Whatever we do to a row of $\mathbf{A}$, we also do to the corresponding row of $\mathbf{I}_n$.

Example 4. Let

$$\mathbf{A} = \begin{bmatrix} 1 & 1 & 1 \\ 0 & 2 & 3 \\ 5 & 5 & 1 \end{bmatrix}.$$

Then

$$\left[\mathbf{A} \mid \mathbf{I}_3\right] = \left[\begin{array}{ccc|ccc} 1 & 1 & 1 & 1 & 0 & 0 \\ 0 & 2 & 3 & 0 & 1 & 0 \\ 5 & 5 & 1 & 0 & 0 & 1 \end{array}\right].$$

We now arrange our computations as follows:

$$\begin{array}{ccc|ccc} & \mathbf{A} & & & \mathbf{I}_3 & \\ 1 & 1 & 1 & 1 & 0 & 0 \\ 0 & 2 & 3 & 0 & 1 & 0 \\ 5 & 5 & 1 & 0 & 0 & 1 \end{array}$$

Subtract 5 times the first row from the third row to obtain

$$\begin{array}{ccc|ccc} 1 & 1 & 1 & 1 & 0 & 0 \\ 0 & 2 & 3 & 0 & 1 & 0 \\ 0 & 0 & -4 & -5 & 0 & 1 \end{array}$$

Divide the second row by 2 to obtain

$$\begin{array}{ccc|ccc} 1 & 1 & 1 & 1 & 0 & 0 \\ 0 & 1 & \frac{3}{2} & 0 & \frac{1}{2} & 0 \\ 0 & 0 & -4 & -5 & 0 & 1 \end{array}$$

Subtract the second row from the first row to obtain

$$\begin{array}{ccc|ccc} 1 & 0 & -\frac{1}{2} & 1 & -\frac{1}{2} & 0 \\ 0 & 1 & \frac{3}{2} & 0 & \frac{1}{2} & 0 \\ 0 & 0 & -4 & -5 & 0 & 1 \end{array}$$

Divide the third row by -4 to obtain

$$\begin{array}{ccc|ccc} 1 & 0 & -\frac{1}{2} & 1 & -\frac{1}{2} & 0 \\ 0 & 1 & \frac{3}{2} & 0 & \frac{1}{2} & 0 \\ 0 & 0 & 1 & \frac{5}{4} & 0 & -\frac{1}{4} \end{array}$$

Add $-\frac{3}{2}$ times the third row to the second row to obtain

$$\begin{array}{ccc|ccc} 1 & 0 & -\frac{1}{2} & 1 & -\frac{1}{2} & 0 \\ 0 & 1 & 0 & -\frac{15}{8} & \frac{1}{2} & \frac{3}{8} \\ 0 & 0 & 1 & \frac{5}{4} & 0 & -\frac{1}{4} \end{array}$$

Add $\frac{1}{2}$ times the third row to the first row to obtain

$$\begin{array}{ccc|ccc} 1 & 0 & 0 & \frac{13}{8} & -\frac{1}{2} & -\frac{1}{8} \\ 0 & 1 & 0 & -\frac{15}{8} & \frac{1}{2} & \frac{3}{8} \\ 0 & 0 & 1 & \frac{5}{4} & 0 & -\frac{1}{4} \end{array}.$$

Hence

$$\mathbf{A}^{-1} = \begin{bmatrix} \frac{13}{8} & -\frac{1}{2} & -\frac{1}{8} \\ -\frac{15}{8} & \frac{1}{2} & \frac{3}{8} \\ \frac{5}{4} & 0 & -\frac{1}{4} \end{bmatrix}.$$

It is easy to verify that $\mathbf{AA}^{-1} = \mathbf{A}^{-1}\mathbf{A} = \mathbf{I}_3$.

If the reduced row echelon matrix under **A** has a row of zeros, then **A** is singular. Since each matrix under **A** is row equivalent to **A**, once a matrix under **A** has a row of zeros, every subsequent matrix that is row equivalent to **A** will have a row of zeros. Thus we can stop as soon as we obtain a matrix **F** that is row equivalent to **A** and has a row of zeros, thereby concluding that $\mathbf{A}^{-1}$ does not exist.

Example 5. Let

$$\mathbf{A} = \begin{bmatrix} 1 & 2 & -3 \\ 1 & -2 & 1 \\ 5 & -2 & -3 \end{bmatrix}.$$

To find $\mathbf{A}^{-1}$, we proceed as follows:

$$\begin{array}{rrr|rrr} & \mathbf{A} & & & \mathbf{I}_3 & \\ 1 & 2 & -3 & 1 & 0 & 0 \\ 1 & -2 & 1 & 0 & 1 & 0 \\ 5 & -2 & -3 & 0 & 0 & 1 \end{array}$$

Subtract the first row from the second row to obtain

$$\begin{array}{rrr|rrr} 1 & 2 & -3 & 1 & 0 & 0 \\ 0 & -4 & 4 & -1 & 1 & 0 \\ 5 & -2 & -3 & 0 & 0 & 1 \end{array}$$

Subtract 5 times the first row from the third row to obtain

$$\begin{array}{rrr|rrr} 1 & 2 & -3 & 1 & 0 & 0 \\ 0 & -4 & 4 & -1 & 1 & 0 \\ 0 & -12 & 12 & -5 & 0 & 1 \end{array}$$

Subtract 3 times the second row from the third row to obtain

$$\begin{array}{rrr|rrr} 1 & 2 & -3 & 1 & 0 & 0 \\ 0 & -4 & 4 & -1 & 1 & 0 \\ 0 & 0 & 0 & -2 & -3 & 1. \end{array}$$

At this point **A** is row equivalent to

$$\mathbf{F} = \begin{bmatrix} 1 & -2 & -3 \\ 0 & -4 & 4 \\ 0 & 0 & 0 \end{bmatrix}.$$

Since **F** has a row of zeros, we stop and conclude that **A** is a singular matrix.

Observe that to find $\mathbf{A}^{-1}$ we do not have to determine, in advance, whether or not it exists. We merely start the method given above and either obtain $\mathbf{A}^{-1}$ or find out that **A** is singular.

The above discussion for the practical method of obtaining $\mathbf{A}^{-1}$ has actually established the following theorem.

THEOREM 1.10. *An $n \times n$ matrix is nonsingular if and only if it is row equivalent to $\mathbf{I}_n$.*

Earlier, we defined a matrix **B** to be the inverse of **A** if $\mathbf{AB} = \mathbf{BA} = \mathbf{I}_n$. We now observe that one of these equations follows from the other in the following theorem, whose proof we omit.

THEOREM 1.11. *Suppose that* **A** *and* **B** *are $n \times n$ matrices.*

(a) *If*

$$\mathbf{AB} = \mathbf{I}_n,$$

then $\mathbf{BA} = \mathbf{I}_n$.

(b) *If*

$$\mathbf{BA} = \mathbf{I}_n,$$

then $\mathbf{AB} = \mathbf{I}_n$.

Linear Systems and Inverses

If **A** is an $n \times n$ matrix, then the linear system $\mathbf{AX} = \mathbf{B}$ is a system of n equations in n unknowns. Suppose that **A** is nonsingular. Then $\mathbf{A}^{-1}$ exists and we can multiply $\mathbf{AX} = \mathbf{B}$ by $\mathbf{A}^{-1}$ on both sides, obtaining

$$\mathbf{A}^{-1}(\mathbf{AX}) = \mathbf{A}^{-1}\mathbf{B},$$

or

$$\mathbf{I}_n\mathbf{X} = \mathbf{X} = \mathbf{A}^{-1}\mathbf{B}.$$

Moreover, $\mathbf{X} = \mathbf{A}^{-1}\mathbf{B}$ is clearly a solution to the given linear system. Thus, if **A** is nonsingular, we have a unique solution.

This observation is useful in industrial problems. Many physical models are described by linear systems. This means that if n values are used as inputs (which can be arranged as the $n \times 1$ matrix $\mathbf{X}$), then m values are obtained as outputs (which can be arranged as the $m \times 1$ matrix $\mathbf{B}$) by the rule $\mathbf{AX} = \mathbf{B}$. The matrix $\mathbf{A}$ is inherently tied to the process. Thus suppose that a chemical process has a certain matrix $\mathbf{A}$ associated with it. Any change in the process may result in a new matrix. The problem frequently encountered in systems analysis is that of determining the input to be used to obtain a desired output. That is, we want to solve the linear system $\mathbf{AX} = \mathbf{B}$ for $\mathbf{X}$ as we vary $\mathbf{B}$. If $\mathbf{A}$ is a nonsingular square matrix, an efficient way of handling this is as follows. Compute $\mathbf{A}^{-1}$ once; then whenever we change $\mathbf{B}$, we find the corresponding solution $\mathbf{X}$ by forming $\mathbf{A}^{-1}\mathbf{B}$.

Example 6. Consider the linear system

$$\begin{bmatrix} 1 & 1 & 1 \\ 0 & 2 & 3 \\ 5 & 5 & 1 \end{bmatrix} \begin{bmatrix} x_1 \\ x_2 \\ x_3 \end{bmatrix} = \begin{bmatrix} 2 \\ 3 \\ 4 \end{bmatrix}. \tag{5}$$

Writing (5) in matrix form as $\mathbf{AX} = \mathbf{B}$, we observe that $\mathbf{A}$ is the matrix of Example 4. Then

$$\mathbf{X} = \mathbf{A}^{-1}\mathbf{B} = \begin{bmatrix} \frac{13}{8} & -\frac{1}{2} & -\frac{1}{8} \\ -\frac{15}{8} & \frac{1}{2} & \frac{3}{8} \\ \frac{5}{4} & 0 & -\frac{1}{4} \end{bmatrix} \begin{bmatrix} 2 \\ 3 \\ 4 \end{bmatrix} = \begin{bmatrix} \frac{5}{4} \\ -\frac{3}{4} \\ \frac{3}{2} \end{bmatrix}.$$

On the other hand, the linear system with $\mathbf{B}$ replaced by

$$\begin{bmatrix} 5 \\ -2 \\ 3 \end{bmatrix}$$

has the solution

$$\mathbf{X} = \mathbf{A}^{-1} \begin{bmatrix} 5 \\ -2 \\ 3 \end{bmatrix} = \begin{bmatrix} \frac{35}{4} \\ -\frac{37}{4} \\ \frac{11}{2} \end{bmatrix}.$$

THEOREM 1.12. *If $\mathbf{A}$ is an $n \times n$ matrix, the homogeneous system*

$$\mathbf{AX} = \mathbf{0} \tag{6}$$

has a nontrivial solution if and only if $\mathbf{A}$ is singular.

proof. Suppose that $\mathbf{A}$ is nonsingular. Then $\mathbf{A}^{-1}$ exists, and multiplying (6) by $\mathbf{A}^{-1}$, we have

$$\mathbf{A}^{-1}(\mathbf{AX}) = \mathbf{A}^{-1}\mathbf{0} = \mathbf{0} = (\mathbf{A}^{-1}\mathbf{A})\mathbf{X} = \mathbf{I}_n\mathbf{X} = \mathbf{X}.$$

Hence the only solution to (6) is $\mathbf{X} = \mathbf{0}$.

We leave the proof of the converse—if $\mathbf{A}$ is singular, then (6) has a nontrivial solution—as an exercise (Exercise T.2).

Example 7. Consider the homogeneous system $\mathbf{AX} = \mathbf{0}$, where $\mathbf{A}$ is the matrix of Example 4. Since $\mathbf{A}$ is nonsingular,

$$\mathbf{X} = \mathbf{A}^{-1}\mathbf{0} = \mathbf{0}.$$

We could also solve the given system by Gauss–Jordan reduction. In this case we find that the matrix in reduced row echelon form that is row equivalent to the augmented matrix of the given system,

$$\left[\begin{array}{ccc|c} 1 & 1 & 1 & 0 \\ 0 & 2 & 3 & 0 \\ 5 & 5 & 1 & 0 \end{array}\right],$$

is

$$\left[\begin{array}{ccc|c} 1 & 0 & 0 & 0 \\ 0 & 1 & 0 & 0 \\ 0 & 0 & 1 & 0 \end{array}\right].$$

which again shows that the solution is

$$\mathbf{X} = \mathbf{0}.$$

Example 8. Consider the homogeneous system $\mathbf{AX} = \mathbf{0}$, where $\mathbf{A}$ is the matrix of Example 5. In this case the matrix in reduced row echelon form that is row equivalent to the augmented matrix of the given system,

$$\left[\begin{array}{ccc|c} 1 & 2 & -3 & 0 \\ 1 & -2 & 1 & 0 \\ 5 & -2 & -3 & 0 \end{array}\right],$$

is

$$\left[\begin{array}{ccc|c} 1 & 0 & -1 & 0 \\ 0 & 1 & -1 & 0 \\ 0 & 0 & 0 & 0 \end{array}\right],$$

which implies that

$$x = r$$
$$y = r$$
$$z = r,$$

where r is any real number. Thus the given system has nontrivial solutions.

1.4 Exercises

In Exercises 1 through 4 use the method of Examples 2 and 3.

1. Show that

$$\begin{bmatrix} 2 & 1 \\ -2 & 3 \end{bmatrix}$$

is nonsingular.

2. Show that

$$\begin{bmatrix} 2 & 1 \\ -4 & -2 \end{bmatrix}$$

is singular.

3. Is the matrix

$$\begin{bmatrix} 1 & 1 \\ 3 & 4 \end{bmatrix}$$

singular or nonsingular? If it is nonsingular, find its inverse.

4. Is the matrix

$$\begin{bmatrix} 1 & 2 & -1 \\ 3 & 2 & 3 \\ 2 & 2 & 1 \end{bmatrix}$$

singular or nonsingular? If it is nonsingular, find its inverse.

In Exercises 5 through 10 find the inverses of the given matrices, if possible:

5. (a) $\begin{bmatrix} 1 & 3 \\ -2 & 6 \end{bmatrix}$. (b) $\begin{bmatrix} 1 & 2 & 3 \\ 1 & 1 & 2 \\ 0 & 1 & 2 \end{bmatrix}$. (c) $\begin{bmatrix} 1 & 1 & 1 & 1 \\ 1 & 2 & -1 & 2 \\ 1 & -1 & 2 & 1 \\ 1 & 3 & 3 & 2 \end{bmatrix}$.

6. (a) $\begin{bmatrix} 1 & 3 \\ 2 & 6 \end{bmatrix}$. (b) $\begin{bmatrix} 1 & 2 & 3 \\ 0 & 2 & 3 \\ 1 & 2 & 4 \end{bmatrix}$. (c) $\begin{bmatrix} 1 & 1 & 2 & 1 \\ 0 & -2 & 0 & 0 \\ 1 & 2 & 1 & -2 \\ 0 & 3 & 2 & 1 \end{bmatrix}$.

7. (a) $\begin{bmatrix} 1 & 3 \\ 2 & 4 \end{bmatrix}$. (b) $\begin{bmatrix} 1 & 1 & 1 & 1 \\ 1 & 3 & 1 & 2 \\ 1 & 2 & -1 & 1 \\ 5 & 9 & 1 & 6 \end{bmatrix}$. (c) $\begin{bmatrix} 1 & 2 & 1 \\ 1 & 3 & 2 \\ 1 & 0 & 1 \end{bmatrix}$.

8. (a) $\begin{bmatrix} 1 & 1 & 1 \\ 1 & 2 & 3 \\ 0 & 1 & 1 \end{bmatrix}$. (b) $\begin{bmatrix} 1 & 2 & 2 \\ 1 & 3 & 1 \\ 1 & 3 & 2 \end{bmatrix}$. (c) $\begin{bmatrix} 1 & 2 & 3 \\ 1 & 1 & 2 \\ 0 & 1 & 1 \end{bmatrix}$.

9. (a) $\begin{bmatrix} 1 & 2 & -3 & 1 \\ -1 & 3 & -3 & -2 \\ 2 & 0 & 1 & 5 \\ 3 & 1 & -2 & 5 \end{bmatrix}$. (b) $\begin{bmatrix} 3 & 1 & 2 \\ 2 & 1 & 2 \\ 1 & 2 & 2 \end{bmatrix}$. (c) $\begin{bmatrix} 1 & 2 & 3 \\ 1 & 1 & 2 \\ 1 & 1 & 0 \end{bmatrix}$.

10. (a) $\begin{bmatrix} 2 & 1 & 3 \\ 0 & 1 & 2 \\ 1 & 0 & 3 \end{bmatrix}$. (b) $\begin{bmatrix} 1 & -1 & 2 & 3 \\ 4 & 1 & 2 & 0 \\ 2 & -1 & 3 & 1 \\ 4 & 2 & 1 & -5 \end{bmatrix}$. (c) $\begin{bmatrix} 2 & 1 & -2 \\ 3 & 4 & 6 \\ 7 & 6 & 2 \end{bmatrix}$.

11. Which of the following linear systems have a nontrivial solution?

(a)
$$\begin{aligned} x + 2y + 3z &= 0 \\ 2y + 2z &= 0 \\ x + 2y + 3z &= 0. \end{aligned}$$

(b)
$$\begin{aligned} 2x + y - z &= 0 \\ x - 2y - 3z &= 0 \\ -3x - y + 2z &= 0. \end{aligned}$$

12. Which of the following linear systems have a nontrivial solution?

(a)
$$\begin{aligned} x + y + 2z &= 0 \\ 2x + y + z &= 0 \\ 3x - y + z &= 0. \end{aligned}$$

(b)
$$\begin{aligned} x - y + z &= 0 \\ 2x + y &= 0 \\ 2x - 2y + 2z &= 0. \end{aligned}$$

(c)
$$\begin{aligned} 2x - y + 5z &= 0 \\ 3x + 2y - 3z &= 0 \\ x - y + 4z &= 0. \end{aligned}$$

13. If

$$\mathbf{A}^{-1} = \begin{bmatrix} 2 & 3 \\ 1 & 4 \end{bmatrix},$$

find **A**.

14. If

$$\mathbf{A}^{-1} = \begin{bmatrix} 3 & 4 \\ -1 & -1 \end{bmatrix},$$

find **A**.

Theoretical Exercises

T.1. Prove Corollary 1.2.

T.2. Let **A** be an $n \times n$ matrix. Prove that if **A** is singular, then the homogeneous system $\mathbf{AX} = \mathbf{0}$ has a nontrivial solution.

T.3. Prove that the matrix

$$\mathbf{A} = \begin{bmatrix} a & b \\ c & d \end{bmatrix}$$

is nonsingular if and only if $ad - bc \neq 0$. If this condition holds, show that

$$\mathbf{A}^{-1} = \begin{bmatrix} \dfrac{d}{ad - bc} & \dfrac{-b}{ad - bc} \\ \dfrac{-c}{ad - bc} & \dfrac{a}{ad - bc} \end{bmatrix}.$$

2 Determinants

2.1 Definition and Properties

In this section we define the notion of a determinant and study some of its properties. Determinants arose first in the solution of linear systems. Although the method given in Chapter 1 for solving such systems is much more efficient than those involving determinants, determinants are useful in other aspects of linear algebra; some of these areas will be considered in Chapter 5. First, we deal rather briefly with permutations, which are used in our definition of determinant. In this chapter, by matrix we mean a square matrix, and every matrix considered in the chapter is a square one.

DEFINITION Let $S = \{1, 2, \ldots, n\}$ be the set of integers from 1 to n, arranged in ascending order. A rearrangement $j_1 j_2 \cdots j_n$ of the elements of S is called a **permutation** of S.

We can put any one of the n elements of S in first position, any one of the remaining $n - 1$ elements in second position, any one of the remaining

$n - 2$ elements in third position, and so on, until the nth position can only be filled by the last remaining element. Thus there are

$$n(n-1)(n-2) \cdot \cdots \cdot 2 \cdot 1 \tag{1}$$

permutations of S; we denote the set of all permutations of S by S_n.

The expression in Equation (1) is denoted by

$$n!, \qquad \textbf{\textit{n} factorial}.$$

We have

$$\begin{aligned}
1! &= 1 \\
2! &= 2 \cdot 1 = 2 \\
3! &= 3 \cdot 2 \cdot 1 = 6 \\
4! &= 4 \cdot 3 \cdot 2 \cdot 1 = 24 \\
5! &= 5 \cdot 4 \cdot 3 \cdot 2 \cdot 1 = 120 \\
6! &= 6 \cdot 5 \cdot 4 \cdot 3 \cdot 2 \cdot 1 = 720 \\
7! &= 7 \cdot 6 \cdot 5 \cdot 4 \cdot 3 \cdot 2 \cdot 1 = 5040 \\
8! &= 8 \cdot 7 \cdot 6 \cdot 5 \cdot 4 \cdot 3 \cdot 2 \cdot 1 = 40{,}320 \\
9! &= 9 \cdot 8 \cdot 7 \cdot 6 \cdot 5 \cdot 4 \cdot 3 \cdot 2 \cdot 1 = 362{,}880.
\end{aligned}$$

Example 1. S_1 consists of only $1! = 1$ permutation of the set $\{1\}$, namely 1; S_2 consists of $2! = 2$ permutations of the set $\{1, 2\}$, namely, 12 and 21; and S_3 consists of $3! = 3 \cdot 2 \cdot 1 = 6$ permutations of the set $\{1, 2, 3\}$, namely, 123, 231, 312, 132, 213, and 321.

A permutation $j_1 j_2 \cdots j_n$ of $S = \{1, 2, \ldots, n\}$ is said to have an **inversion** if a larger integer j_s precedes a smaller one j_r. A permutation is called **even** or **odd** according to whether the total number of inversions in it is even or odd.

If $n \geqslant 2$, S_n has $n!/2$ even permutations and an equal number of odd permutations.

Example 2. In S_2, the permutation 12 is even, since it has no inversions; the permutation 21 is odd, since it has one inversion.

Example 3. The even permutations in S_3 are: 123 (no inversions), 231 (two inversions: 21 and 31); 312 (two inversions: 31 and 32). The odd permutations in S_3 are: 132 (one inversion: 32); 213 (one inversion: 21); 321 (three inversions: 32, 31, and 21).

DEFINITION Let $\mathbf{A} = [a_{ij}]$ be an $n \times n$ matrix. We define the **determinant** of **A** (written $|\mathbf{A}|$) by

$$|\mathbf{A}| = \sum (\pm) a_{1j_1} a_{2j_2} \cdots a_{nj_n}, \tag{2}$$

where the summation ranges over all permutations $j_1 j_2 \cdots j_n$ of the set $S = \{1, 2, \ldots, n\}$. The sign is taken as + or − according to whether the permutation $j_1 j_2 \cdots j_n$ is even or odd.

In each term $(\pm) a_{1j_1} a_{2j_2} \cdots a_{nj_n}$ of $|\mathbf{A}|$, the row subscripts are in their natural order, whereas the column subscripts are in the order $j_1 j_2 \cdots j_n$. Since the permutation $j_1 j_2 \cdots j_n$ is merely a rearrangement of the numbers from 1 to n, it has no repeats. Thus each term in $|\mathbf{A}|$ is a product of n elements of **A** each with its appropriate sign, with exactly one element from each row and exactly one element from each column. Since we sum over all the permutations of the set, $S = \{1, 2, \ldots, n\}$, $|\mathbf{A}|$ has $n!$ terms in the sum.

Example 4. If $A = [a_{11}]$ is a 1×1 matrix, then S_1 has only one permutation in it, the identity permutation 1, which is even. Thus $|\mathbf{A}| = a_{11}$.

Example 5. If

$$\mathbf{A} = \begin{bmatrix} a_{11} & a_{12} \\ a_{21} & a_{22} \end{bmatrix}$$

is a 2×2 matrix, then to obtain $|\mathbf{A}|$ we write down the terms

$$a_{1-}\, a_{2-} \qquad \text{and} \qquad a_{1-}\, a_{2-},$$

and fill in the blanks with all possible elements of S_2; these are 12 and 21. Since 12 is an even permutation, the term $a_{11}a_{22}$ has a + sign associated with it; since 21 is an odd permutation, the term $a_{12}a_{21}$ has a − sign associated with it. Hence

$$|\mathbf{A}| = a_{11}a_{22} - a_{12}a_{21}.$$

We can also obtain $|\mathbf{A}|$ by forming the product of the entries on the line from left to right and subtracting from this number the product of the entries on the line from right to left.

$$\begin{array}{cc} a_{11} & a_{12} \\ a_{21} & a_{22} \end{array}.$$

Example 6. If

$$\mathbf{A} = \begin{bmatrix} a_{11} & a_{12} & a_{13} \\ a_{21} & a_{22} & a_{23} \\ a_{31} & a_{32} & a_{33} \end{bmatrix},$$

then, to compute $|\mathbf{A}|$, we write down the six terms

$$a_{1\_}\, a_{2\_}\, a_{3\_}, \qquad a_{1\_}\, a_{2\_}\, a_{3\_}, \qquad a_{1\_}\, a_{2\_}\, a_{3\_}, \qquad a_{1\_}\, a_{2\_}\, a_{3\_},$$
$$a_{1\_}\, a_{2\_}\, a_{3\_}, \qquad \text{and} \qquad a_{1\_}\, a_{2\_}\, a_{3\_}.$$

All the elements of S_3 are used to fill in the blanks, and if we prefix each term by + or − according to whether the permutation used is even or odd, we find that

$$\begin{aligned} |\mathbf{A}| = a_{11}a_{22}a_{33} + a_{12}a_{23}a_{31} + a_{13}a_{21}a_{32} - a_{11}a_{23}a_{32} \\ - a_{12}a_{21}a_{33} - a_{13}a_{22}a_{31}. \end{aligned} \tag{3}$$

We can also obtain $|\mathbf{A}|$ as follows. Repeat the first and second columns of $\mathbf{A}$ as shown below. Form the sum of the products of the entries on the lines from left to right, and subtract from this number the products of the entries on the lines from right to left (verify).

$$\begin{array}{ccccc} a_{11} & a_{12} & a_{13} & a_{11} & a_{12} \\ a_{21} & a_{22} & a_{23} & a_{21} & a_{22} \\ a_{31} & a_{32} & a_{33} & a_{31} & a_{32}. \end{array}$$

It should be emphasized that, for $n \geqslant 4$, there is no "easy" method for evaluating $|A|$ as in Examples 5 and 6.

Example 7. Let

$$\mathbf{A} = \begin{bmatrix} 1 & 2 & 3 \\ 2 & 1 & 3 \\ 3 & 1 & 2 \end{bmatrix}.$$

Substituting in (3), we find that

$$|\mathbf{A}| = (1)(1)(2) + (2)(3)(3) + (3)(2)(1)$$
$$- (1)(3)(1) - (2)(2)(2) - (3)(1)(3) = 6.$$

We could obtain the same result by using the illustration of lines above (verify).

It may already have struck the reader that this is an extremely tedious way of computing determinants. In fact, $10! = 3.6288 \times 10^6$ and $20! = 2.4329 \times 10^{18}$, enormous numbers. We shall soon develop a number of properties satisfied by determinants, which will greatly reduce the computational effort.

Permutations are studied to some depth in abstract algebra courses or in courses dealing with group theory. We shall not make use of permutations in our methods for computing determinants. We require the following property of permutations. If we interchange two numbers in the permutation $j_1 j_2 \cdots j_n$, then the number of inversions is either increased or decreased by an odd number (Exercise T.1).

Example 8. The number of inversions in the permutation 54132 is 8. The number of inversions in 52134 is 5. The permutation 52134 was obtained from 54132 by interchanging 2 and 4. The number of inversions differs by 3, an odd number.

Properties of Determinants

THEOREM 2.1. *The determinants of a matrix and its transpose are equal.*

proof. Let $\mathbf{A} = [a_{ij}]$ and let $\mathbf{A}^T = [b_{ij}]$, where $b_{ij} = a_{ji}$ $(1 \leqslant i \leqslant n, 1 \leqslant j \leqslant n)$. Then from (2) we have

$$|\mathbf{A}^T| = \sum (\pm) b_{1j_1} b_{2j_2} \cdots b_{nj_n} = \sum (\pm) a_{j_1 1} a_{j_2 2} \cdots a_{j_n n}. \tag{4}$$

We can now rearrange the factors in the term $a_{j_1 1} a_{j_2 2} \cdots a_{j_n n}$ so that the row indices are in their natural order. Thus

$$b_{1j_1} b_{2j_2} \cdots b_{nj_n} = a_{j_1 1} a_{j_2 2} \cdots a_{j_n n} = a_{1k_1} a_{2k_2} \cdots a_{nk_n}.$$

We shall state, without proof, that the permutations $k_1 k_2 \cdots k_n$, which determines the sign associated with $a_{1k_1}a_{2k_2}\cdots a_{nk_n}$, and $j_1j_2\cdots j_n$, which determines the sign associated with $a_{1j_1}a_{2j_2}\cdots a_{nj_n}$, are both even or both odd. As an example,

$$b_{13}b_{24}b_{35}b_{41}b_{52} = a_{31}a_{42}a_{53}a_{14}a_{25} = a_{14}a_{25}a_{31}a_{42}a_{53};$$

the number of inversions in the permutation 45123 is 6 and the number of inversions in the permutation 34512 is also 6. Since the terms and corresponding signs in (2) and (4) agree, we conclude that $|\mathbf{A}| = |\mathbf{A}^T|$.

Example 9. Let $\mathbf{A}$ be the matrix of Example 7. Then

$$\mathbf{A}^T = \begin{bmatrix} 1 & 2 & 3 \\ 2 & 1 & 1 \\ 3 & 3 & 2 \end{bmatrix}.$$

Substituting in (3), we find that

$$\begin{aligned} |\mathbf{A}^T| &= (1)(1)(2) + (2)(1)(3) + (3)(2)(3) - (1)(1)(3) \\ &\quad - (2)(2)(2) - (3)(1)(3) \\ &= 6 = |\mathbf{A}|. \end{aligned}$$

Theorem 2.1 will enable us to replace "row" by "column" in many of the additional properties of determinants; we see how to do this in the following theorem.

THEOREM 2.2. *If matrix* $\mathbf{B}$ *results from matrix* $\mathbf{A}$ *by interchanging two rows (columns) of* $\mathbf{A}$, *then* $|\mathbf{B}| = -|\mathbf{A}|$.

proof. Suppose that $\mathbf{B}$ arises from $\mathbf{A}$ by interchanging rows r and s of $\mathbf{A}$, say $r < s$. Then we have $b_{rj} = a_{sj}$, $b_{sj} = a_{rj}$, and $b_{ij} = a_{ij}$ for $i \neq r$, $i \neq s$. Now

$$\begin{aligned} |\mathbf{B}| &= \sum (\pm) b_{1j_1}b_{2j_2}\cdots b_{rj_r}\cdots b_{sj_s}\cdots b_{nj_n} \\ &= \sum (\pm) a_{1j_1}a_{2j_2}\cdots a_{sj_r}\cdots a_{rj_s}\cdots a_{nj_n} \\ &= \sum (\pm) a_{1j_1}a_{2j_2}\cdots a_{rj_s}\cdots a_{sj_r}\cdots a_{nj_n}. \end{aligned}$$

The permutation $j_1j_2\cdots j_s\cdots j_r\cdots j_n$ results from the permutation

$j_1 j_2 \cdots j_r \cdots j_s \cdots j_n$ by an interchange of two numbers; the number of inversions in the former differs by an odd number from the number of inversions in the latter. This means that the sign of each term in $|\mathbf{B}|$ is the negative of the corresponding term in $|\mathbf{A}|$. Hence $|\mathbf{B}| = -|\mathbf{A}|$.

Now suppose that $\mathbf{B}$ is obtained from $\mathbf{A}$ by interchanging two columns of $\mathbf{A}$. Then $\mathbf{B}^T$ is obtained from $\mathbf{A}^T$ by interchanging two rows of $\mathbf{A}^T$. So $|\mathbf{B}^T| = -|\mathbf{A}^T|$, but $|\mathbf{B}^T| = |\mathbf{B}|$ and $|\mathbf{A}^T| = |\mathbf{A}|$. Hence $|\mathbf{B}| = -|\mathbf{A}|$.

In the results to follow, proofs will be given only for the rows of $\mathbf{A}$; the proofs for the corresponding column case proceed as at the end of the proof of Theorem 2.2.

Example 10. We have

$$\begin{vmatrix} 2 & -1 \\ 3 & 2 \end{vmatrix} = 7 \quad \text{and} \quad \begin{vmatrix} 3 & 2 \\ 2 & -1 \end{vmatrix} = -7.$$

THEOREM 2.3. *If two rows (columns) of* $\mathbf{A}$ *are equal, then* $|\mathbf{A}| = 0$.

proof. Suppose that rows r and s of $\mathbf{A}$ are equal. Interchange rows r and s of $\mathbf{A}$ to obtain a matrix $\mathbf{B}$. Then $|\mathbf{B}| = -|\mathbf{A}|$. On the other hand, $\mathbf{B} = \mathbf{A}$, so $|\mathbf{B}| = |\mathbf{A}|$. Thus $|\mathbf{A}| = -|\mathbf{A}|$, and so $|\mathbf{A}| = 0$.

Example 11. Using Equation (3), it follows that

$$\begin{vmatrix} 1 & 2 & 3 \\ -1 & 0 & 7 \\ 1 & 2 & 3 \end{vmatrix} = 0.$$

THEOREM 2.4. *If a row (column) of* $\mathbf{A}$ *consists entirely of zeros, then* $|\mathbf{A}| = 0$.

proof. Let the rth row of $\mathbf{A}$ consist entirely of zeros. Since each term in the definition for the determinant of $\mathbf{A}$ contains a factor from the rth row, each term in $|\mathbf{A}|$ is zero. Hence $|\mathbf{A}| = 0$.

Example 12. Using Equation (3), it follows that

$$\begin{vmatrix} 1 & 2 & 3 \\ 4 & 5 & 6 \\ 0 & 0 & 0 \end{vmatrix} = 0.$$

THEOREM 2.5. *If* $\mathbf{B}$ *is obtained from* $\mathbf{A}$ *by multiplying a row (column) of* $\mathbf{A}$ *by a real number* c, *then* $|\mathbf{B}| = c|\mathbf{A}|$.

proof. Suppose that the rth row of $\mathbf{A} = [a_{ij}]$ is multiplied by c to obtain $\mathbf{B} = [b_{ij}]$. Then $b_{ij} = a_{ij}$ if $i \neq r$ and $b_{rj} = ca_{rj}$. We obtain $|\mathbf{B}|$ from Equation (2) as

$$\begin{aligned}
|\mathbf{B}| &= \sum (\pm) b_{1j_1} b_{2j_2} \cdots b_{rj_r} \cdots b_{nj_n} \\
&= \sum (\pm) a_{1j_1} a_{2j_2} \cdots (ca_{rj_r}) \cdots a_{nj_n} \\
&= c\left(\sum (\pm) a_{1j_1} a_{2j_2} \cdots a_{rj_r} \cdots a_{nj_n}\right) = c|\mathbf{A}|.
\end{aligned}$$

Example 13. We have

$$\begin{vmatrix} 2 & 6 \\ 3 & 5 \end{vmatrix} = 2\begin{vmatrix} 1 & 3 \\ 3 & 5 \end{vmatrix} = 2(-4) = -8.$$

We can use Theorem 2.5 to simplify the computation of $|\mathbf{A}|$ by factoring out common factors from rows and columns of $\mathbf{A}$.

Example 14. We have

$$\begin{vmatrix} 1 & 2 & 3 \\ 1 & 5 & 3 \\ 2 & 8 & 6 \end{vmatrix} = 2\begin{vmatrix} 1 & 2 & 3 \\ 1 & 5 & 3 \\ 1 & 4 & 3 \end{vmatrix} = (2)(3)\begin{vmatrix} 1 & 2 & 1 \\ 1 & 5 & 1 \\ 1 & 4 & 1 \end{vmatrix} = (2)(3)(0) = 0.$$

Here we first factored out 2 from the third row, then 3 from the third column, and then used Theorem 2.3, since the first and third columns are equal.

THEOREM 2.6. *If* $\mathbf{B} = [b_{ij}]$ *is obtained from* $\mathbf{A} = [a_{ij}]$ *by adding to each element of the* rth *row (column) of* $\mathbf{A}$ *a constant* c *times the corresponding element of its* sth *row (column)* $r \neq s$, *then* $|\mathbf{B}| = |\mathbf{A}|$.

proof. We prove the theorem for rows. We have $b_{ij} = a_{ij}$ for $i \neq r$, and $b_{rj} = a_{rj} + ca_{sj}$, $r \neq s$, say $r < s$. Then

$$
\begin{aligned}
|\mathbf{B}| &= \sum (\pm) b_{1j_1} b_{2j_2} \cdots b_{rj_r} \cdots b_{nj_n} \\
&= \sum (\pm) a_{1j_1} a_{2j_2} \cdots (a_{rj_r} + ca_{sj_r}) \cdots a_{sj_s} \cdots a_{nj_n} \\
&= \sum (\pm) a_{1j_1} a_{2j_2} \cdots a_{rj_r} \cdots a_{sj_s} \cdots a_{nj_n} \\
&\quad + \sum (\pm) a_{1j_1} a_{2j_2} \cdots (ca_{sj_r}) \cdots a_{sj_s} \cdots a_{nj_n}.
\end{aligned}
$$

The first sum in this last expression is $|\mathbf{A}|$; the second sum can be written as $c[\Sigma(\pm) a_{1j_1} a_{2j_2} \cdots a_{sj_r} \cdots a_{sj_s} \cdots a_{nj_n}]$. Note that

$$
\sum (\pm) a_{1j_1} a_{2j_2} \cdots a_{sj_r} \cdots a_{sj_s} \cdots a_{nj_n}
= \begin{vmatrix}
a_{11} & a_{12} & \cdots & a_{1n} \\
a_{21} & a_{22} & \cdots & a_{2n} \\
 & \vdots & & \\
a_{s1} & a_{s2} & \cdots & a_{sn} \\
 & \vdots & & \\
a_{s1} & a_{s2} & \cdots & a_{sn} \\
 & \vdots & & \\
a_{n1} & a_{n2} & \cdots & a_{nn}
\end{vmatrix}
\begin{matrix} \\ \\ \\ \cdots r\text{th row} \\ \\ \cdots s\text{th row} \\ \\ \\ \end{matrix}
= 0,
$$

because there are two equal rows.

Hence $|\mathbf{B}| = |\mathbf{A}| + 0 = |\mathbf{A}|$.

Example 15. We have

$$
\begin{vmatrix}
1 & 2 & 3 \\
2 & -1 & 3 \\
1 & 0 & 1
\end{vmatrix}
=
\begin{vmatrix}
5 & 0 & 9 \\
2 & -1 & 3 \\
1 & 0 & 1
\end{vmatrix}
= 4,
$$

obtained by adding twice the second row to its first row.

THEOREM 2.7. *If a matrix* $\mathbf{A} = [a_{ij}]$ *is upper (lower) triangular (see Exercise T.18, Section 1.2), then*

$$|\mathbf{A}| = a_{11}a_{22} \cdots a_{nn};$$

that is, the determinant of a triangular matrix is the product of the elements on the main diagonal.

proof. Let $\mathbf{A} = [a_{ij}]$ be upper triangular (that is, $a_{ij} = 0$ for $i > j$). Then a term $a_{1j_1}a_{2j_2} \cdots a_{nj_n}$ in the expression for $|\mathbf{A}|$ can be nonzero only for $1 \leqslant j_1, 2 \leqslant j_2, \ldots, n \leqslant j_n$. Now $j_1 j_2 \cdots j_n$ must be a permutation, or rearrangement, of $\{1, 2, \ldots, n\}$. Hence we must have $j_1 = 1, j_2 = 2, \ldots, j_n = n$. Thus the only term of $|\mathbf{A}|$ that can be nonzero is the product of the elements on the main diagonal of $\mathbf{A}$. Hence $|\mathbf{A}| = a_{11}a_{22} \cdots a_{nn}$.

We leave the proof of the lower triangular case to the reader (Exercise T.2).

Theorems 2.2, 2.5, 2.6, and 2.7 are very useful in the evaluation of $|\mathbf{A}|$. What we do is transform $\mathbf{A}$ by means of our elementary row operations to a triangular matrix. Of course, we must keep track of how the determinant of the resulting matrices changes as we perform the elementary row operations.

Example 16. We have

$$\begin{vmatrix} 4 & 3 & 2 \\ 3 & -2 & 5 \\ 2 & 4 & 6 \end{vmatrix}_{\frac{1}{2}\mathbf{r}_3 \to \mathbf{r}_3} = 2\begin{vmatrix} 4 & 3 & 2 \\ 3 & -2 & 5 \\ 1 & 2 & 3 \end{vmatrix}_{\mathbf{r}_1 \leftrightarrow \mathbf{r}_3} = -2\begin{vmatrix} 1 & 2 & 3 \\ 3 & -2 & 5 \\ 4 & 3 & 2 \end{vmatrix}_{\mathbf{r}_2 - 3\mathbf{r}_1 \to \mathbf{r}_2}$$

$$= -2\begin{vmatrix} 1 & 2 & 3 \\ 0 & -8 & -4 \\ 4 & 3 & 2 \end{vmatrix}_{\mathbf{r}_3 - 4\mathbf{r}_1 \to \mathbf{r}_3} = -2\begin{vmatrix} 1 & 2 & 3 \\ 0 & -8 & -4 \\ 0 & -5 & -10 \end{vmatrix}_{\frac{1}{4}\mathbf{r}_2 \to \mathbf{r}_2}$$

$$= (-2)(4)\begin{vmatrix} 1 & 2 & 3 \\ 0 & -2 & -1 \\ 0 & -5 & -10 \end{vmatrix}_{\frac{1}{5}\mathbf{r}_3 \to \mathbf{r}_3} = (-2)(4)(5)\begin{vmatrix} 1 & 2 & 3 \\ 0 & -2 & -1 \\ 0 & -1 & -2 \end{vmatrix}_{\mathbf{r}_3 - \frac{1}{2}\mathbf{r}_2 \to \mathbf{r}_3}$$

$$= (-2)(4)(5)\begin{vmatrix} 1 & 2 & 3 \\ 0 & -2 & -1 \\ 0 & 0 & -\frac{3}{2} \end{vmatrix} = (-2)(4)(5)(1)(-2)(-\tfrac{3}{2}) = -120.$$

Here $\frac{1}{2}\mathbf{r}_3 \to \mathbf{r}_3$ means that $\frac{1}{2}$ times the third row $\mathbf{r}_3$ replaces the third row; $\mathbf{r}_1 \leftrightarrow \mathbf{r}_3$ means that the first row $\mathbf{r}_1$ and the third rows are interchanged; $\mathbf{r}_2 - 3\mathbf{r}_1 \to \mathbf{r}_2$ means that the second row $\mathbf{r}_2$ minus three times the first row replaces the second row.

We shall omit the proof of the following theorem.

THEOREM 2.8. *The determinant of a product of two matrices is the product of their determinants; that is,*

$$|\mathbf{AB}| = |\mathbf{A}|\,|\mathbf{B}|.$$

Example 17. Let

$$\mathbf{A} = \begin{bmatrix} 1 & 2 \\ 3 & 4 \end{bmatrix} \quad \text{and} \quad \mathbf{B} = \begin{bmatrix} 2 & -1 \\ 1 & 2 \end{bmatrix}.$$

Then

$$|\mathbf{A}| = -2 \quad \text{and} \quad |\mathbf{B}| = 5.$$

Also,

$$\mathbf{AB} = \begin{bmatrix} 4 & 3 \\ 10 & 5 \end{bmatrix}$$

and

$$|\mathbf{AB}| = -10 = |\mathbf{A}|\,|\mathbf{B}|.$$

COROLLARY 2.1. *If* $\mathbf{A}$ *is nonsingular, then* $|\mathbf{A}| \neq 0$ *and*

$$|\mathbf{A}^{-1}| = \frac{1}{|\mathbf{A}|}.$$

proof. Exercise T.4.

Example 18. Let

$$\mathbf{A} = \begin{bmatrix} 1 & 2 \\ 3 & 4 \end{bmatrix}.$$

Then $|\mathbf{A}| = -2$ and

$$\mathbf{A}^{-1} = \begin{bmatrix} -2 & 1 \\ \frac{3}{2} & -\frac{1}{2} \end{bmatrix}.$$

Now

$$|\mathbf{A}^{-1}| = -\frac{1}{2} = \frac{1}{|\mathbf{A}|}.$$

2.1 Exercises

1. Find the number of inversions in each of the following permutations of $S = \{1, 2, 3, 4, 5\}$.
(a) 52134. (b) 45213. (c) 42135.
(d) 13542. (e) 35241. (f) 12345.

2. Determine whether each of the following permutations of $S = \{1, 2, 3, 4\}$ is even or odd.
(a) 4213. (b) 1243. (c) 1234.
(d) 3214. (e) 1423. (f) 2431.

3. Determine the sign associated with each of the following permutations of $S = \{1, 2, 3, 4, 5\}$.
(a) 25431. (b) 31245. (c) 21345.
(d) 52341. (e) 34125. (f) 41253.

4. In each of the following pairs of permutations of $S = \{1, 2, 3, 4, 5, 6\}$, verify that the number of inversions differ by an odd number.
(a) 436215 and 416235. (b) 623415 and 523416.
(c) 321564 and 341562. (d) 123564 and 423561.

In Exercises 5 and 6 compute the determinants using Equation (2).

5. (a) $\begin{vmatrix} 2 & -1 \\ 3 & 2 \end{vmatrix}$. (b) $\begin{vmatrix} 4 & 2 & 0 \\ 0 & -2 & 5 \\ 0 & 0 & 3 \end{vmatrix}$. (c) $\begin{vmatrix} 4 & 2 & 2 & 0 \\ 2 & 0 & 0 & 0 \\ 3 & 0 & 0 & 1 \\ 0 & 0 & 1 & 0 \end{vmatrix}$.

6. (a) $\begin{vmatrix} 2 & 1 \\ 4 & 3 \end{vmatrix}$. (b) $\begin{vmatrix} 3 & 4 & 2 \\ 2 & 5 & 0 \\ 3 & 0 & 0 \end{vmatrix}$. (c) $\begin{vmatrix} -4 & 2 & 0 & 0 \\ 2 & 3 & 1 & 0 \\ 3 & 1 & 0 & 2 \\ 1 & 3 & 0 & 3 \end{vmatrix}$.

7. Let $\mathbf{A} = [a_{ij}]$ be a 4×4 matrix. Write the general expression for $|\mathbf{A}|$ using Equation (2).

8. If

$$|\mathbf{A}| = \begin{vmatrix} a_1 & a_2 & a_3 \\ b_1 & b_2 & b_3 \\ c_1 & c_2 & c_3 \end{vmatrix} = -4,$$

find the determinants of the following matrices:

$$\mathbf{B} = \begin{bmatrix} a_3 & a_2 & a_1 \\ b_3 & b_2 & b_1 \\ c_3 & c_2 & c_1 \end{bmatrix}, \qquad \mathbf{C} = \begin{bmatrix} a_1 & a_2 & a_3 \\ b_1 & b_2 & b_3 \\ 2c_1 & 2c_2 & 2c_3 \end{bmatrix},$$

and

$$\mathbf{D} = \begin{bmatrix} a_1 & a_2 & a_3 \\ b_1 + 4c_1 & b_2 + 4c_2 & b_3 + 4c_3 \\ c_1 & c_2 & c_3 \end{bmatrix}.$$

9. If

$$|\mathbf{A}| = \begin{vmatrix} a_1 & a_2 & a_3 \\ b_1 & b_2 & b_3 \\ c_1 & c_2 & c_3 \end{vmatrix} = 3,$$

find the determinants of the following matrices:

$$\mathbf{B} = \begin{bmatrix} a_1 + 2b_1 - 3c_1 & a_2 + 2b_2 - 3c_2 & a_3 + 2b_3 - 3c_3 \\ b_1 & b_2 & b_3 \\ c_1 & c_2 & c_3 \end{bmatrix},$$

$$\mathbf{C} = \begin{bmatrix} a_1 & 3a_2 & a_3 \\ b_1 & 3b_2 & b_3 \\ c_1 & 3c_2 & c_3 \end{bmatrix}, \quad \text{and} \quad \mathbf{D} = \begin{bmatrix} a_1 & a_2 & a_3 \\ c_1 & c_2 & c_3 \\ b_1 & b_2 & b_3 \end{bmatrix}.$$

10. If

$$\mathbf{A} = \begin{bmatrix} 1 & -1 & 2 \\ 3 & 4 & 1 \\ 2 & 5 & 1 \end{bmatrix},$$

verify that

$$|\mathbf{A}| = |\mathbf{A}^T|.$$

In Exercises 11 through 14 evaluate the given determinants.

11. (a) $\begin{vmatrix} 4 & -3 & 5 \\ 5 & 2 & 0 \\ 2 & 0 & 4 \end{vmatrix}$. (b) $\begin{vmatrix} 2 & 0 & 1 & 4 \\ 3 & 2 & -4 & -2 \\ 2 & 3 & -1 & 0 \\ 11 & 8 & -4 & 6 \end{vmatrix}$. (c) $\begin{vmatrix} 4 & 1 & 2 \\ 0 & 2 & 3 \\ 0 & 0 & -3 \end{vmatrix}$.

12. (a) $\begin{vmatrix} 4 & 0 & 0 & 0 \\ -1 & 2 & 0 & 0 \\ 1 & 2 & -3 & 0 \\ 1 & 5 & 3 & 5 \end{vmatrix}$. (b) $\begin{vmatrix} 4 & 1 & 3 \\ 2 & 3 & 0 \\ 1 & 3 & 2 \end{vmatrix}$. (c) $\begin{vmatrix} 1 & 2 & 3 \\ 2 & 1 & 0 \\ -3 & 1 & 2 \end{vmatrix}$.

13. (a) $\begin{vmatrix} 4 & 2 & 3 & -4 \\ 3 & -2 & 1 & 5 \\ -2 & 0 & 1 & -3 \\ 8 & -2 & 6 & 4 \end{vmatrix}$. (b) $\begin{vmatrix} 1 & 3 & -4 \\ -2 & 1 & 2 \\ -9 & 15 & 0 \end{vmatrix}$.

(c) $\begin{vmatrix} 1 & 1 & 2 \\ 0 & 2 & -2 \\ 0 & 0 & 3 \end{vmatrix}$.

14. (a) $\begin{vmatrix} 1 & 0 & 1 \\ 1 & 1 & 0 \\ 2 & 1 & 0 \end{vmatrix}$. (b) $\begin{vmatrix} 2 & 0 & 0 & 0 \\ -5 & 3 & 0 & 0 \\ 3 & 2 & 4 & 0 \\ 4 & 2 & 1 & -5 \end{vmatrix}$. (c) $\begin{vmatrix} 1 & 2 & -1 \\ 3 & 2 & 0 \\ 1 & 4 & 3 \end{vmatrix}$.

15. Verify that $|\mathbf{AB}| = |\mathbf{A}|\,|\mathbf{B}|$ for the following:

(a) $\mathbf{A} = \begin{bmatrix} 1 & -2 & 3 \\ -2 & 3 & 1 \\ 0 & 1 & 0 \end{bmatrix}$, $\mathbf{B} = \begin{bmatrix} 1 & 0 & 2 \\ 3 & -2 & 5 \\ 2 & 1 & 3 \end{bmatrix}$.

(b) $\mathbf{A} = \begin{bmatrix} 2 & 3 & 6 \\ 0 & 3 & 2 \\ 0 & 0 & -4 \end{bmatrix}$, $\mathbf{B} = \begin{bmatrix} 3 & 0 & 0 \\ 4 & 5 & 0 \\ 2 & 1 & -2 \end{bmatrix}$.

16. If $|\mathbf{A}| = -4$, find

(a) $|\mathbf{A}^2|$. (b) $|\mathbf{A}^4|$. (c) $|\mathbf{A}^{-1}|$.

Theoretical Exercises

T.1. Prove that if we interchange two numbers in the permutation $j_1 j_2 \cdots j_n$, then the number of inversions is either increased or decreased by an odd number.

(*Hint:* First show that if two adjacent numbers are interchanged, the number of inversions is either increased or decreased by 1. Then prove that an interchange of any two numbers can be achieved by an odd number of successive interchanges of adjacent numbers.)

T.2. Prove Theorem 2.7 for the lower triangular case.

T.3. Show that if c is a real number and $\mathbf{A}$ is $n \times n$, then $|c\mathbf{A}| = c^n|\mathbf{A}|$.

T.4. Prove Corollary 2.1.

T.5. Show that if $|\mathbf{AB}| = 0$, then $|\mathbf{A}| = 0$ or $|\mathbf{B}| = 0$.

T.6. Is $|\mathbf{AB}| = |\mathbf{BA}|$? Justify your answer.

T.7. Show that if $\mathbf{A}$ is a matrix such that in each row and in each column one and only one element is $\neq 0$, then $|\mathbf{A}| \neq 0$.

T.8. Show that if $\mathbf{AB} = \mathbf{I}_n$, then $|\mathbf{A}| \neq 0$ and $|\mathbf{B}| \neq 0$.

T.9. (a) Show that if $\mathbf{A} = \mathbf{A}^{-1}$, then $|\mathbf{A}| = \pm 1$.
(b) Show that if $\mathbf{A}^T = \mathbf{A}^{-1}$, then $|\mathbf{A}| = \pm 1$.

T.10. Show that if $\mathbf{A}$ is a nonsingular matrix such that $\mathbf{A}^2 = \mathbf{A}$, then $|\mathbf{A}| = 0$ or $|\mathbf{A}| = 1$.

T.11. Show that

$$|\mathbf{A}^T\mathbf{B}^T| = |\mathbf{A}||\mathbf{B}^T| = |\mathbf{A}^T||\mathbf{B}|.$$

T.12. Show that

$$\begin{vmatrix} 1 & a & a^2 \\ 1 & b & b^2 \\ 1 & c & c^2 \end{vmatrix} = (b - a)(c - a)(c - b).$$

This determinant is called a *Vandermonde determinant*.

T.13. Let $\mathbf{A} = [a_{ij}]$ be an upper triangular matrix. Prove that $\mathbf{A}$ is nonsingular if and only if $a_{ii} \neq 0$, $1 \leqslant i \leqslant n$.

2.2. Cofactor Expansion and Applications

So far, we have been evaluating determinants by using Equation (2) and have been aided by the properties established in Section 2.1. We now develop a different method for evaluating the determinant of an $n \times n$ matrix, which reduces the problem to the evaluation of determinants of matrices of the order $n - 1$. We can then repeat the process for these $(n - 1) \times (n - 1)$ matrices until we get to 2×2 matrices.

DEFINITION Let $\mathbf{A} = [a_{ij}]$ be an $n \times n$ matrix. Let $\mathbf{M}_{ij}$ be the $(n - 1) \times (n - 1)$ submatrix of $\mathbf{A}$ obtained by deleting the ith row and jth column of

$\mathbf{A}$. The determinant $|\mathbf{M}_{ij}|$ is called the **minor** of a_{ij}. The **cofactor** A_{ij} of a_{ij} is defined as

$$A_{ij} = (-1)^{i+j}|\mathbf{M}_{ij}|.$$

Example 1. Let

$$\mathbf{A} = \begin{bmatrix} 3 & -1 & 2 \\ 4 & 5 & 6 \\ 7 & 1 & 2 \end{bmatrix}.$$

Then

$$|\mathbf{M}_{12}| = \begin{vmatrix} 4 & 6 \\ 7 & 2 \end{vmatrix} = 8 - 42 = -34, \qquad |\mathbf{M}_{23}| = \begin{vmatrix} 3 & -1 \\ 7 & 1 \end{vmatrix} = 3 + 7 = 10,$$

and

$$|\mathbf{M}_{31}| = \begin{vmatrix} -1 & 2 \\ 5 & 6 \end{vmatrix} = -6 - 10 = -16.$$

Also,

$$A_{12} = (-1)^{1+2}|\mathbf{M}_{12}| = (-1)(-34) = 34,$$

$$A_{23} = (-1)^{2+3}|\mathbf{M}_{23}| = (-1)(10) = -10,$$

and

$$A_{31} = (-1)^{1+3}|\mathbf{M}_{31}| = (1)(-16) = -16.$$

If we think of the sign $(-1)^{i+j}$ as being located in position (i, j) of an $n \times n$ matrix, then the signs form a checkerboard pattern that has a + in the (1, 1) position. The patterns for $n = 3$ and $n = 4$ are as follows:

$$\begin{matrix} + & - & + \\ - & + & - \\ + & - & + \end{matrix} \qquad\qquad \begin{matrix} + & - & + & - \\ - & + & - & + \\ + & - & + & - \\ - & + & - & + \end{matrix}$$

$$n = 3 \qquad\qquad\qquad n = 4$$

The following theorem, whose proof is omitted, gives a more practical way of evaluating determinants.

THEOREM 2.9. *Let* $\mathbf{A} = [a_{ij}]$ *be an* $n \times n$ *matrix. Then*

$$|\mathbf{A}| = a_{i1}A_{i1} + a_{i2}A_{i2} + \cdots + a_{in}A_{in} \tag{1}$$

(expansion of $|\mathbf{A}|$ *about the* ith *row).*

and

$$|\mathbf{A}| = a_{1j}A_{1j} + a_{2j}A_{2j} + \cdots + a_{nj}A_{nj} \tag{2}$$

(*expansion of* $|\mathbf{A}|$ *about the* jth *column*).

Example 2. To evaluate the determinant

$$\begin{vmatrix} 1 & 2 & -3 & 4 \\ -4 & 2 & 1 & 3 \\ 3 & 0 & 0 & -3 \\ 2 & 0 & -2 & 3 \end{vmatrix},$$

we note that it is best to expand about either the second column or the third row because they each have two zeros. Obviously, the optimal course of action is to expand about a row or column having the largest number of zeros because in that case the cofactors A_{ij} of those a_{ij} which are zero do not have to be evaluated, since $a_{ij}A_{ij} = (0)(A_{ij}) = 0$. Thus, expanding about the third row, we have

$$\begin{vmatrix} 1 & 2 & -3 & 4 \\ -4 & 2 & 1 & 3 \\ 3 & 0 & 0 & -3 \\ 2 & 0 & -2 & 3 \end{vmatrix} = (-1)^{3+1}(3)\begin{vmatrix} 2 & -3 & 4 \\ 2 & 1 & 3 \\ 0 & -2 & 3 \end{vmatrix}$$

$$+ (-1)^{3+2}(0)\begin{vmatrix} 1 & -3 & 4 \\ -4 & 1 & 3 \\ 2 & -2 & 3 \end{vmatrix} + (-1)^{3+3}(0)\begin{vmatrix} 1 & 2 & 4 \\ -4 & 2 & 3 \\ 2 & 0 & 3 \end{vmatrix} \tag{3}$$

$$+ (-1)^{3+4}(-3)\begin{vmatrix} 1 & 2 & -3 \\ -4 & 2 & 1 \\ 2 & 0 & -2 \end{vmatrix}.$$

We now evaluate

$$\begin{vmatrix} 2 & -3 & 4 \\ 2 & 1 & 3 \\ 0 & -2 & 3 \end{vmatrix}$$

by expanding about the first column, obtaining

$$(-1)^{1+1}(2)\begin{vmatrix} 1 & 3 \\ -2 & 3 \end{vmatrix} + (-1)^{1+2}(2)\begin{vmatrix} -3 & 4 \\ -2 & 3 \end{vmatrix}$$

$$= (1)(2)(9) + (-1)(2)(-1) = 20.$$

Similarly, we evaluate

$$\begin{vmatrix} 1 & 2 & -3 \\ -4 & 2 & 1 \\ 2 & 0 & -2 \end{vmatrix}$$

by expanding about the third row, obtaining

$$(-1)^{1+3}(2)\begin{vmatrix} 2 & -3 \\ 2 & 1 \end{vmatrix} + (-1)^{3+3}(-2)\begin{vmatrix} 1 & 2 \\ -4 & 2 \end{vmatrix}$$

$$= (1)(2)(8) + (1)(-2)(10) = -4.$$

Substituting in Equation (3), we find the value of the given determinant as

$$(+1)(3)(20) + 0 + 0 + (-1)(-3)(-4) = 48.$$

On the other hand, expanding about the first column we have

$$(-1)^{1+1}(1)\begin{vmatrix} 2 & 1 & 3 \\ 0 & 0 & -3 \\ 0 & -2 & 3 \end{vmatrix} + (-1)^{1+2}(-4)\begin{vmatrix} 2 & -3 & 4 \\ 0 & 0 & -3 \\ 0 & -2 & 3 \end{vmatrix}$$

$$+ (-1)^{1+3}(3)\begin{vmatrix} 2 & -3 & 4 \\ 2 & 1 & 3 \\ 0 & -2 & 3 \end{vmatrix} + (-1)^{1+4}(2)\begin{vmatrix} 2 & -3 & 4 \\ 2 & 1 & 3 \\ 0 & 0 & -3 \end{vmatrix}$$

$$= (1)(1)(-12) + (-1)(-4)(-12) + (1)(3)(20) + (-1)(2)(-24) = 48.$$

We can use the properties of Section 2.1 to introduce many zeros in a given row or column and then expand about that row or column. Consider the following example.

Example 3

$$\begin{vmatrix} 1 & 2 & -3 & 4 \\ -4 & 2 & 1 & 3 \\ 1 & 0 & 0 & -3 \\ 2 & 0 & -2 & 3 \end{vmatrix}_{c_4+3c_1\to c_4} = \begin{vmatrix} 1 & 2 & -3 & 7 \\ -4 & 2 & 1 & -9 \\ 1 & 0 & 0 & 0 \\ 2 & 0 & -2 & 9 \end{vmatrix}$$

$$= (-1)^{3+1}(1)\begin{vmatrix} 2 & -3 & 7 \\ 2 & 1 & -9 \\ 0 & -2 & 9 \end{vmatrix}_{r_1-r_2\to r_1} = (-1)^4(1)\begin{vmatrix} 0 & -4 & 16 \\ 2 & 1 & -9 \\ 0 & -2 & 9 \end{vmatrix}$$

$$= (-1)^4(1)(8) = 8.$$

Here $\mathbf{c}_4 + 3\mathbf{c}_1 \to \mathbf{c}_4$ means that the fourth column $\mathbf{c}_4$ of the matrix plus three times the first column $\mathbf{c}_1$ replaces the fourth column.

The Inverse of a Matrix

It is interesting to ask what $a_{i1}A_{k1} + a_{i2}A_{k2} + \cdots + a_{in}A_{kn}$ is for $i \neq k$ because, as soon as we answer this question, we shall obtain another method for finding the inverse of a nonsingular matrix.

THEOREM 2.10. *If* $\mathbf{A} = [a_{ij}]$ *is an* $n \times n$ *matrix, then*

$$a_{i1}A_{k1} + a_{i2}A_{k2} + \cdots + a_{in}A_{kn} = 0 \qquad \text{for } i \neq k; \tag{4}$$

$$a_{1j}A_{1k} + a_{2j}A_{2k} + \cdots + a_{nj}A_{nk} = 0 \qquad \text{for } j \neq k. \tag{5}$$

proof. We only prove the first formula. The second one follows from the first one by Theorem 2.1.

Consider the matrix $\mathbf{B}$ obtained from $\mathbf{A}$ by replacing the kth row of $\mathbf{A}$ by its ith row. Thus $\mathbf{B}$ is a matrix having two identical rows—the ith and kth rows. Then $|\mathbf{B}| = 0$. Now expand $|\mathbf{B}|$ about the kth row. The elements of the kth row of $\mathbf{B}$ are $a_{i1}, a_{i2}, \ldots, a_{in}$. The cofactors of the kth row are $A_{k1}, A_{k2}, \ldots, A_{kn}$. Thus from Equation (1) we have

$$0 = |\mathbf{B}| = a_{i1}A_{k1} + a_{i2}A_{k2} + \cdots + a_{in}A_{kn},$$

which is what we wanted to show.

Example 4. Let

$$\mathbf{A} = \begin{bmatrix} 1 & 2 & 3 \\ -2 & 3 & 1 \\ 4 & 5 & -2 \end{bmatrix}.$$

Then

$$A_{21} = (-1)^{2+1}\begin{vmatrix} 2 & 3 \\ 5 & -2 \end{vmatrix} = 19, \qquad A_{22} = (-1)^{2+2}\begin{vmatrix} 1 & 3 \\ 4 & -2 \end{vmatrix} = -14,$$

$$A_{23} = (-1)^{2+3}\begin{vmatrix} 1 & 2 \\ 4 & 5 \end{vmatrix} = 3.$$

Now $a_{31}A_{21} + a_{32}A_{22} + a_{33}A_{23} = (4)(19) + (5)(-14) + (-2)(3) = 0.$

We may combine (1) and (4) as

$$\begin{aligned} a_{i1}A_{k1} + a_{i2}A_{k2} + \cdots + a_{in}A_{kn} &= |\mathbf{A}| \qquad \text{if } i = k \\ &= 0 \qquad \text{if } i \neq k. \end{aligned} \tag{6}$$

Similarly, we may combine (2) and (5) as

$$\begin{aligned} a_{1j}A_{1k} + a_{2j}A_{2k} + \cdots + a_{nj}A_{nk} &= |\mathbf{A}| \qquad \text{if } j = k \\ &= 0 \qquad \text{if } j \neq k. \end{aligned} \tag{7}$$

Let $\mathbf{A} = [a_{ij}]$ be an $n \times n$ matrix. The $n \times n$ matrix adj $\mathbf{A}$, called the **adjoint** of $\mathbf{A}$, is the matrix whose i, jth element is the cofactor A_{ji} of a_{ji}. Thus

$$\operatorname{adj} \mathbf{A} = \begin{bmatrix} A_{11} & A_{21} & \cdots & A_{n1} \\ A_{12} & A_{22} & \cdots & A_{n2} \\ \vdots & & & \vdots \\ A_{1n} & A_{2n} & \cdots & A_{nn} \end{bmatrix}.$$

Example 5. Let

$$\mathbf{A} = \begin{bmatrix} 3 & -2 & 1 \\ 5 & 6 & 2 \\ 1 & 0 & -3 \end{bmatrix}.$$

The cofactors of A are

$$A_{11} = (-1)^{1+1}\begin{vmatrix} 6 & 2 \\ 0 & -3 \end{vmatrix} = -18; \qquad A_{12} = (-1)^{1+2}\begin{vmatrix} 5 & 2 \\ 1 & -3 \end{vmatrix} = 17;$$

$$A_{13} = (-1)^{1+3}\begin{vmatrix} 5 & 6 \\ 1 & 0 \end{vmatrix} = -6;$$

$$A_{21} = (-1)^{2+1}\begin{vmatrix} -2 & 1 \\ 0 & -3 \end{vmatrix} = -6; \qquad A_{22} = (-1)^{2+2}\begin{vmatrix} 3 & 1 \\ 1 & -3 \end{vmatrix} = -10;$$

$$A_{23} = (-1)^{2+3}\begin{vmatrix} 3 & -2 \\ 1 & 0 \end{vmatrix} = -2;$$

$$A_{31} = (-1)^{3+1}\begin{vmatrix} -2 & 1 \\ 6 & 2 \end{vmatrix} = -10; \qquad A_{32} = (-1)^{3+2}\begin{vmatrix} 3 & 1 \\ 5 & 2 \end{vmatrix} = -1;$$

$$A_{33} = (-1)^{3+3}\begin{vmatrix} 3 & -2 \\ 5 & 6 \end{vmatrix} = 28.$$

Then

$$\operatorname{adj} \mathbf{A} = \begin{bmatrix} -18 & -6 & -10 \\ 17 & -10 & -1 \\ -6 & -2 & 28 \end{bmatrix}.$$

THEOREM 2.11. *If* $\mathbf{A} = [a_{ij}]$ *is an* $n \times n$ *matrix, then*

$$\mathbf{A}(\operatorname{adj} \mathbf{A}) = (\operatorname{adj} \mathbf{A})\mathbf{A} = |\mathbf{A}|\mathbf{I}_n.$$

proof. We have $\mathbf{A}(\operatorname{adj} \mathbf{A}) =$

$$\begin{bmatrix} a_{11} & a_{12} & \cdots & a_{1n} \\ a_{21} & a_{22} & \cdots & a_{2n} \\ \vdots & \vdots & & \vdots \\ a_{i1} & a_{i2} & \cdots & a_{in} \\ \vdots & \vdots & & \vdots \\ a_{n1} & a_{n2} & \cdots & a_{nn} \end{bmatrix} \begin{bmatrix} A_{11} & A_{21} & \cdots & A_{j1} & \cdots & A_{n1} \\ A_{12} & A_{22} & \cdots & A_{j2} & \cdots & A_{n2} \\ \vdots & \vdots & & \vdots & & \vdots \\ A_{1n} & A_{2n} & \cdots & A_{jn} & \cdots & A_{nn} \end{bmatrix}.$$

The i, jth element in the product matrix $\mathbf{A}(\operatorname{adj} \mathbf{A})$ is, by (6),

$$\begin{aligned} a_{i1}A_{j1} + a_{i2}A_{j2} + \cdots + a_{in}A_{jn} &= |\mathbf{A}| \quad \text{if } i = j \\ &= 0 \quad \text{if } i \neq j. \end{aligned}$$

This means that

$$\mathbf{A}(\operatorname{adj} \mathbf{A}) = \begin{bmatrix} |\mathbf{A}| & 0 & \cdots & 0 \\ 0 & |\mathbf{A}| & & \vdots \\ \vdots & & \ddots & 0 \\ 0 & \cdots & 0 & |\mathbf{A}| \end{bmatrix} = |\mathbf{A}|\mathbf{I}_n.$$

The i, jth element in the product matrix $(\operatorname{adj} \mathbf{A})\mathbf{A}$ is, by (7),

$$\begin{aligned} A_{1i}a_{1j} + A_{2i}a_{2j} + \cdots + A_{ni}a_{nj} &= |\mathbf{A}| \quad \text{if } i = j \\ &= 0 \quad \text{if } i \neq j. \end{aligned}$$

Thus $(\operatorname{adj} \mathbf{A})\mathbf{A} = |\mathbf{A}|\mathbf{I}_n$.

Example 6. Consider the matrix of Example 5. Then

$$\begin{bmatrix} 3 & -2 & 1 \\ 5 & 6 & 2 \\ 1 & 0 & -3 \end{bmatrix}\begin{bmatrix} -18 & -6 & -10 \\ 17 & -10 & -1 \\ -6 & -2 & 28 \end{bmatrix}$$

$$= \begin{bmatrix} -94 & 0 & 0 \\ 0 & -94 & 0 \\ 0 & 0 & -94 \end{bmatrix} = -94\begin{bmatrix} 1 & 0 & 0 \\ 0 & 1 & 0 \\ 0 & 0 & 1 \end{bmatrix}$$

and

$$\begin{bmatrix} -18 & -6 & -10 \\ 17 & -10 & -1 \\ -6 & -2 & 28 \end{bmatrix}\begin{bmatrix} 3 & -2 & 1 \\ 5 & 6 & 2 \\ 1 & 0 & -3 \end{bmatrix} = -94\begin{bmatrix} 1 & 0 & 0 \\ 0 & 1 & 0 \\ 0 & 0 & 1 \end{bmatrix}.$$

We now have a new method for finding the inverse of a nonsingular matrix, and we state this result as the following corollary.

COROLLARY 2.2. *If* $\mathbf{A}$ *is an* $n \times n$ *matrix and* $|\mathbf{A}| \neq 0$, *then*

$$\mathbf{A}^{-1} = \frac{1}{|\mathbf{A}|}(\text{adj } \mathbf{A}) = \begin{bmatrix} \dfrac{A_{11}}{|\mathbf{A}|} & \dfrac{A_{21}}{|\mathbf{A}|} & \cdots & \dfrac{A_{n1}}{|\mathbf{A}|} \\ \dfrac{A_{12}}{|\mathbf{A}|} & \dfrac{A_{22}}{|\mathbf{A}|} & \cdots & \dfrac{A_{n2}}{|\mathbf{A}|} \\ \vdots & & & \vdots \\ \dfrac{A_{1n}}{|\mathbf{A}|} & \dfrac{A_{2n}}{|\mathbf{A}|} & \cdots & \dfrac{A_{nn}}{|\mathbf{A}|} \end{bmatrix}.$$

proof. By Theorem 2.11, $\mathbf{A}(\text{adj } \mathbf{A}) = |\mathbf{A}|\mathbf{I}_n$, so if $|\mathbf{A}| \neq 0$, then

$$\mathbf{A}\frac{1}{|\mathbf{A}|}(\text{adj } \mathbf{A}) = \frac{1}{|\mathbf{A}|}[\mathbf{A}(\text{adj } \mathbf{A})] = \frac{1}{|\mathbf{A}|}(|\mathbf{A}|\mathbf{I}_n) = \mathbf{I}_n.$$

Hence

$$\mathbf{A}^{-1} = \frac{1}{|\mathbf{A}|}(\text{adj } \mathbf{A}).$$

Example 7. Again consider the matrix of Example 5. Then $|\mathbf{A}| = -94$, and

$$\mathbf{A}^{-1} = \frac{1}{|\mathbf{A}|}(\text{adj }\mathbf{A}) = \begin{bmatrix} \frac{18}{94} & \frac{6}{94} & \frac{10}{94} \\ -\frac{17}{94} & \frac{10}{94} & \frac{1}{94} \\ \frac{6}{94} & \frac{2}{94} & -\frac{28}{94} \end{bmatrix}.$$

THEOREM 2.12. *A matrix* $\mathbf{A}$ *is nonsingular if and only if* $|\mathbf{A}| \neq 0$.

proof. If $|\mathbf{A}| \neq 0$, then Corollary 2.2 gives an expression for $\mathbf{A}^{-1}$, so $\mathbf{A}$ is nonsingular.

The converse has already been established in Corollary 2.1, whose proof was left to the reader as Exercise T.4 of Section 2.1. We now prove the converse here. Suppose that $\mathbf{A}$ is nonsingular. Then

$$\mathbf{A}\mathbf{A}^{-1} = \mathbf{I}_n.$$

From Theorem 2.8 we have

$$|\mathbf{A}\mathbf{A}^{-1}| = |\mathbf{A}||\mathbf{A}^{-1}| = |\mathbf{I}_n| = 1,$$

which implies that $|\mathbf{A}| \neq 0$.

This completes the proof.

COROLLARY 2.3. *If* $\mathbf{A}$ *is an* $n \times n$ *matrix, then the homogeneous system* $\mathbf{AX} = \mathbf{0}$ *has a nontrivial solution if and only if* $|\mathbf{A}| = 0$.

proof. If $|\mathbf{A}| \neq 0$, then, by Theorem 2.12, $\mathbf{A}$ is nonsingular and thus $\mathbf{AX} = \mathbf{0}$ has only the trivial solution (Theorem 1.12 in Section 1.4).

Conversely, if $|\mathbf{A}| = 0$, then $\mathbf{A}$ is singular (Theorem 2.12). Suppose that $\mathbf{A}$ is row equivalent to a matrix $\mathbf{B}$ in reduced row echelon form. It then follows from Theorem 1.10 in Section 1.4 and Exercise T.7 in Section 1.3 that $\mathbf{B}$ has a row of zeros. The system $\mathbf{BX} = \mathbf{0}$ has the same solutions as the system $\mathbf{AX} = \mathbf{0}$. Let $\mathbf{C}_1$ be the matrix obtained by deleting the zero row of $\mathbf{B}$. Then the system $\mathbf{BX} = \mathbf{0}$ has the same solutions as the system $\mathbf{CX} = \mathbf{0}$. Since the latter is a homogeneous system of $n - 1$ equations in n

unknowns, it has a nontrivial solution (Theorem 1.7 in Section 1.3). Hence the given system $\mathbf{AX} = \mathbf{0}$ has a nontrivial solution. We might note that the proof of the converse is essentially the proof of Exercise T.2 in Section 1.4.

In Section 1.4 we developed a practical method for finding $\mathbf{A}^{-1}$. In describing the situation showing that a matrix is singular and has no inverse, we used the fact that if we start with the identity matrix $\mathbf{I}_n$ and use only elementary row operations on $\mathbf{I}_n$, we can never obtain a matrix having a row consisting entirely of zeros. We can now justify this statement as follows. If matrix $\mathbf{B}$ results from $\mathbf{I}_n$ by interchanging two rows of $\mathbf{I}_n$, then $|\mathbf{B}| = -|\mathbf{I}_n| = -1$ (Theorem 2.2); if $\mathbf{C}$ results from $\mathbf{I}_n$ by multiplying a row of $\mathbf{I}_n$ by $c \neq 0$, then $|\mathbf{C}| = c|\mathbf{I}_n| = c$ (Theorem 2.5), and if $\mathbf{D}$ results from $\mathbf{I}_n$ by adding a multiple of a row of $\mathbf{I}_n$ to another row of $\mathbf{I}_n$, then $|\mathbf{D}| = |\mathbf{I}_n| = 1$ (Theorem 2.6). Performing elementary row operations on $\mathbf{I}_n$ will thus never yield a matrix with zero determinant. Now suppose that $\mathbf{F}$ is obtained from $\mathbf{I}_n$ by a sequence of elementary row operations and $\mathbf{F}$ has a row of zeros. Then $|\mathbf{F}| = 0$ (Theorem 2.4). This contradiction justifies the statement used in Section 1.4.

We might note that the method of inverting a nonsingular matrix given in Corollary 2.2 is much less efficient than the method given in Chapter 1. In fact, the computation of $\mathbf{A}^{-1}$ using determinants, as given in Corollary 2.2, becomes too expensive for $n > 4$ from a computing point of view. We discuss these matters in Section 2.3, where we deal with determinants from a computational point of view. However, Corollary 2.2 is still a useful result on other grounds.

Cramer's Rule

We can use the result of Theorem 2.11 to obtain another method, known as Cramer's rule, for solving a linear system of n equations in n unknowns with a nonsingular coefficient matrix.

THEOREM 2.13 (Cramer's Rule). *Let*

$$\begin{aligned}
a_{11}x_1 + a_{12}x_2 + \cdots + a_{1n}x_n &= b_1 \\
a_{21}x_1 + a_{22}x_2 + \cdots + a_{2n}x_n &= b_2 \\
&\vdots \\
a_{n1}x_1 + a_{n2}x_2 + \cdots + a_{nn}x_n &= b_n
\end{aligned}$$

be a linear system of n equations in n unknowns and let $\mathbf{A} = [a_{ij}]$ *be the coefficient matrix so that we can write the given system as* $\mathbf{AX} = \mathbf{B}$, *where*

$$\mathbf{B} = \begin{bmatrix} b_1 \\ b_2 \\ \vdots \\ b_n \end{bmatrix}.$$

If $|\mathbf{A}| \neq 0$, *then the system has the unique solution*

$$x_1 = \frac{|\mathbf{A}_1|}{|\mathbf{A}|}, \quad x_2 = \frac{|\mathbf{A}_2|}{|\mathbf{A}|}, \quad \ldots, \quad x_n = \frac{|\mathbf{A}_n|}{|\mathbf{A}|},$$

where $\mathbf{A}_i$ *is the matrix obtained from* $\mathbf{A}$ *by replacing the* ith *column of* $\mathbf{A}$ *by* $\mathbf{B}$.

proof. If $|\mathbf{A}| \neq 0$, then, by Theorem 2.12, $\mathbf{A}$ is nonsingular. Hence

$$\mathbf{X} = \begin{bmatrix} x_1 \\ x_2 \\ \vdots \\ x_n \end{bmatrix} = \mathbf{A}^{-1}\mathbf{B} = \begin{bmatrix} \frac{A_{11}}{|\mathbf{A}|} & \frac{A_{21}}{|\mathbf{A}|} & \cdots & \frac{A_{n1}}{|\mathbf{A}|} \\ \frac{A_{12}}{|\mathbf{A}|} & \frac{A_{22}}{|\mathbf{A}|} & \cdots & \frac{A_{n2}}{|\mathbf{A}|} \\ \vdots & \vdots & & \vdots \\ \frac{A_{1i}}{|\mathbf{A}|} & \frac{A_{2i}}{|\mathbf{A}|} & \cdots & \frac{A_{ni}}{|\mathbf{A}|} \\ \vdots & \vdots & & \vdots \\ \frac{A_{1n}}{|\mathbf{A}|} & \frac{A_{2n}}{|\mathbf{A}|} & \cdots & \frac{A_{nn}}{|\mathbf{A}|} \end{bmatrix} \begin{bmatrix} b_1 \\ b_2 \\ \vdots \\ b_n \end{bmatrix}.$$

This means that

$$x_i = \frac{A_{1i}}{|\mathbf{A}|} b_1 + \frac{A_{2i}}{|\mathbf{A}|} b_2 + \cdots + \frac{A_{ni}}{|\mathbf{A}|} b_n \qquad (1 \leqslant i \leqslant n).$$

Now let

$$\mathbf{A}_i = \begin{bmatrix} a_{11} & a_{12} & \cdots & a_{1i-1} & b_1 & a_{1i+1} & \cdots & a_{1n} \\ a_{21} & a_{22} & \cdots & a_{2i-1} & b_2 & a_{2i+1} & \cdots & a_{2n} \\ \vdots & \vdots & & \vdots & \vdots & \vdots & & \vdots \\ a_{n1} & a_{n2} & \cdots & a_{ni-1} & b_n & a_{ni+1} & \cdots & a_{nn} \end{bmatrix}.$$

If we evaluate $|\mathbf{A}_i|$ by expanding about the ith column, we find that

$$|\mathbf{A}_i| = A_{1i}b_1 + A_{2i}b_2 + \cdots + A_{ni}b_n.$$

Hence

$$x_i = \frac{|\mathbf{A}_i|}{|\mathbf{A}|}$$

for $i = 1, 2, \ldots, n$. In this expression for x_i, the determinant $|\mathbf{A}_i|$ of $\mathbf{A}_i$ can be calculated by any method. It was only in the derivation of the expression for x_i that we had to evaluate it by expanding about the ith column.

Example 8. Consider the following linear system:

$$\begin{aligned} -2x_1 + 3x_2 - x_3 &= 1 \\ x_1 + 2x_2 - x_3 &= 4 \\ -2x_1 - x_2 + x_3 &= -3. \end{aligned}$$

Then

$$|\mathbf{A}| = \begin{vmatrix} -2 & 3 & -1 \\ 1 & 2 & -1 \\ -2 & -1 & 1 \end{vmatrix} = -2.$$

Hence

$$x_1 = \frac{\begin{vmatrix} 1 & 3 & -1 \\ 4 & 2 & -1 \\ -3 & -1 & 1 \end{vmatrix}}{|\mathbf{A}|} = \frac{-4}{-2} = 2;$$

$$x_2 = \frac{\begin{vmatrix} -2 & 1 & -1 \\ 1 & 4 & -1 \\ -2 & -3 & 1 \end{vmatrix}}{|\mathbf{A}|} = \frac{-6}{-2} = 3;$$

$$x_3 = \frac{\begin{vmatrix} -2 & 3 & 1 \\ 1 & 2 & 4 \\ -2 & -1 & -3 \end{vmatrix}}{|\mathbf{A}|} = \frac{-8}{-2} = 4.$$

We note that Cramer's rule is only applicable to the case where we have n equations in n unknowns and where the coefficient matrix $\mathbf{A}$ is nonsingular. Cramer's rule becomes computationally inefficient for $n > 4$, and it is better to use the Gauss–Jordan reduction method discussed in Section 1.3.

2.2 Exercises

1. Let

$$\mathbf{A} = \begin{bmatrix} 1 & 0 & -2 \\ 3 & 1 & 4 \\ 5 & 2 & -3 \end{bmatrix}.$$

Compute all the cofactors.

2. Let

$$\mathbf{A} = \begin{bmatrix} 1 & 0 & 3 & 0 \\ 2 & 1 & 4 & -1 \\ 3 & 2 & 4 & 0 \\ 0 & 3 & -1 & 0 \end{bmatrix}.$$

Compute all the cofactors.

In Exercises 3 through 6 compute the determinants using Theorem 2.9.

3. (a) $\begin{bmatrix} 1 & 2 & 3 \\ -1 & 5 & 2 \\ 3 & 2 & 0 \end{bmatrix}$. (b) $\begin{bmatrix} 4 & -4 & 2 & 1 \\ 1 & 2 & 0 & 3 \\ 2 & 0 & 3 & 4 \\ 0 & -3 & 2 & 1 \end{bmatrix}$.

(c) $\begin{bmatrix} 4 & -2 & 0 \\ 0 & 2 & 4 \\ -1 & -1 & -3 \end{bmatrix}$.

4. (a) $\begin{bmatrix} 2 & 2 & -3 & 1 \\ 0 & 1 & 2 & -1 \\ 3 & -1 & 4 & 1 \\ 2 & 3 & 0 & 0 \end{bmatrix}$. (b) $\begin{bmatrix} 0 & 1 & -2 \\ -1 & 3 & 1 \\ 2 & -2 & 3 \end{bmatrix}$.

(c) $\begin{bmatrix} 2 & 1 & -3 \\ 0 & 1 & 2 \\ -4 & 2 & 1 \end{bmatrix}$.

5. (a) $\begin{bmatrix} 3 & 1 & 2 & -1 \\ 2 & 0 & 3 & -7 \\ 1 & 3 & 4 & -5 \\ 0 & -1 & 1 & -5 \end{bmatrix}$. (b) $\begin{bmatrix} 3 & 1 & 0 \\ 3 & 2 & 1 \\ 0 & 1 & -1 \end{bmatrix}$.

(c) $\begin{bmatrix} 3 & -3 & 0 \\ 2 & 0 & 2 \\ 2 & 1 & -3 \end{bmatrix}$.

6. (a) $\begin{bmatrix} 0 & 0 & -1 & 3 \\ 0 & 1 & 2 & 1 \\ 2 & -2 & 5 & 2 \\ 3 & 3 & 0 & 0 \end{bmatrix}$. (b) $\begin{bmatrix} 4 & 2 & 0 \\ 1 & 1 & 2 \\ -1 & 3 & 4 \end{bmatrix}$.

(c) $\begin{bmatrix} -1 & 2 & -1 \\ 3 & 2 & 1 \\ 1 & 4 & 2 \end{bmatrix}$.

7. Verify Theorem 2.10 for the matrix $\mathbf{A} = \begin{bmatrix} -2 & 3 & 0 \\ 4 & 1 & -3 \\ 2 & 0 & 1 \end{bmatrix}$ by computing

$$a_{11}A_{12} + a_{21}A_{22} + a_{31}A_{32}.$$

8. Let $\mathbf{A} = \begin{bmatrix} 2 & 1 & 3 \\ -1 & 2 & 0 \\ 3 & -2 & 1 \end{bmatrix}$.

(a) Find adj $\mathbf{A}$.
(b) Compute $|\mathbf{A}|$.
(c) Verify Theorem 2.11; that is, show that $\mathbf{A}(\text{adj } \mathbf{A}) = |\mathbf{A}|\mathbf{I}_3$.

9. Let $\mathbf{A} = \begin{bmatrix} 6 & 2 & 8 \\ -3 & 4 & 1 \\ 4 & -4 & 5 \end{bmatrix}$.

(a) Find adj $\mathbf{A}$.
(b) Compute $|\mathbf{A}|$.
(c) Verify Theorem 2.11; that is, show that $(\text{adj } \mathbf{A})\mathbf{A} = |\mathbf{I}|\mathbf{I}_3$.

In Exercises 10 through 13 compute the inverses of the given matrices, if they exist, using Corollary 2.2.

10. (a) $\begin{bmatrix} 3 & 2 \\ -3 & 4 \end{bmatrix}$. (b) $\begin{bmatrix} 4 & 2 & 2 \\ 0 & 1 & 2 \\ 1 & 0 & 3 \end{bmatrix}$. (c) $\begin{bmatrix} 2 & 0 & -1 \\ 3 & 7 & 2 \\ 1 & 1 & 0 \end{bmatrix}$.

11. (a) $\begin{bmatrix} 1 & 2 & -3 \\ -4 & -5 & 2 \\ -1 & 1 & -7 \end{bmatrix}$. (b) $\begin{bmatrix} 2 & 3 \\ -1 & 2 \end{bmatrix}$.

(c) $\begin{bmatrix} 4 & 0 & 2 \\ 0 & 3 & 4 \\ 0 & 1 & -2 \end{bmatrix}$.

12. (a) $\begin{bmatrix} 2 & 0 & 1 \\ 3 & 2 & -1 \\ 1 & 0 & 1 \end{bmatrix}$. (b) $\begin{bmatrix} 5 & -1 \\ 2 & -1 \end{bmatrix}$. (c) $\begin{bmatrix} 1 & 2 & 4 \\ 1 & -5 & 6 \\ 3 & -1 & 2 \end{bmatrix}$.

13. (a) $\begin{bmatrix} -3 & 1 \\ 2 & 0 \end{bmatrix}$. (b) $\begin{bmatrix} 4 & 0 & 0 \\ 0 & -3 & 0 \\ 0 & 0 & 2 \end{bmatrix}$. (c) $\begin{bmatrix} 0 & 2 & 1 & 3 \\ 2 & -1 & 3 & 4 \\ -2 & 1 & 5 & 2 \\ 0 & 1 & 0 & 2 \end{bmatrix}$.

Use Theorem 2.12 to determine which of the following matrices are nonsingular.

14. (a) $\begin{bmatrix} 1 & 2 & 3 \\ 0 & 1 & 2 \\ 2 & -3 & 1 \end{bmatrix}$. (b) $\begin{bmatrix} 1 & 2 \\ 3 & 4 \end{bmatrix}$. (c) $\begin{bmatrix} 1 & 3 & 2 \\ 2 & 1 & 4 \\ 1 & -7 & 2 \end{bmatrix}$.

(d) $\begin{bmatrix} 1 & 2 & 0 & 5 \\ 3 & 4 & 1 & 7 \\ -2 & 5 & 2 & 0 \\ 0 & 1 & 2 & -7 \end{bmatrix}$.

Use Theorem 2.12 to determine which of the following matrices are nonsingular.

15. (a) $\begin{bmatrix} 4 & 3 & -5 \\ -2 & -1 & 3 \\ 4 & 6 & -2 \end{bmatrix}$. (b) $\begin{bmatrix} 1 & 3 & -1 & 2 \\ 2 & -6 & 4 & 1 \\ 3 & 5 & -1 & 3 \\ 4 & -6 & 5 & 2 \end{bmatrix}$.

(c) $\begin{bmatrix} 2 & 2 & -4 \\ 1 & 5 & 2 \\ 3 & 7 & -2 \end{bmatrix}$. (d) $\begin{bmatrix} 0 & 1 & 2 \\ 1 & 2 & 0 \\ 1 & 3 & 4 \end{bmatrix}$.

16. Use Corollary 2.3 to find out whether the following homogeneous systems have nontrivial solutions.

(a)
$$\begin{aligned} x - 2y + z &= 0 \\ 2x + 3y + z &= 0 \\ 3x + y + 2z &= 0. \end{aligned}$$

(b)
$$\begin{aligned} x + 2y + w &= 0 \\ x + 2y + 3z &= 0 \\ z + 2w &= 0 \\ y + 2z - w &= 0. \end{aligned}$$

17. Repeat Exercise 16 for the following homogeneous systems.

(a)
$$\begin{aligned} x + y - z &= 0 \\ 2x + y + 2z &= 0 \\ 3x - y + z &= 0. \end{aligned}$$

(b)
$$\begin{aligned} x + y + 2z + w &= 0 \\ 2x - y + z - w &= 0 \\ 3x + y + 2z + 3w &= 0 \\ 2x - y - z + w &= 0. \end{aligned}$$

In Exercises 18 through 21, if possible, solve the given linear system by Cramer's rule.

18.
$$\begin{aligned} 2x + 4y + 6z &= 2 \\ x + 2z &= 0 \\ 2x + 3y - z &= -5. \end{aligned}$$

19.
$$\begin{aligned} x + y + z - 2w &= -4 \\ 2y + z + 3w &= 4 \\ 2x + y - z + 2w &= 5 \\ x - y + w &= 4. \end{aligned}$$

20.
$$\begin{aligned} 2x + y + z &= 6 \\ 3x + 2y - 2z &= -2 \\ x + y + 2z &= 4. \end{aligned}$$

21.
$$\begin{aligned} 2x + 3y + 7z &= 2 \\ -2x \qquad - 4z &= 0 \\ x + 2y + 4z &= 0. \end{aligned}$$

Theoretical Exercises

T.1. Show by a column (row) expansion that if $\mathbf{A} = [a_{ij}]$ is upper (lower) triangular, then $|\mathbf{A}| = a_{11}a_{22} \cdots a_{nn}$.

T.2. If $\mathbf{A} = [a_{ij}]$ is a 3×3 matrix, develop the general expression for $|\mathbf{A}|$ by expanding (a) about the second column, and (b) about the third row. Compare these answers with those obtained for Example 6 in Section 2.1.

T.3. Prove that if $\mathbf{A}$ is a nonsingular matrix that is symmetric, then $\mathbf{A}^{-1}$ is symmetric.

T.4. Prove that if $\mathbf{A}$ is a nonsingular upper triangular matrix, then $\mathbf{A}^{-1}$ is also upper triangular.

T.5. Prove that

$$\mathbf{A} = \begin{bmatrix} a & b \\ c & d \end{bmatrix}$$

is nonsingular if and only if $ad - bc \neq 0$. If this condition is satisfied, use Corollary 2.2 to find $\mathbf{A}^{-1}$.

T.6. Using Corollary 2.2, find the inverse of

$$\mathbf{A} = \begin{bmatrix} 1 & a & a^2 \\ 1 & b & b^2 \\ 1 & c & c^2 \end{bmatrix}.$$

(*Hint:* See Exercise T.12 in Section 2.1, where $|\mathbf{A}|$ was computed.)

T.7. Prove that $|\text{adj } \mathbf{A}| = |\mathbf{A}|^{n-1}$.

T.8. Prove that if $\mathbf{A}$ is symmetric, then adj $\mathbf{A}$ is also symmetric.

T.9. Do Exercise T.8 in Section 1.3 using determinants.

2.3 Determinants from a Computational Point of View

In this book we have, by now, developed two methods for solving a linear system of n equations in n unknowns: Gauss–Jordan reduction and Cramer's rule. We also have two methods for finding the inverse of a nonsingular matrix: the method involving determinants and the method developed in Section 1.3. We must then develop some criteria for choosing one or the other method depending upon our needs.

Most sizable problems in linear algebra are solved on computers so that it is natural to compare two methods by estimating their computing time for the same problem. Since addition is so much faster than multiplication, the number of multiplications is often used as a basis of comparison for two numerical procedures.

Consider the linear system $\mathbf{AX} = \mathbf{B}$, where $\mathbf{A}$ is 25×25. If we find $\mathbf{X}$ by Cramer's rule, we must first obtain $|\mathbf{A}|$. We can find $|\mathbf{A}|$ by cofactor expansion, say $|\mathbf{A}| = a_{11}A_{11} + a_{21}A_{21} + \cdots + a_{n1}A_{n1}$, where we have expanded $|\mathbf{A}|$ about the first column. Note that if each cofactor is available, we require 25 multiplications. Now each cofactor A_{ij} is the determinant of a 24×24 matrix, and it can be expanded about a given row or column, requiring 24 multiplications. Thus the computation of $|\mathbf{A}|$ requires $25 \times 24 \times \cdots \times 2 \times 1 = 25!$ multiplications. Even if we use the fastest computer in existence today, it would take *one billion years* to evaluate $|\mathbf{A}|$. However, Gauss–Jordan reduction takes about 25^3 multiplications, and we would find the solution in less than *one second*. Of course, we can compute $|\mathbf{A}|$ in a much more efficient way by using elementary row operations to reduce $\mathbf{A}$ to triangular form and then use Theorem 2.7 (see Example 16 in Section 2.1). When implemented this way, Cramer's rule will require approximately n^4 multiplications. Thus, Gauss–Jordan reduction is still much faster.

In general, if we are seeking numerical answers, then any method involving determinants can be used for $n \leqslant 4$. For $n > 5$, determinant-dependent methods are much less efficient than Gauss–Jordan reduction or the method of Section 1.3 for inverting a matrix.

The importance of determinants obviously does not lie in their computational usage. Note that methods involving determinants enable one to express the inverse of a matrix and the solution to a linear system of n equations in n unknowns by means of expressions or formulas. Gauss–Jordan reduction and the method for finding $\mathbf{A}^{-1}$ given in Section 1.3 do not yield a formula for the answer; we must proceed numerically to obtain the answer. Sometimes we do not need numerical answers but an expression for the answer because we may wish to further manipulate the answer. Another important reason for studying determinants is that they play a key role in the study of eigenvalues and eigenvectors, which will be undertaken in Chapter 5.

3

Vectors and Vector Spaces

3.1 Vectors in the Plane

Coordinate Systems

In many applications we deal with measurable quantities, such as pressure, mass, and speed, which can be described completely by giving their magnitude. There are also many other measurable quantities, such as velocity, force, and acceleration, which require for their description not only magnitude, but also a sense of direction. These are called **vectors**, and their study comprises this chapter. Vectors will be denoted by boldface capital letters, such as **O**, **U**, **V**, **X**, **Y**, and **Z**. The real numbers will be called **scalars** and will be denoted by lowercase italic letters.

We recall that the real number system may be visualized as a straight line L, which is usually taken in a horizontal position. A point O, called the **origin**, is chosen on L; O corresponds to the number 0. A point A is chosen to the right of O, thereby fixing the length of OA as 1 and specifying a positive direction. Thus the positive real numbers lie to the right of O; the negative real numbers lie to the left of O (Figure 3.1).

The real number x corresponding to the point P is called the **coordinate of** P and the point P whose coordinate is x is denoted by $P(x)$. The line L is called a **coordinate axis.** If P is to the right of O, then its coordinate is the length of the segment OP. If Q is to the left of O, then its coordinate is the negative of the length of the segment OQ. The distance between the points P and Q with respective coordinates a and b is $|b - a|$.

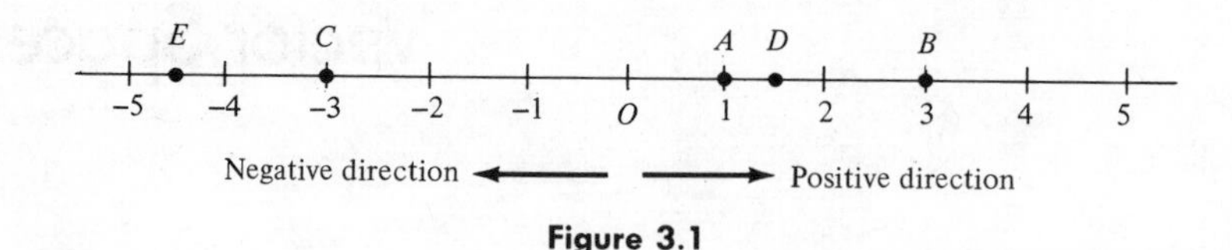

Figure 3.1

Example 1. Referring to Figure 3.1, the coordinates of the points B, C, D, and E are, respectively, 3, -3, 1.5, and -4.5. The distance between B and C is $|-3-3| = 6$. The distance between A and B is $|3 - 1| = 2$. The distance between C and E is $|-4.5-(-3)| = 1.5$.

We shall now turn to the analogous situation for the plane. We draw a pair of perpendicular lines intersecting at a point O, called the **origin**. One of the lines, the ***x*-axis**, is usually taken in a horizontal position. The other line, the ***y*-axis**, is then taken in a vertical position. We now choose a point on the x-axis to the right of O and a point on the y-axis above O to fix the units of length and positive directions on the x- and y-axes. Frequently, but not always, these points are chosen so that they are both equidistant from O, that is, so that the same unit of length is used for both axes. The x and y axes together are called **coordinate axes** (Figure 3.2). The **projection** of a point P in the plane on a line L is the point Q obtained by intersecting L with the line L' passing through Q and perpendicular to L [Figure 3.3(a) and (b)].

Let P be a point in the plane and let Q be its projection onto the x-axis. The coordinate of Q on the x-axis is called the ***x*-coordinate** of P. Similarly, let Q' be the projection of P onto the y-axis. The coordinate of Q' is called the ***y*-coordinate** of P. Thus with every point in the plane we associate an ordered pair (x, y) of real numbers, its coordinates. The point P with coordinates x and y is denoted by $P(x, y)$. Conversely, it is easy to see (Exercise T.1) how we can associate a point in the plane with each ordered pair (x, y) of real numbers (Figure 3.4). The correspondence given above

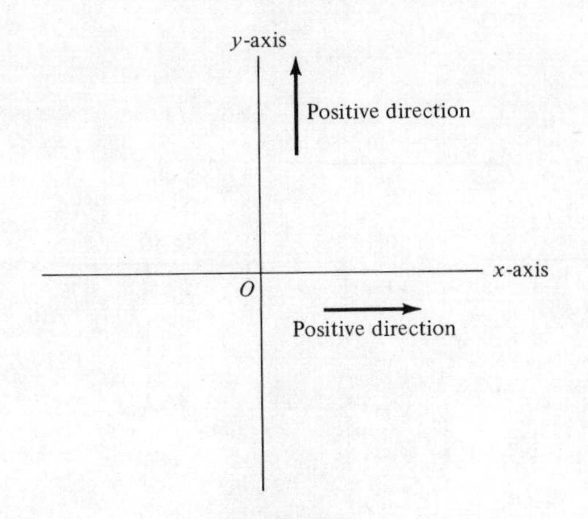

Figure 3.2

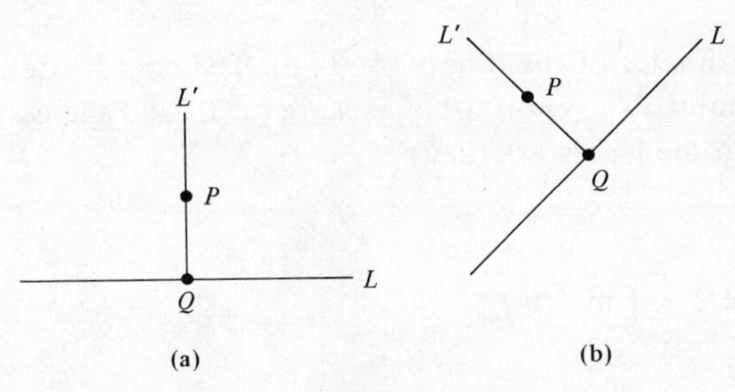

Figure 3.3

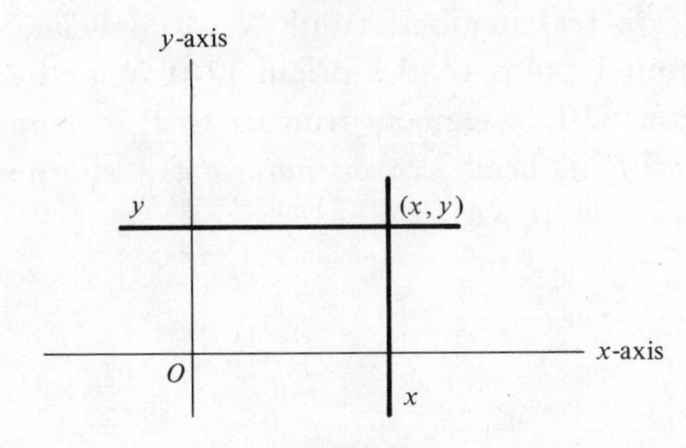

Figure 3.4

between points in the plane and ordered pairs of real numbers is called a **rectangular coordinate system.** The set of all points in the plane is denoted by R^2. It is also called **2-space.**

Example 2. In Figure 3.5 we show a number of points and their coordinates.

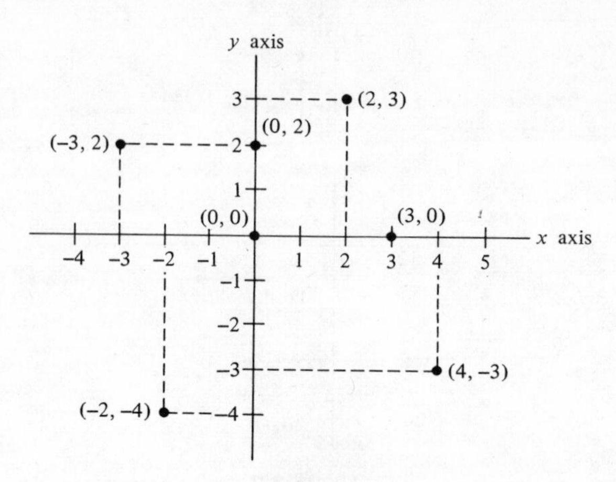

Figure 3.5

The coordinates of the origin are (0, 0). The coordinates of the projection of the point $P(x, y)$ on the x-axis are $(x, 0)$ and the coordinates of its projection on the y-axis are $(0, y)$.

Vectors

Consider the 2×1 matrix

$$\mathbf{X} = \begin{bmatrix} x \\ y \end{bmatrix},$$

where x and y are real numbers. With $\mathbf{X}$ we associate the directed line segment with initial point at the origin $O(0, 0)$ and terminal point at $P(x, y)$. The directed line segment from O to P is denoted by $\overrightarrow{OP}$; O is called its **tail** and P its **head**. We distinguish tail and head by placing an arrow at the head (Figure 3.6).

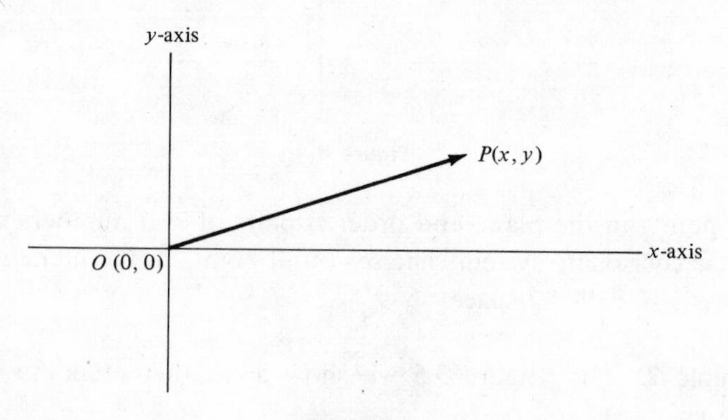

Figure 3.6

A directed line segment has a **direction**, which is the angle made with the positive axis, indicated by the arrow at its head. The **magnitude** of a directed line segment is its length.

Example 3. Let

$$\mathbf{X} = \begin{bmatrix} 2 \\ 3 \end{bmatrix}.$$

With $\mathbf{X}$ we can associate the directed line segment with tail $O(0, 0)$ and head $P(2, 3)$, shown in Figure 3.7.

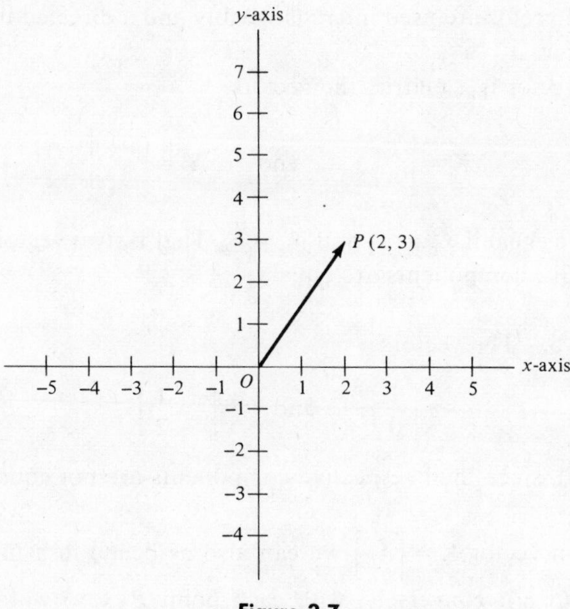

Figure 3.7

Conversely, with a directed line segment $\overrightarrow{OP}$ with tail $O(0, 0)$ and head $P(x, y)$, we can associate the matrix

$$\begin{bmatrix} x \\ y \end{bmatrix}.$$

Example 4. With the directed line segment $\overrightarrow{OP}$ with head $P(4, 5)$, we can associate the matrix

$$\begin{bmatrix} 4 \\ 5 \end{bmatrix}.$$

DEFINITION A **vector in the plane** is a 2×1 matrix

$$\mathbf{X} = \begin{bmatrix} x \\ y \end{bmatrix},$$

where x and y are real numbers, called the **components of X**. We refer to a vector in the plane merely as a **vector**.

Thus with every vector we can associate a directed line segment, and conversely, with every directed line segment we can associate a vector. The magnitude and direction of a vector are the magnitude and direction of its associated directed line segment. Frequently, the notions of directed line segment and vector are used interchangeably and a directed line segment is called a **vector**.

Since a vector is a matrix, the vectors

$$\mathbf{X} = \begin{bmatrix} x_1 \\ y_1 \end{bmatrix} \quad \text{and} \quad \mathbf{Y} = \begin{bmatrix} x_2 \\ y_2 \end{bmatrix}$$

are said to be **equal** if $x_1 = x_2$ and $y_1 = y_2$. That is, two vectors are equal if their respective components are equal.

Example 5. The vectors

$$\begin{bmatrix} 1 \\ 0 \end{bmatrix} \quad \text{and} \quad \begin{bmatrix} 1 \\ -2 \end{bmatrix}$$

are not equal, since their respective components are not equal.

With each vector $\mathbf{X} = \begin{bmatrix} x \\ y \end{bmatrix}$ we can also associate in a unique manner the point $P(x, y)$; conversely, with each point $P(x, y)$ we associate in a unique manner the vector $\begin{bmatrix} x \\ y \end{bmatrix}$. Thus we also write the vector $\mathbf{X}$ as

$$\mathbf{X} = (x, y).$$

Of course, this association is obtained by means of the directed line segment $\overrightarrow{OP}$, where O is the origin (Figure 3.6).

Thus the plane may be viewed both as the set of all points or as the set of all vectors. For this reason and, depending upon the context, we sometimes take R^2 as the set of all ordered pairs (x, y) and sometimes as the set of all 2×1 matrices $\begin{bmatrix} x \\ y \end{bmatrix}$.

Frequently, in physical applications it is necessary to deal with a directed line segment $\overrightarrow{PQ}$, from the point $P(x, y)$ (not the origin) to the point $Q(x', y')$, as shown in Figure 3.8(a). Such a directed line segment will also be called a **vector in the plane**, or simply a **vector** with **tail** at $P(x, y)$ and **head** at $Q(x', y')$. The components of such a vector are $x' - x$ and $y' - y$. Two such vectors will be called **equal** if they have the same directions and equal components. Consider the vectors $\overrightarrow{P_1Q_1}$, $\overrightarrow{P_2Q_2}$, and $\overrightarrow{P_3Q_3}$ joining the points $P_1(3, 2)$ and $Q_1(5, 5)$, $P_2(0, 0)$ and $Q_2(2, 3)$, $P_3(-3, 1)$ and $Q_3(-1, 4)$, respectively, as shown in Figure 3.8(b). Since they all have the same directions and since their components are also equal, the three vectors are equal. Moreover, the head $Q_4(x'_4, y'_4)$ of the vector $\overrightarrow{P_4Q_4} = \begin{bmatrix} 2 \\ 3 \end{bmatrix} = \overrightarrow{P_2Q_2}$, with tail $P_4(-5, 2)$ can be determined as follows. We must have $x'_4 - (-5) = 2$ and $y'_4 - 2 = 3$ so that $x'_4 = 2 - 5 = -3$ and $y'_4 = 3 + 2 = 5$. Similarly, the tail $P_5(x_5, y_5)$ of the vector $\overrightarrow{P_5Q_5} = \begin{bmatrix} 2 \\ 3 \end{bmatrix}$ with head $Q_5(8, 6)$ is determined as follows. We must have $8 - x_5 = 2$ and $6 - y_5 = 3$ so that $x_5 = 8 - 2 = 6$ and $y_5 = 6 - 3 = 3$.

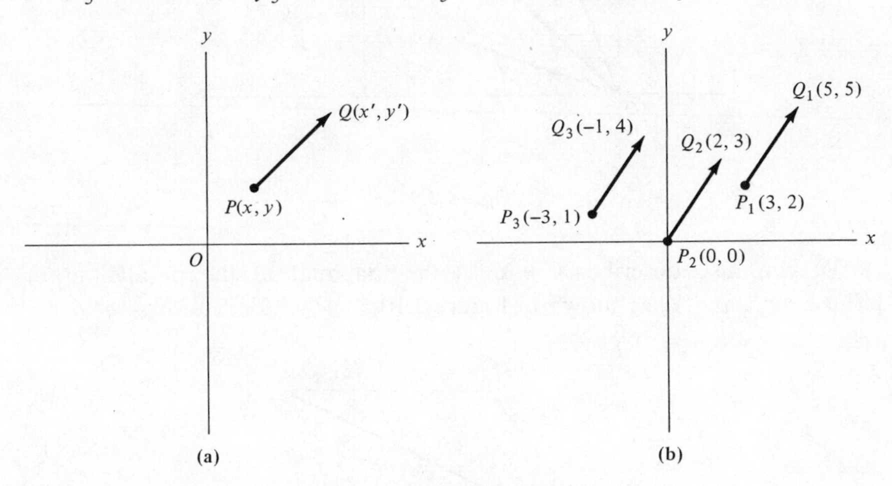

Figure 3.8

Vector Operations

DEFINITION Let $\mathbf{X} = (x_1, y_1)$ and $\mathbf{Y} = (x_2, y_2)$ be two vectors in the plane. The **sum** of the vectors $\mathbf{X}$ and $\mathbf{Y}$ is the vector

$$(x_1 + x_2, y_1 + y_2)$$

and is denoted by **X** + **Y**. Thus vectors are added by adding their components.

Example 6. Let **X** = (1, 2) and Y = (3, − 4). Then

$$\mathbf{X} + \mathbf{Y} = (1 + 3, 2 + (-4)) = (4, -2).$$

We can interpret vector addition geometrically as follows. In Figure 3.9, the vector **Z**, which is parallel to **Y**, is of the same length as **Y** and its tail is the head (x_1, y_1) of **X**; so its head is $(x_1 + x_2, y_1 + y_2)$. Thus the vector with tail at O and head at $(x_1 + x_2, y_1 + y_2)$ is **X** + **Y**.

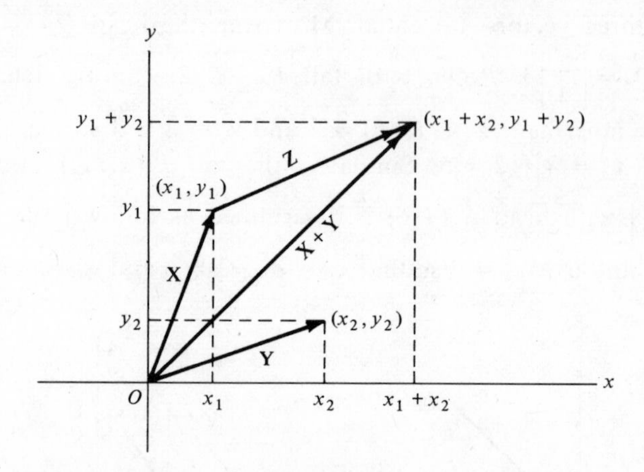

Figure 3.9

We can also describe **X** + **Y** as the diagonal of the parallelogram defined by **X** and **Y**, as shown in Figure 3.10.

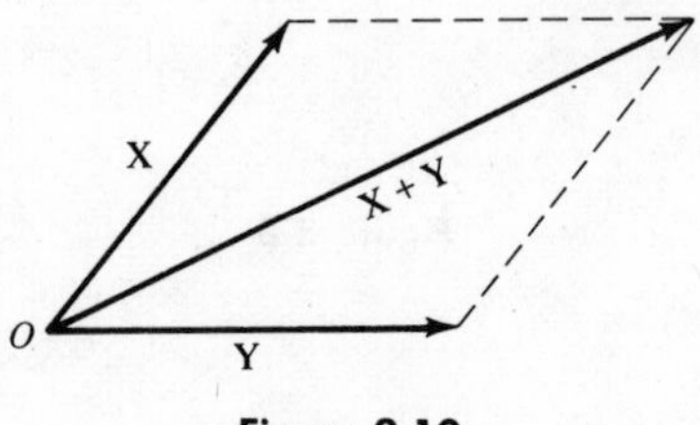

Figure 3.10

Example 7. If **X** and **Y** are as in Example 6, then **X** + **Y** is shown in Figure 3.11.

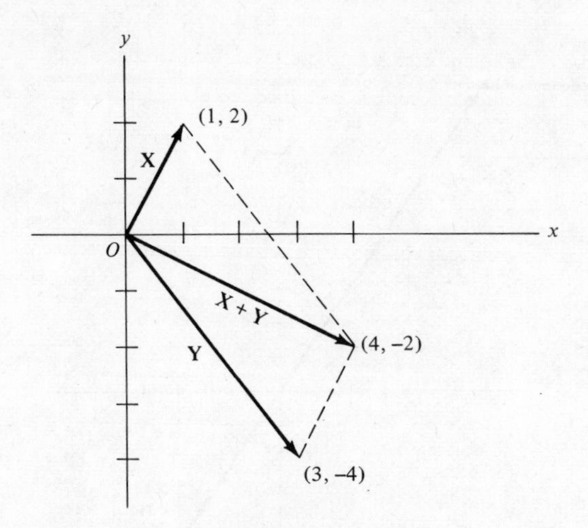

Figure 3.11

DEFINITION If $\mathbf{X} = (x, y)$ and c is a scalar (a real number), then the **scalar multiple $c\mathbf{X}$** of $\mathbf{X}$ by c is the vector (cx, cy). Thus the scalar multiple $c\mathbf{X}$ of $\mathbf{X}$ by c is obtained by multiplying each component of $\mathbf{X}$ by c.

If $c > 0$, then $c\mathbf{X}$ is in the same direction as $\mathbf{X}$, whereas if $d < 0$, then $d\mathbf{X}$ is in the opposite direction (Figure 3.12).

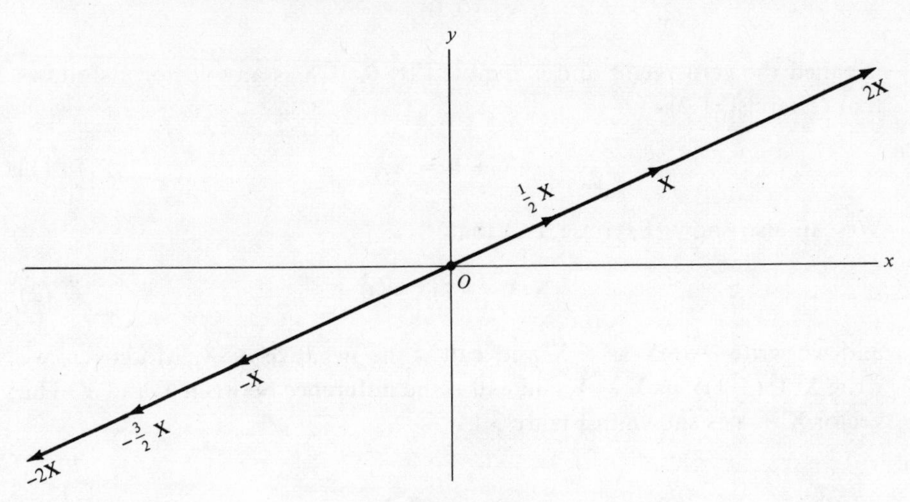

Figure 3.12

Example 8. If $c = 2$, $d = -3$, and $\mathbf{X} = (2, -3)$ then

$$c\mathbf{X} = 2(2, -3) = (4, -6) \quad \text{and} \quad d\mathbf{X} = -3(2, -3) = (-6, 9),$$

which are shown in Figure 3.13.

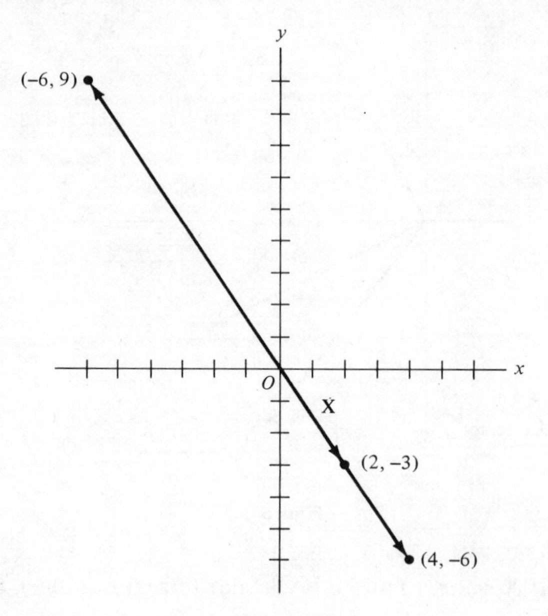

Figure 3.13

The vector

$$(0,\ 0)$$

is called the **zero vector** and is denoted by **0**. If **X** is any vector, it follows that (Exercise T.2)

$$\mathbf{X} + \mathbf{0} = \mathbf{X}. \tag{1}$$

We can also show (Exercise T.3) that

$$\mathbf{X} + (-1)\mathbf{X} = \mathbf{0}, \tag{2}$$

and we write $(-1)\mathbf{X}$ as $-\mathbf{X}$ and call it the **negative** of **X**. Moreover, we write $\mathbf{X} + (-1)\mathbf{Y}$ as $\mathbf{X} - \mathbf{Y}$ and call it the **difference between X and Y**. The vector $\mathbf{X} - \mathbf{Y}$ is shown in Figure 3.14.

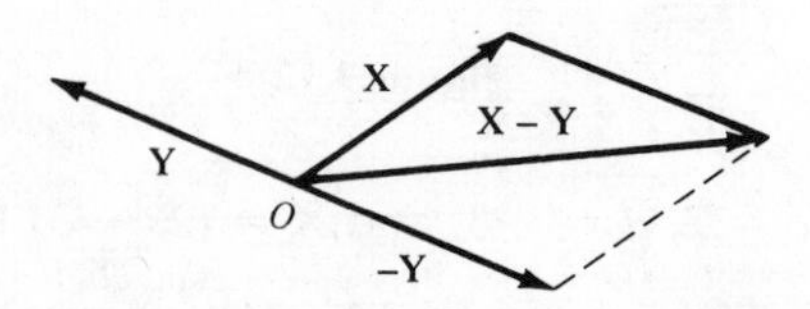

Figure 3.14

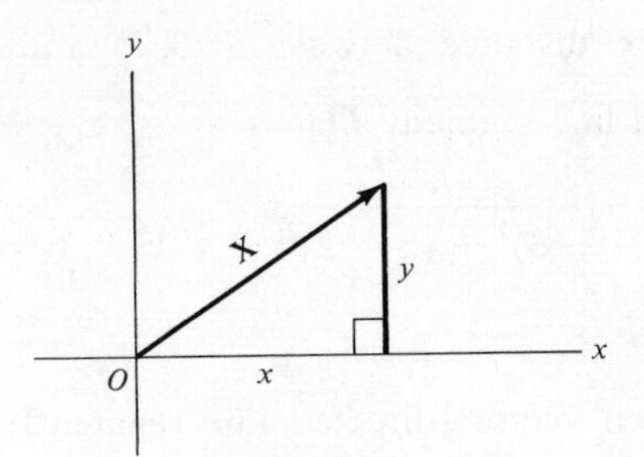

Figure 3.15

Length

By the Pythagorean Theorem (Figure 3.15) the **length** or **magnitude** of the vector $\mathbf{X} = (x, y)$ is

$$\|\mathbf{X}\| = \sqrt{x^2 + y^2}\,. \tag{3}$$

It also follows, by the Pythagorean Theorem, that the length of the directed line segment with initial point $P_1(x_1, y_1)$ and terminal point $P_2(x_2, y_2)$ is (Figure 3.16)

$$\|\overrightarrow{P_1P_2}\| = \sqrt{(x_2 - x_1)^2 + (y_2 - y_1)^2}\,. \tag{4}$$

Equation (4) also gives the distance between the points P_1 and P_2.

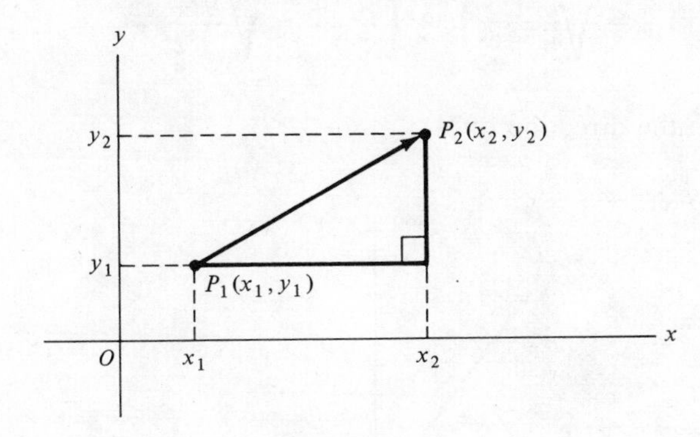

Figure 3.16

Example 9. If $\mathbf{X} = (2, -5)$, then

$$\|\mathbf{X}\| = \sqrt{(2)^2 + (-5)^2} = \sqrt{4 + 25} = \sqrt{29}\,.$$

Example 10. The distance between $P(3, 2)$ and $Q(-1, 5)$, or the length of the directed line segment $\overrightarrow{PQ}$, is

$$\|\overrightarrow{PQ}\| = \sqrt{(-1-3)^2 + (5-2)^2} = \sqrt{4^2 + 3^2} = \sqrt{25} = 5.$$

The length of each vector (directed line segment) $\overrightarrow{P_1Q_1}$, $\overrightarrow{P_2Q_2}$, and $\overrightarrow{P_3Q_3}$ in Figure 3.8(b) is $\sqrt{13}$ (verify).

A **unit vector** is a vector whose length is 1. If $\mathbf{X}$ is any nonzero vector, then the vector

$$\mathbf{U} = \frac{1}{\|\mathbf{X}\|}\mathbf{X} \tag{5}$$

is a unit vector in the direction of $\mathbf{X}$ (Exercise T.5).

Example 11. Let $\mathbf{X} = (-3, 4)$. Then

$$\|\mathbf{X}\| = \sqrt{(-3)^2 + 4^2} = 5.$$

Hence the vector $\mathbf{U} = \frac{1}{5}(-3, 4) = (-\frac{3}{5}, \frac{4}{5})$ is a unit vector, since

$$\|\mathbf{U}\| = \sqrt{\left(-\frac{3}{5}\right)^2 + \left(\frac{4}{5}\right)^2} = \sqrt{\frac{9+15}{25}} = 1.$$

Also, $\mathbf{U}$ lies in the direction of $\mathbf{X}$ (Figure 3.17).

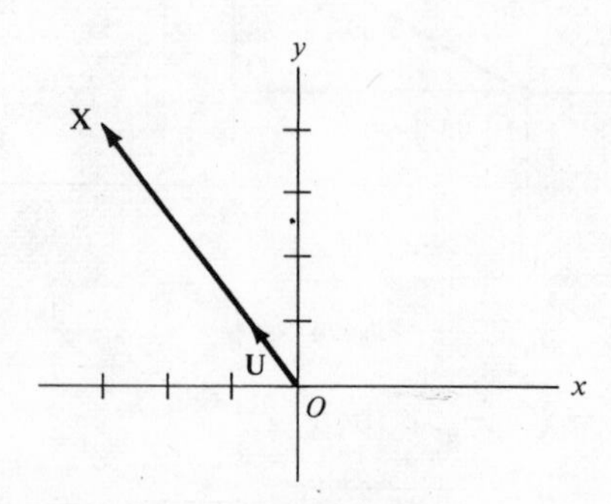

Figure 3.17

Angle Between Two Vectors

The angle between the nonzero vectors $\mathbf{X} = (x_1, y_1)$ and $\mathbf{Y} = (x_2, y_2)$ is the angle θ, $0 \leqslant \theta \leqslant 180°$, shown in Figure 3.18. Applying the law of cosines to the triangle in Figure 3.18, we obtain

$$\|\mathbf{X} - \mathbf{Y}\|^2 = \|\mathbf{X}\|^2 + \|\mathbf{Y}\|^2 - 2\|\mathbf{X}\|\,\|\mathbf{Y}\|\cos\theta. \tag{6}$$

From (3),

$$\begin{aligned} \|\mathbf{X} - \mathbf{Y}\|^2 &= (x_1 - x_2)^2 + (y_1 - y_2)^2 \\ &= x_1^2 + x_2^2 + y_1^2 + y_2^2 - 2(x_1x_2 + y_1y_2) \\ &= \|\mathbf{X}\|^2 + \|\mathbf{Y}\|^2 - 2(x_1x_2 + y_1y_2). \end{aligned}$$

If we substitute this expression in (6) and solve for $\cos\theta$ (recall that since $\mathbf{X}$ and $\mathbf{Y}$ are nonzero vectors, then $\|\mathbf{X}\| \neq 0$ and $\|\mathbf{Y}\| \neq 0$), we obtain

$$\cos\theta = \frac{x_1x_2 + y_1y_2}{\|\mathbf{X}\|\,\|\mathbf{Y}\|}. \tag{7}$$

DEFINITION The **inner product** or **dot product** of the vectors $\mathbf{X} = (x_1, y_1)$ and $\mathbf{Y} = (x_2, y_2)$ is defined by

$$\mathbf{X} \cdot \mathbf{Y} = x_1x_2 + y_1y_2.$$

Thus we can rewrite (7) as

$$\cos\theta = \frac{\mathbf{X} \cdot \mathbf{Y}}{\|\mathbf{X}\|\,\|\mathbf{Y}\|} \qquad 0 \leqslant \theta \leqslant \pi. \tag{8}$$

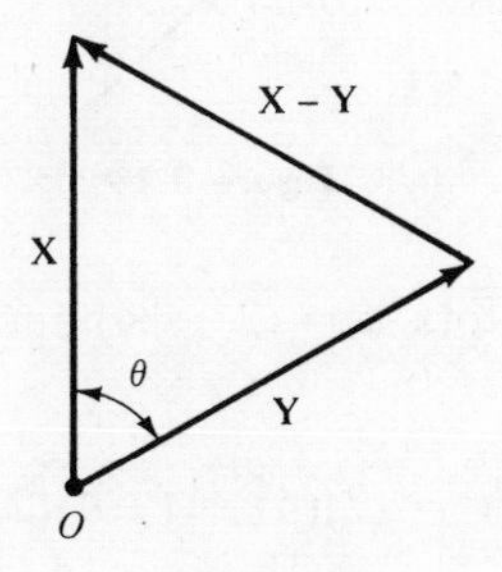

Figure 3.18

Example 12. If $\mathbf{X} = (2, 4)$ and $\mathbf{Y} = (-1, 2)$, then

$$\mathbf{X} \cdot \mathbf{Y} = (2)(-1) + (4)(2) = 6.$$

Also,

$$\|\mathbf{X}\| = \sqrt{2^2 + 4^2} = \sqrt{20}$$

and

$$\|\mathbf{Y}\| = \sqrt{(-1)^2 + 2^2} = \sqrt{5}\ .$$

Hence

$$\cos\theta = \frac{6}{\sqrt{20}\cdot\sqrt{5}} = 0.6.$$

We can obtain the approximate angle by using a table of cosines; we find that θ is approximately $53°8'$.

If the nonzero vectors **X** and **Y** are at right angles (Figure 3.19), then the cosine of the angle θ between them is zero. Hence, from (8), we have $\mathbf{X} \cdot \mathbf{Y} = 0$. Conversely, if $\mathbf{X} \cdot \mathbf{Y} = 0$, then $\cos\theta = 0$ and the vectors are at right angles. Thus the nonzero vectors **X** and **Y** are **perpendicular** or **orthogonal** if and only if $\mathbf{X} \cdot \mathbf{Y} = 0$.

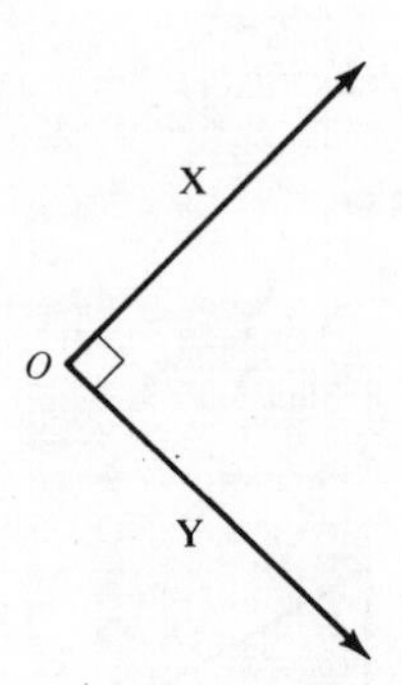

Figure 3.19

Example 13. The vectors $\mathbf{X} = (2, -4)$ and $\mathbf{Y} = (4, 2)$ are orthogonal, since

$$\mathbf{X} \cdot \mathbf{Y} = (2)(4) + (-4)(2) = 0$$

(Figure 3.20).

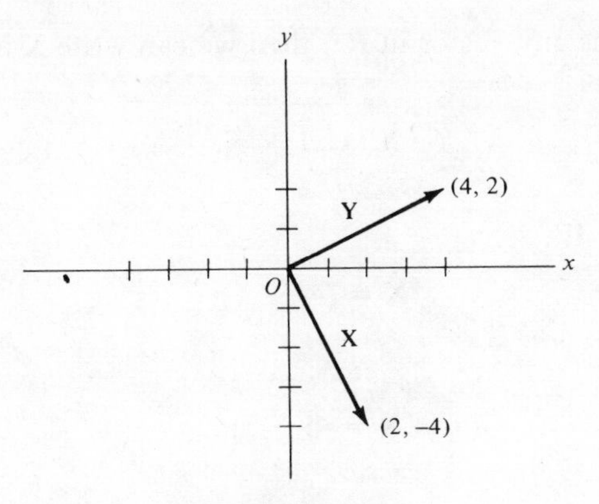

Figure 3.20

It is not too difficult to verify that the vectors (directed line segments) $\overrightarrow{P_1Q_1}$, $\overrightarrow{P_2Q_2}$, and $\overrightarrow{P_3Q_3}$ in Figure 3.8(b) are parallel (verify).

The dot product satisfies the following properties.

THEOREM 3.1. *If* $\mathbf{X}$, $\mathbf{Y}$, *and* $\mathbf{Z}$ *are vectors and* c *is a scalar, then*:

(a) $\mathbf{X}\cdot\mathbf{X} = \|\mathbf{X}\|^2 \geqslant 0$, *with equality if and only if* $\mathbf{X} = 0$.

(b) $\mathbf{X}\cdot\mathbf{Y} = \mathbf{Y}\cdot\mathbf{X}$.

(c) $(\mathbf{X}+\mathbf{Y})\cdot\mathbf{Z} = \mathbf{X}\cdot\mathbf{Z} + \mathbf{Y}\cdot\mathbf{Z}$.

(d) $(c\mathbf{X})\cdot\mathbf{Y} = \mathbf{X}\cdot(c\mathbf{Y}) = c(\mathbf{X}\cdot\mathbf{Y})$.

proof. Exercise T.7.

There are two unit vectors in R^2 that are of special importance. They are $\mathbf{i} = (1, 0)$ and $\mathbf{j} = (0, 1)$, the unit vectors along the positive x- and y-axes, respectively (Figure 3.21).

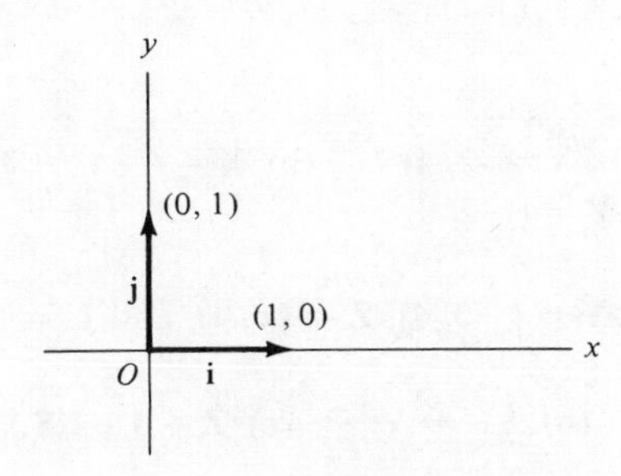

Figure 3.21

If $\mathbf{X} = (x, y)$ is any vector in R^2, then we can write $\mathbf{X}$ in terms of $\mathbf{i}$ and $\mathbf{j}$ as

$$\mathbf{X} = x\mathbf{i} + y\mathbf{j}.$$

Example 14. If

$$\mathbf{X} = (4, -5),$$

then

$$\mathbf{X} = 4\mathbf{i} - 5\mathbf{j}.$$

3.1 Exercises

1. Plot the following points in R^2: (a) $(2, -1)$. (b) $(-1, 2)$. (c) $(3, 4)$. (d) $(-3, -2)$. (e) $(0, 2)$. (f) $(-3, 0)$.

2. Sketch a directed line segment in R^2 representing each of the following vectors.
(a) $\mathbf{X}_1 = \begin{bmatrix} -2 \\ 3 \end{bmatrix}$. (b) $\mathbf{X}_2 = \begin{bmatrix} 3 \\ 4 \end{bmatrix}$. (c) $\mathbf{X}_3 = \begin{bmatrix} -3 \\ -3 \end{bmatrix}$.
(d) $\mathbf{X}_4 = \begin{bmatrix} 0 \\ -3 \end{bmatrix}$.

3. Determine the head of the vector $\begin{bmatrix} -2 \\ 5 \end{bmatrix}$ whose tail is at $(3, 2)$. Make a sketch.

4. Determine the tail of the vector $\begin{bmatrix} 2 \\ 5 \end{bmatrix}$ whose head is at $(1, 2)$. Make a sketch.

5. Find $\mathbf{X} + \mathbf{Y}$, $\mathbf{X} - \mathbf{Y}$, $2\mathbf{X}$, and $3\mathbf{X} - 2\mathbf{Y}$ if
(a) $\mathbf{X} = (2, 3)$, $\mathbf{Y} = (-2, 5)$. (b) $\mathbf{X} = (0, 3)$, $\mathbf{Y} = (3, 2)$.
(c) $\mathbf{X} = (2, 6)$, $\mathbf{Y} = (3, 2)$.

6. Repeat Exercise 5 for
(a) $\mathbf{X} = (-1, 3)$, $\mathbf{Y} = (2, 4)$. (b) $\mathbf{X} = (-4, -3)$, $\mathbf{Y} = (5, 2)$.
(c) $\mathbf{X} = (3, 2)$, $\mathbf{Y} = (-2, 0)$.

7. Let $\mathbf{X} = (1, 2)$, $\mathbf{Y} = (-3, 4)$, $\mathbf{Z} = (x, 4)$, and $\mathbf{U} = (-2, y)$. Find x and y so that
(a) $\mathbf{Z} = 2\mathbf{X}$. (b) $\frac{3}{2}\mathbf{U} = \mathbf{Y}$. (c) $\mathbf{Z} + \mathbf{U} = \mathbf{X}$.

8. Let $\mathbf{X} = (-4, 3)$, $\mathbf{Y} = (2, -5)$, and $\mathbf{Z} = (x, y)$. Find x and y so that
(a) $\mathbf{Z} = 2\mathbf{X} + 3\mathbf{Y}$. (b) $\mathbf{X} + \mathbf{Z} = 2\mathbf{X} - \mathbf{Y}$. (c) $\mathbf{Z} = \frac{5}{2}\mathbf{Y}$.

9. Find the length of the following vectors.
(a) (1, 2). (b) (−3, −4). (c) (0, 2). (d) (−4, 3).

10. Find the length of the following vectors.
(a) (−2, 3). (b) (3, 0). (c) (−4, −5). (d) 3, 2).

11. Find the distance between the following pairs of points.
(a) (2, 3), (4, 5). (b) (0, 0), (3, 4).
(c) (−3, 2), (0, 1). (d) (0, 3), (2, 0).

12. Find the distance between the following pairs of points.
(a) (4, 2), (1, 2). (b) (−2, −3), (0, 1).
(c) (2, 4), (−1, 1). (d) (2, 0), (3, 2).

13. Find a unit vector in the direction of **X**.
(a) $\mathbf{X} = (3, 4)$. (b) $\mathbf{X} = (-2, -3)$. (c) $\mathbf{X} = (5, 0)$.

14. Find a unit vector in the direction of X.
(a) $\mathbf{X} = (2, 4)$. (b) $\mathbf{X} = (0, -2)$. (c) $\mathbf{X} = (-1, -3)$.

15. Find $\mathbf{X} \cdot \mathbf{Y}$.
(a) $\mathbf{X} = (1, 2)$, $\mathbf{Y} = (2, -3)$. (b) $\mathbf{X} = (1, 0)$, $\mathbf{Y} = (0, 1)$.
(c) $\mathbf{X} = (-3, -4)$, $\mathbf{Y} = (4, -3)$. (d) $\mathbf{X} = (2, 1)$, $\mathbf{Y} = (-2, -1)$.

16. Find $\mathbf{X} \cdot \mathbf{Y}$.
(a) $\mathbf{X} = (0, -1)$, $\mathbf{Y} = (1, 0)$. (b) $\mathbf{X} = (2, 2)$, $\mathbf{Y} = (4, -5)$.
(c) $\mathbf{X} = (2, -1)$, $\mathbf{Y} = (-3, -2)$. (d) $\mathbf{X} = (0, 2)$, $\mathbf{Y} = (3, -3)$.

17. Find the cosine of the angle between each pair of vectors **X** and **Y** in Exercise 15.

18. Find the cosine of the angle between each pair of vectors **X** and **Y** in Exercise 16.

19. Show that
(a) $\mathbf{i} \cdot \mathbf{i} = \mathbf{j} \cdot \mathbf{j} = 1$. (b) $\mathbf{i} \cdot \mathbf{j} = 0$.

20. Which of the vectors $\mathbf{X}_1 = (1, 2)$, $\mathbf{X}_2 = (0, 1)$, $\mathbf{X}_3 = (-2, -4)$, $\mathbf{X}_4 = (-2, 1)$, $\mathbf{X}_5 = (2, 4)$, and $\mathbf{X}_6 = (-6, 3)$ are
(a) Orthogonal? (b) In the same direction?
(c) In opposite directions?

21. Write each of the following vectors in terms of **i** and **j**.
(a) (1, 3). (b) (−2, −3). (c) (−2, 0). (d) 0, 3).

22. Write each of the following vectors as a 2 × 1 matrix.
(a) $3\mathbf{i} - 2\mathbf{j}$. (b) $2\mathbf{i}$. (c) $-2\mathbf{i} - 3\mathbf{j}$.

Theoretical Exercises

T.1. Show how we can associate a point in the plane with each ordered pair (x, y) of numbers.

T.2. Show that $\mathbf{X} + \mathbf{0} = \mathbf{X}$.

T.3. Show that $\mathbf{X} + (-1)\mathbf{X} = \mathbf{0}$.

T.4. Show that if c is a scalar, then $\|c\mathbf{X}\| = |c|\ \|\mathbf{X}\|$.

T.5. Show that if $\mathbf{X}$ is a nonzero vector, then $\mathbf{U} = (1/\|\mathbf{X}\|)\mathbf{X}$ is a unit vector in the direction of $\mathbf{X}$.

T.6. Show that

(a) $1\mathbf{X} = \mathbf{X}$.

(b) $(rs)\mathbf{X} = r(s\mathbf{X})$, where r and s are scalars.

T.7. Prove Theorem 3.1.

T.8. Show that if $\mathbf{Z}$ is orthogonal to $\mathbf{X}$ and $\mathbf{Y}$, then $\mathbf{Z}$ is orthogonal to $r\mathbf{X} + s\mathbf{Y}$, where r and s are scalars.

3.2 n-Vectors

We shall now generalize the notion of a vector in the plane, discussed in the previous section, to n-vectors. For $n = 2$, we obtain the vectors in the plane. The case of $n = 3$ will also be of special interest, and we shall discuss it in some detail.

DEFINITION An n-**vector** is an $n \times 1$ matrix

$$\mathbf{X} = \begin{bmatrix} x_1 \\ x_2 \\ \cdot \\ \vdots \\ x_n \end{bmatrix},$$

where $x_1, x_2, \ldots, x_n$ are real numbers, which are called the **components** of $\mathbf{X}$.

Since an n-vector is a matrix, the n-vectors

$$\mathbf{X} = \begin{bmatrix} x_1 \\ x_2 \\ \cdot \\ \vdots \\ x_n \end{bmatrix} \quad \text{and} \quad \mathbf{Y} = \begin{bmatrix} y_1 \\ y_2 \\ \cdot \\ \vdots \\ y_n \end{bmatrix}$$

are said to be **equal** if $x_i = y_i$ $(1 \leqslant i \leqslant n)$.

Example 1. The 4-vectors

$$\begin{bmatrix} 1 \\ -2 \\ 3 \\ 4 \end{bmatrix} \quad \text{and} \quad \begin{bmatrix} 1 \\ -2 \\ 3 \\ -4 \end{bmatrix}$$

are not equal, since their fourth components are not equal.

The set of all n-vectors is denoted by R^n and is called n-**space**. When the actual value of n need not be specified, we refer to n-vectors simply as **vectors**. The real numbers are also called **scalars**.

Vector Operations

DEFINITION Let

$$\mathbf{X} = \begin{bmatrix} x_1 \\ x_2 \\ \cdot \\ \vdots \\ x_n \end{bmatrix} \quad \text{and} \quad \mathbf{Y} = \begin{bmatrix} y_1 \\ y_2 \\ \vdots \\ \vdots \\ y_n \end{bmatrix}$$

be two vectors in R^n. The **sum** of the vectors **X** and **Y** is the vector

$$\begin{bmatrix} x_1 + y_1 \\ x_2 + y_2 \\ \vdots \\ \vdots \\ x_n + y_n \end{bmatrix},$$

and it is denoted by $\mathbf{X} + \mathbf{Y}$.

Example 2. If

$$\mathbf{X} = \begin{bmatrix} 1 \\ -2 \\ 3 \end{bmatrix} \quad \text{and} \quad \mathbf{Y} = \begin{bmatrix} 2 \\ 3 \\ -3 \end{bmatrix}$$

are vectors in R^3, then

$$\mathbf{X} + \mathbf{Y} = \begin{bmatrix} 1 + 2 \\ -2 + 3 \\ 3 + (-3) \end{bmatrix} = \begin{bmatrix} 3 \\ 1 \\ 0 \end{bmatrix}.$$

DEFINITION If

$$\mathbf{X} = \begin{bmatrix} x_1 \\ x_2 \\ \cdot \\ \cdot \\ \cdot \\ x_n \end{bmatrix}$$

is a vector in R^n and c is a scalar, then the **scalar multiple** $c\mathbf{X}$ of $\mathbf{X}$ by c is the vector

$$\begin{bmatrix} cx_1 \\ cx_2 \\ \cdot \\ \cdot \\ \cdot \\ cx_n \end{bmatrix}.$$

Example 3. If

$$\mathbf{X} = \begin{bmatrix} 2 \\ 3 \\ -1 \\ 2 \end{bmatrix}$$

is a vector in R^4 and $c = -2$, then

$$c\mathbf{X} = (-2)\begin{bmatrix} 2 \\ 3 \\ -1 \\ 2 \end{bmatrix} = \begin{bmatrix} -4 \\ -6 \\ 2 \\ -4 \end{bmatrix}.$$

The operations of vector addition and scalar multiplication satisfy the following properties.

THEOREM 3.2. *Let* $\mathbf{X}$, $\mathbf{Y}$, *and* $\mathbf{Z}$ *be any vectors in* R^n; *let* c *and* d *be any scalars. Then*

(α) $\mathbf{X} + \mathbf{Y}$ *is a vector in* R^n (*that is,* R^n *is closed under the operation of vector addition*).

(a) $\mathbf{X} + \mathbf{Y} = \mathbf{Y} + \mathbf{X}$.

(b) $\mathbf{X} + (\mathbf{Y} + \mathbf{Z}) = (\mathbf{X} + \mathbf{Y}) + \mathbf{Z}$.

(c) *There is a unique vector* **0** *in* R^n,

$$\mathbf{0} = \begin{bmatrix} 0 \\ 0 \\ \vdots \\ 0 \end{bmatrix},$$

such that $\mathbf{X} + \mathbf{0} = \mathbf{0} + \mathbf{X} = \mathbf{X}$; **0** *is called the* **zero vector**.

(d) *There is a unique vector* $-\mathbf{X}$,

$$-\mathbf{X} = \begin{bmatrix} -x_1 \\ -x_2 \\ \vdots \\ -x_n \end{bmatrix},$$

such that

$$\mathbf{X} + (-\mathbf{X}) = \mathbf{0}.$$

(β) $c\mathbf{X}$ *is a vector in* R^n.

(e) $c(\mathbf{X} + \mathbf{Y}) = c\mathbf{X} + c\mathbf{Y}$.

(f) $(c + d)\mathbf{X} = c\mathbf{X} + d\mathbf{X}$.

(g) $c(d\mathbf{X}) = (cd)\mathbf{X}$.

(h) $1\mathbf{X} = \mathbf{X}$.

proof. (α) and (β) are immediate from the definitions for vector sum and scalar multiple. We verify (f) here and leave the rest of the proof to the reader (Exercise T.1). Thus

$$(c + d)\mathbf{X} = (c + d)\begin{bmatrix} x_1 \\ x_2 \\ \vdots \\ x_n \end{bmatrix} = \begin{bmatrix} (c + d)x_1 \\ (c + d)x_2 \\ \vdots \\ (c + d)x_n \end{bmatrix} = \begin{bmatrix} cx_1 + dx_1 \\ cx_2 + dx_2 \\ \vdots \\ cx_n + dx_n \end{bmatrix}$$

$$= \begin{bmatrix} cx_1 \\ cx_2 \\ \vdots \\ cx_n \end{bmatrix} + \begin{bmatrix} dx_1 \\ dx_2 \\ \vdots \\ dx_n \end{bmatrix} = c\begin{bmatrix} x_1 \\ x_2 \\ \vdots \\ x_n \end{bmatrix} + d\begin{bmatrix} x_1 \\ x_2 \\ \vdots \\ x_n \end{bmatrix} = c\mathbf{X} + d\mathbf{X}.$$

The vector $-\mathbf{X}$ in (d) is called the **negative** of $\mathbf{X}$. It is easy to verify (Exercise T.2) that

$$-\mathbf{X} = (-1)\mathbf{X}.$$

We shall also write $\mathbf{X} + (-\mathbf{Y})$ as $\mathbf{X} - \mathbf{Y}$.

Example 4. If $\mathbf{X}$ and $\mathbf{Y}$ are as in Example 2, then

$$\mathbf{X} - \mathbf{Y} = \begin{bmatrix} 1 - 2 \\ -2 - 3 \\ 3 - (-3) \end{bmatrix} = \begin{bmatrix} -1 \\ -5 \\ 6 \end{bmatrix}.$$

As in the case of R^2, we shall identify the vector

$$\begin{bmatrix} x_1 \\ x_2 \\ \vdots \\ x_n \end{bmatrix}$$

with the point $(x_1, x_2, \ldots, x_n)$ so that points and vectors can be used interchangeably. Thus we may view R^n as consisting of vectors or of points, and we write $\mathbf{X} = (x_1, x_2, \ldots, x_n)$. Moreover, an n-vector is an $n \times 1$ matrix, vector addition is matrix addition, and scalar multiplication is merely the operation of multiplication of a matrix by a real number. Thus R^n can be viewed as the set of all $n \times 1$ matrices with the operations of matrix addition and scalar multiplication. The important point here is that no matter how we view R^n—as n-vectors, points, or $n \times 1$ matrices—its algebraic behavior is always the same.

Visualizing R^3

We cannot draw pictures of R^n for $n > 3$. However, since R^3 is the world we live in, we can visualize it in a manner similar to that used for R^2.

We first fix a **coordinate system** by choosing a point, called the **origin**, and three lines, called the **coordinate axes**, so that each line is perpendicular to the other two. These lines are called the **x-, y-, and z-axes**. On each of these axes we choose a point fixing the units of length and positive directions on the coordinate axes. Frequently, but not always, the same unit of length is used for all the coordinate axes. In Figure 3.22(a) and (b) we show two of the many possible coordinate systems.

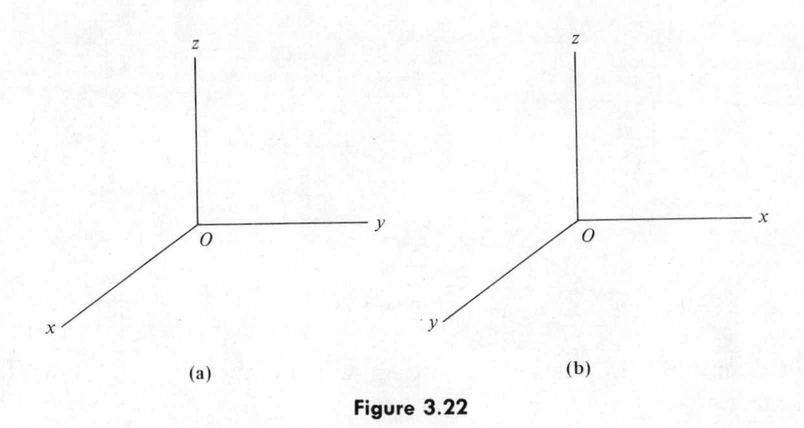

Figure 3.22

The coordinate system shown in Figure 3.22(a) is called a **right-handed coordinate system**; the one shown in Figure 3.22(b) is called **left-handed**. A right-handed system is characterized by the following property. If we rotate the x-axis counterclockwise toward the y-axis, then a right-hand screw will move in the positive z direction (Figure 3.23).

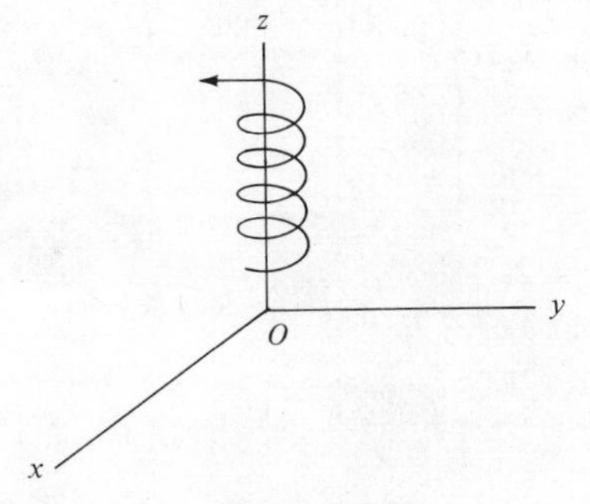

Figure 3.23

The projection of a point P in space on a line L is the point Q obtained by intersecting L with the line L' passing through Q and perpendicular to L (Figure 3.24).

The **x-coordinate** of the point P is the number associated with the projection of P on the x-axis; similarly, for the **y-and z-coordinates**. These three numbers are called the **coordinates** of P. Thus with each point in space we associate an ordered triple (x, y, z) of real numbers, and conversely, with each ordered triple of real numbers, we associate a point in space. This correspondence is called a **rectangular coordinate system**. We write $P(x, y, z)$, or simply (x, y, z).

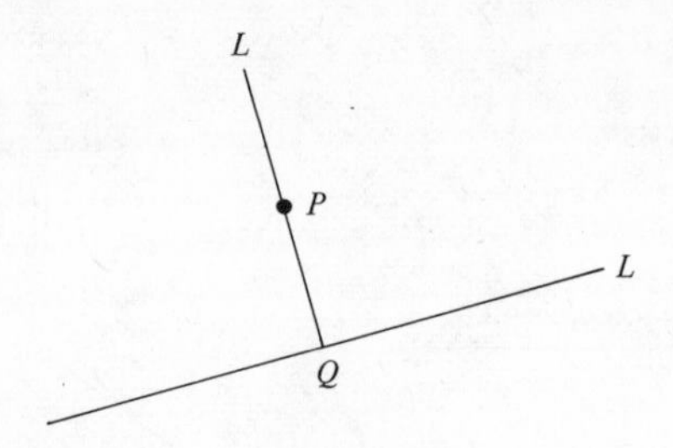

Figure 3.24

Example 5. In Figure 3.25 we show a number of points and their coordinates.

The xy-**plane** is the plane determined by the x- and y-axes. Similarly, we have the xz- and yz-**planes**.

In R^3, the components of a vector $\mathbf{X}$ are denoted by x, y, and z. Thus $\mathbf{X} = (x, y, z)$.

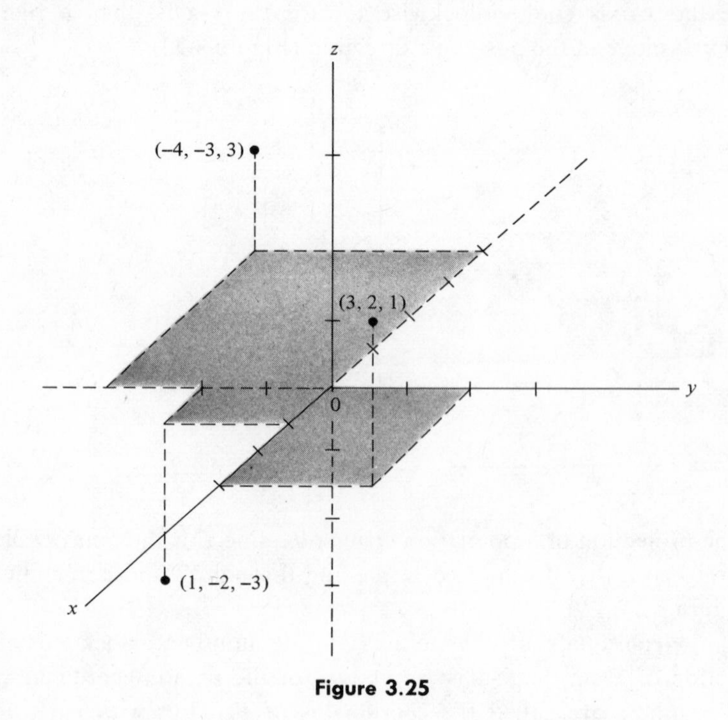

Figure 3.25

As in the plane, with the vector $\mathbf{X} = (x, y, z)$ we associate the directed line segment $\overrightarrow{OP}$, whose tail is $O(0, 0, 0)$ and whose head is $P(x, y, z)$

[Figure 3.26(a)]. Again as in the plane, in physical applications we often deal with a directed line segment $\overrightarrow{PQ}$, from the point $P(x, y, z)$ (not the origin) to the point $Q(x', y', z')$, as shown in Figure 3.26(b). Such a directed line segment will also be called a **vector in** $\mathbf{R}^3$, or simply a **vector** with tail at $P(x, y, z)$ and head at $Q(x', y', z')$. The components of such a vector are $x' - x, y' - y$, and $z' - z$.

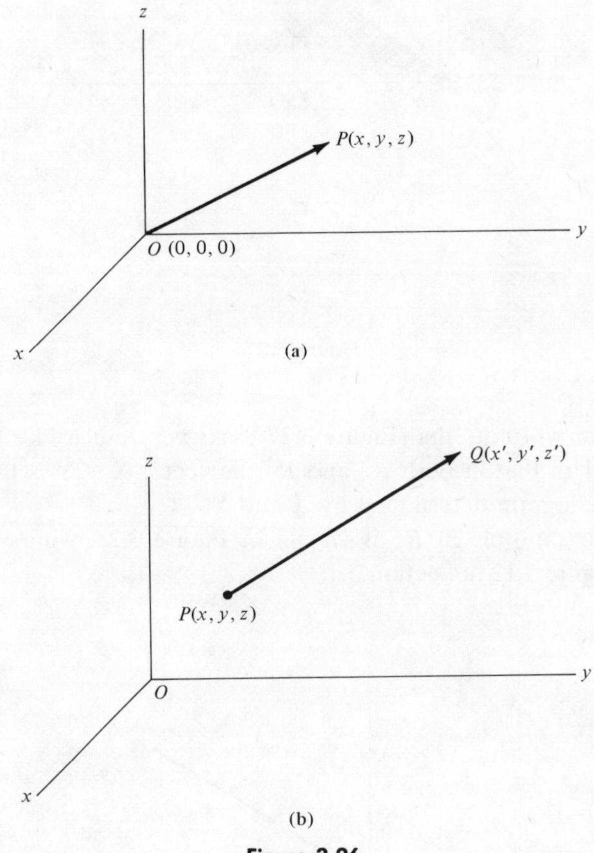

Figure 3.26

It can be shown that two vectors in R^3 are equal if and only if any directed line segments representing them are parallel and of the same length.

The sum $\mathbf{X} + \mathbf{Y}$ of the vectors $\mathbf{X} = (x_1, y_1, z_1)$ and $\mathbf{Y} = (x_2, y_2, z_2)$ in R^3 is the diagonal of the parallelogram determined by X and Y, as shown in Figure 3.27.

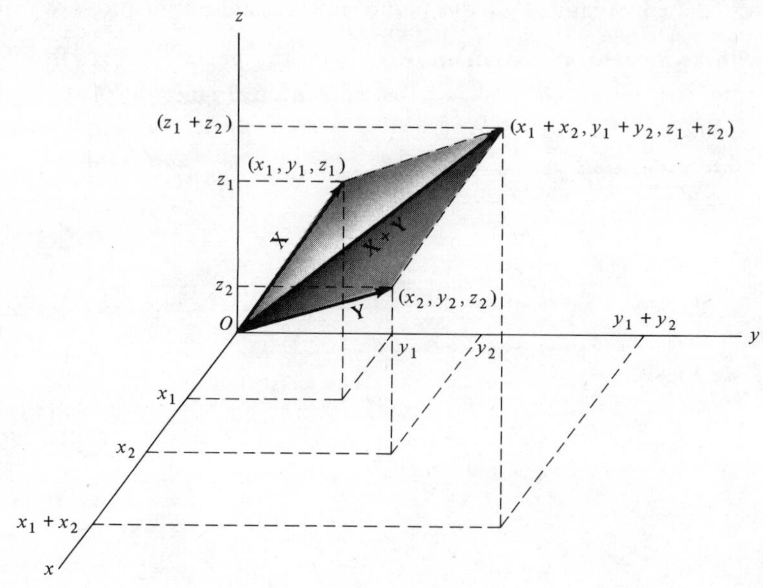

Figure 3.27

The reader will note that Figure 3.27 looks very much like Figure 3.10 in Section 3.1 in that in both R^2 and R^3 the vector $\mathbf{X} + \mathbf{Y}$ is the diagonal of the parallellogram determined by **X** and **Y**.

The scalar multiple in R^3 is shown in Figure 3.28, which looks very much like Figure 3.12 in Section 3.1.

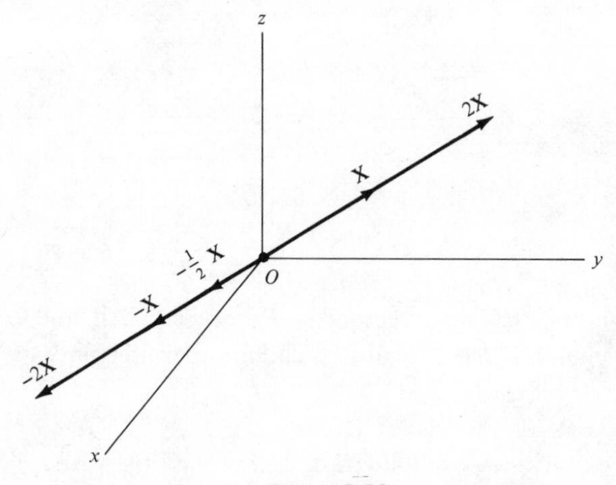

Figure 3.28

Inner Product in R^n

We shall now define the notion of the length of a vector in R^n by generalizing the corresponding idea for R^2.

DEFINITION The **length** or **magnitude** or **norm** of the vector $\mathbf{X} = (x_1, x_2, \ldots, x_n)$ in R^n is

$$\|\mathbf{X}\| = \sqrt{x_1^2 + x_2^2 + \cdots + x_n^2}\,. \tag{1}$$

We also define the distance from the point $(x_1, x_2, \ldots, x_n)$ to the origin by (1). The **distance** between the points $(x_1, x_2, \ldots, x_n)$ and $(y_1, y_2, \ldots, y_n)$ is then defined as the length of the vector $\mathbf{X} - \mathbf{Y}$, where

$$\mathbf{X} = (x_1, x_2, \ldots, x_n) \qquad \text{and} \qquad \mathbf{Y} = (y_1, y_2, \ldots, y_n).$$

Thus this distance is given by

$$\|\mathbf{X} - \mathbf{Y}\| = \sqrt{(x_1 - y_1)^2 + (x_2 - y_2)^2 + \cdots + (x_n - y_n)^2}\,. \tag{2}$$

Example 6. Let $\mathbf{X} = (2, 3, 2, -1)$ and $\mathbf{Y} = (4, 2, 1, 3)$. Then

$$\|\mathbf{X}\| = \sqrt{2^2 + 3^2 + 2^2 + (-1)^2} = \sqrt{18}\,,$$

$$\|\mathbf{Y}\| = \sqrt{4^2 + 2^2 + 1^2 + 3^2} = \sqrt{30}\,.$$

The distance between the points $(2, 3, 2, -1)$ and $(4, 2, 1, 3)$ is the length of the vector $\mathbf{X} - \mathbf{Y}$. Thus, from Equation (2),

$$\|\mathbf{X} - \mathbf{Y}\| = \sqrt{(2-4)^2 + (3-2)^2 + (2-1)^2 + (-1-3)^2} = \sqrt{22}\,.$$

It should be noted that in R^3, Equations (1) and (2), for the length of a vector and the distance between two points, do not have to be defined. They can easily be established by means of two applications of the Pythagorean Theorem (Exercise T.3).

We shall define the cosine of the angle between two vectors in R^n by generalizing the corresponding formula in R^2. However, first we formulate the notion of inner product in R^n.

DEFINITION If $\mathbf{X} = (x_1, x_2, \ldots, x_n)$ and $\mathbf{Y} = (y_1, y_2, \ldots, y_n)$ are vectors in R^n, then their **inner product** is defined by

$$\mathbf{X} \cdot \mathbf{Y} = x_1 y_1 + x_2 y_2 + \cdots + x_n y_n.$$

This is exactly how the dot product was defined in R^2. The inner product in R^n is also called **dot product**.

Example 7. If **X** and **Y** are as in Example 6, then

$$\begin{aligned} \mathbf{X} \cdot \mathbf{Y} &= (2)(4) + (3)(2) + (2)(1) + (-1)(3) \\ &= 13. \end{aligned}$$

The inner product in R^n satisfies the same properties as in R^2. We state these as the following theorem, which reads just like Theorem 3.1.

THEOREM 3.3. *If* **X**, **Y**, *and* **Z** *are vectors in* R^n *and* c *is a scalar, then*:

(a) $\mathbf{X} \cdot \mathbf{X} = \|\mathbf{X}\|^2 \geqslant 0$, *with equality if and only if* $\mathbf{X} = \mathbf{0}$.
(b) $\mathbf{X} \cdot \mathbf{Y} = \mathbf{Y} \cdot \mathbf{X}$.
(c) $(\mathbf{X} + \mathbf{Y}) \cdot \mathbf{Z} = \mathbf{X} \cdot \mathbf{Z} + \mathbf{Y} \cdot \mathbf{Z}$.
(d) $(c\mathbf{X}) \cdot \mathbf{Y} = \mathbf{X} \cdot (c\mathbf{Y}) = c(\mathbf{X} \cdot \mathbf{Y})$.

proof. Exercise T.4.

We shall now prove a result that will enable us to give a worthwhile definition for the cosine of an angle between two nonzero vectors. This result, called the **Cauchy–Schwarz inequality**, has many important applications in mathematics. The proof of this result, although not difficult, is one that is not too natural and does call for a clever start.

THEOREM 3.4 (Cauchy–Schwarz Inequality). *If* **X** *and* **Y** *are vectors in* R^n, *then*

$$|\mathbf{X} \cdot \mathbf{Y}| \leqslant \|\mathbf{X}\| \, \|\mathbf{Y}\|. \tag{3}$$

(*Observe that* $| \quad |$ *on the left stands for the absolute value of a real number*; $\| \quad \|$ *on the right denotes the length of a vector*.)

proof. If $\mathbf{X} = \mathbf{0}$, then $\|\mathbf{X}\| = 0$ and $\mathbf{X} \cdot \mathbf{Y} = 0$, so (3) holds. Now suppose that both **X** and **Y** are nonzero. Let r be a scalar and consider the vector $r\mathbf{X} + \mathbf{Y}$. By Theorem 3.3,

$$0 \leqslant (r\mathbf{X} + \mathbf{Y}) \cdot (r\mathbf{X} + \mathbf{Y}) = r^2\mathbf{X} \cdot \mathbf{X} + 2r\mathbf{X} \cdot \mathbf{Y} + \mathbf{Y} \cdot \mathbf{Y}$$
$$= ar^2 + 2br + c,$$

where

$$a = \mathbf{X} \cdot \mathbf{X}, \qquad b = \mathbf{X} \cdot \mathbf{Y}, \qquad \text{and} \qquad c = \mathbf{Y} \cdot \mathbf{Y}.$$

Now $p(r) = ar^2 + 2br + c$ is a quadratic polynomial in r that is nonnegative for all values of r. This means that either this polynomial has no real roots, or if it has real roots, then both roots are equal. [If $p(r)$ had two distinct roots r_1 and r_2, then it would be negative for some value of r between r_1 and r_2, as seen in Figure 3.29.]

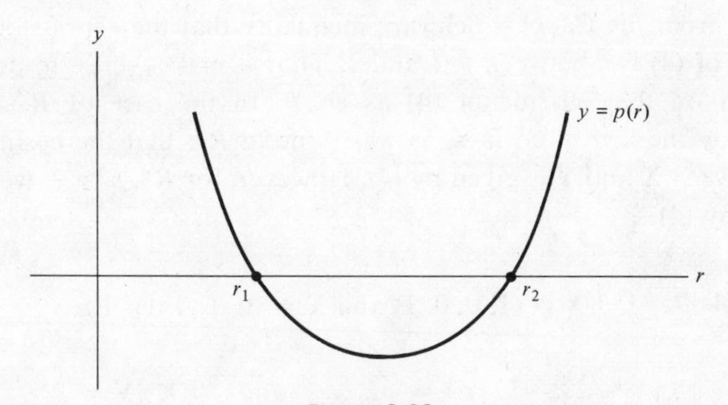

Figure 3.29

Recall that the roots of $p(r)$ are given by the quadratic formula as

$$\frac{-2b + \sqrt{4b^2 - 4ac}}{2a} \qquad \text{and} \qquad \frac{-2b - \sqrt{4b^2 - 4ac}}{2a}$$

($a \neq 0$ since $\mathbf{X} \neq \mathbf{0}$). Thus we must have

$$4b^2 - 4ac \leqslant 0,$$

which means that

$$b^2 \leqslant ac.$$

Taking square roots of both sides and observing that $\sqrt{a} = \sqrt{\mathbf{X} \cdot \mathbf{X}} = \|\mathbf{X}\|$, $\sqrt{c} = \sqrt{\mathbf{Y} \cdot \mathbf{Y}} = \|\mathbf{Y}\|$, we obtain (3).

Example 8. If $\mathbf{X}$ and $\mathbf{Y}$ are as in Example 6, then from Example 7, $\mathbf{X} \cdot \mathbf{Y} = 13$. Hence

$$|\mathbf{X} \cdot \mathbf{Y}| = 13 \leqslant \|\mathbf{X}\| \, \|\mathbf{Y}\| = \sqrt{18} \, \sqrt{30} \, .$$

We can now define the cosine of the angle between two nonzero vectors as follows.

DEFINITION The **cosine** of the angle θ between two nonzero vectors $\mathbf{X}$ and $\mathbf{Y}$ is defined by

$$\cos \theta = \frac{\mathbf{X} \cdot \mathbf{Y}}{\|\mathbf{X}\| \, \|\mathbf{Y}\|} \qquad 0 \leqslant \theta \leqslant \pi \, . \tag{4}$$

It follows from the Cauchy–Schwarz inequality that the expression on the right side of (4) lies between -1 and 1. Thus it makes sense to define the expression on the left-side of (4) as $\cos \theta$. In the case of R^3, we can establish by the law of cosines, as was done in R^2, that the cosine of the angle between $\mathbf{X}$ and $\mathbf{Y}$ is given by (4). However, for R^n, $n > 3$, we have to define it by (4).

Example 9. Let $\mathbf{X} = (1, 0, 0, 1)$ and $\mathbf{Y} = (0, 1, 0, 1)$. Then

$$\|\mathbf{X}\| = \sqrt{2} \, , \qquad \|\mathbf{Y}\| = \sqrt{2} \, , \qquad \text{and} \qquad \mathbf{X} \cdot \mathbf{Y} = 1.$$

Thus

$$\cos \theta = \tfrac{1}{2}$$

and $\theta = 60°$.

It is very useful to talk about orthogonality and parallellism in R^n, and we accordingly formulate the following definitions.

DEFINITION Two vectors $\mathbf{X}$ and $\mathbf{Y}$ in R^n are said to be **orthogonal** if $\mathbf{X} \cdot \mathbf{Y} = 0$. They are said to be **parallel** if $|\mathbf{X} \cdot \mathbf{Y}| = \|\mathbf{X}\| \, \|\mathbf{Y}\|$. They are in the **same direction** if $\mathbf{X} \cdot \mathbf{Y} = \|\mathbf{X}\| \, \|\mathbf{Y}\|$. That is, they are orthogonal if $\cos \theta = 0$, parallel if $\cos \theta = \pm 1$, and in the same direction if $\cos \theta = 1$.

Example 10. Consider the vectors $\mathbf{X} = (1, 0, 0, 1)$, $\mathbf{Y} = (0, 1, 1, 0)$, and $\mathbf{Z} = (3, 0, 0, 3)$. Then

$$\mathbf{X} \cdot \mathbf{Y} = 0 \quad \text{and} \quad \mathbf{Y} \cdot \mathbf{Z} = 0,$$

which implies that $\mathbf{X}$ and $\mathbf{Y}$ are orthogonal and $\mathbf{Y}$ and $\mathbf{Z}$ are orthogonal. Since

$$\mathbf{X} \cdot \mathbf{Z} = 6, \qquad \|\mathbf{X}\| = \sqrt{2}, \qquad \|\mathbf{Z}\| = \sqrt{18}, \qquad \text{and} \qquad \mathbf{X} \cdot \mathbf{Z} = \|\mathbf{X}\|\,\|\mathbf{Z}\|,$$

we conclude that $\mathbf{X}$ and $\mathbf{Z}$ are in the same direction.

An easy consequence of the Cauchy–Schwarz inequality is the triangle inequality, which we prove next.

THEOREM 3.5 (Triangle Inequality). *If* $\mathbf{X}$ *and* $\mathbf{Y}$ *are vectors in* R^n, *then*

$$\|\mathbf{X} + \mathbf{Y}\| \leqslant \|\mathbf{X}\| + \|\mathbf{Y}\|.$$

proof. We have, by Theorem 3.3,

$$\begin{aligned} \|\mathbf{X} + \mathbf{Y}\|^2 &= (\mathbf{X} + \mathbf{Y}) \cdot (\mathbf{X} + \mathbf{Y}) \\ &= \mathbf{X} \cdot \mathbf{X} + 2(\mathbf{X} \cdot \mathbf{Y}) + \mathbf{Y} \cdot \mathbf{Y} \\ &= \|\mathbf{X}\|^2 + 2(\mathbf{X} \cdot \mathbf{Y}) + \|\mathbf{Y}\|^2. \end{aligned}$$

By the Cauchy–Schwarz inequality we have

$$\begin{aligned} \|\mathbf{X} + \mathbf{Y}\|^2 &\leqslant \|\mathbf{X}\|^2 + 2\|\mathbf{X}\|\,\|\mathbf{Y}\| + \|\mathbf{Y}\|^2 \\ &= (\|\mathbf{X}\| + \|\mathbf{Y}\|)^2. \end{aligned}$$

Taking square roots, we obtain the desired result.

The triangle inequality in R^2 and R^3 merely states that the length of a side of a triangle does not exceed the sum of the lengths of the other two sides (Figure 3.30).

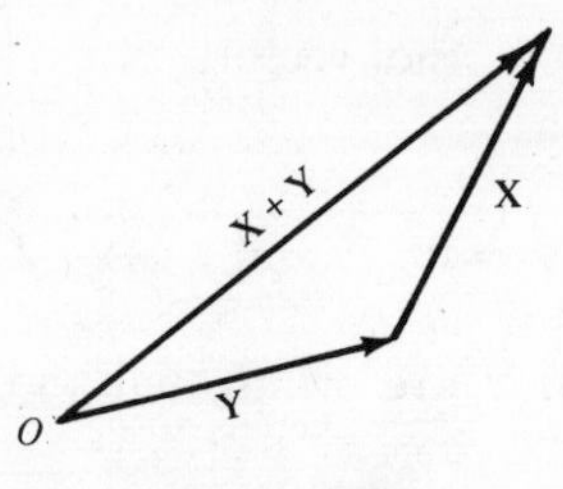

Figure 3.30

Example 11. For $\mathbf{X}$ and $\mathbf{Y}$ as in Example 10, we have

$$\|\mathbf{X} + \mathbf{Y}\| = \sqrt{4} = 2 < \sqrt{2} + \sqrt{2} = \|\mathbf{X}\| + \|\mathbf{Y}\|.$$

A **unit vector** $\mathbf{U}$ in R^n is a vector of unit length. If $\mathbf{X}$ is a nonzero vector, then the vector

$$\mathbf{U} = \left[\frac{1}{\|\mathbf{X}\|}\right]\mathbf{X}$$

is a unit vector in the direction of $\mathbf{X}$.

Example 12. If $\mathbf{X}$ is as in Example 10, then since $\|\mathbf{X}\| = \sqrt{2}$, the vector $\mathbf{U} = (1/\sqrt{2})(1, 0, 0, 1)$ is a unit vector in the direction of $\mathbf{X}$.

In the case of R^3 the unit vectors in the positive directions of the x-, y-, and z-axes are denoted by $\mathbf{i} = (1, 0, 0)$, $\mathbf{j} = (0, 1, 0)$, and $\mathbf{k} = (0, 0, 1)$ and are shown in Figure 3.31. If $\mathbf{X} = (x, y, z)$ is any vector in R^3, then we can

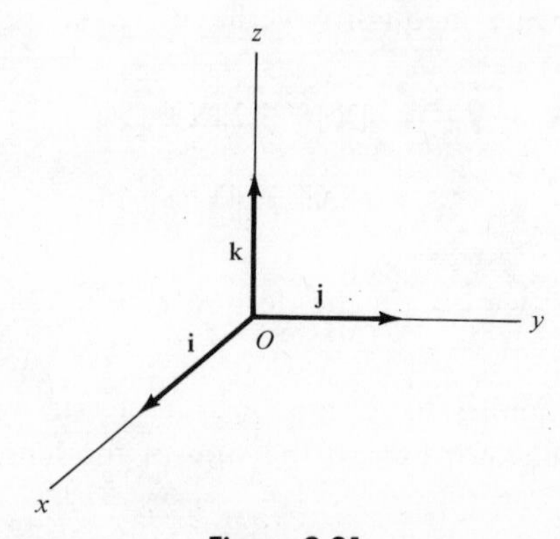

Figure 3.31

write **X** in terms of **i**, **j**, and **k** as

$$\mathbf{X} = x\mathbf{i} + y\mathbf{j} + z\mathbf{k}.$$

Example 13. If $\mathbf{X} = (2, -1, 3)$, then

$$\mathbf{X} = 2\mathbf{i} - \mathbf{j} + 3\mathbf{k}.$$

3.2 Exercises

1. Find $\mathbf{X} + \mathbf{Y}$, $\mathbf{X} - \mathbf{Y}$, $2\mathbf{X}$, and $3\mathbf{X} - 2\mathbf{Y}$ if
(a) $\mathbf{X} = (1, 2, -3)$, $\mathbf{Y} = (0, 1, -2)$.
(b) $\mathbf{X} = (4, -2, 1, 3)$, $\mathbf{Y} = (-1, 2, 5, -4)$.

2. Repeat Exercise 1 for
(a) $\mathbf{X} = (2, 0, -4)$, $\mathbf{Y} = (3, 2, 1)$.
(b) $\mathbf{X} = (-3, 5, -3, 0)$, $\mathbf{Y} = (2, 1, 5, -2)$.

3. Let $\mathbf{X} = (1, -2, 3)$, $\mathbf{Y} = (-3, -1, 3)$, $\mathbf{Z} = (x, -1, y)$, and $\mathbf{U} = (3, u, 2)$. Find x, y, and u so that
(a) $\mathbf{Z} = \frac{1}{2}\mathbf{X}$. (b) $\mathbf{Z} + \mathbf{Y} = \mathbf{X}$. (c) $\mathbf{Z} + \mathbf{U} = \mathbf{Y}$.

4. Let $\mathbf{X} = (4, -1, -2, 3)$, $\mathbf{Y} = (3, -2, -4, 1)$, $\mathbf{Z} = (x, -3, -6, y)$, and $\mathbf{U} = (2, u, v, 4)$. Find x, y, u, and v so that
(a) $\mathbf{Z} = \frac{1}{3}\mathbf{X}$. (b) $\mathbf{Z} + \mathbf{U} = \mathbf{X}$. (c) $\mathbf{Z} - \mathbf{X} = \mathbf{Y}$.

5. Let $\mathbf{X} = (4, 5, -2, 3)$, $\mathbf{Y} = (3, -2, 0, 1)$, $\mathbf{Z} = (-3, 2, -5, 3)$, $c = 2$, and $d = 3$. Verify properties (a) through (h) in Theorem 3.2.

6. Plot the following points in R^3:
(a) $(3, -1, 2)$. (b) $(1, 0, 2)$. (c) $(0, 0, -4)$.
(d) $(1, 0, 0)$. (e) $(0, -2, 0)$.

7. Sketch a directed line segment in R^3 representing each of the following vectors.
(a) $\mathbf{X}_1 = (2, -3, -1)$. (b) $\mathbf{X}_2 = (0, 1, 4)$. (c) $\mathbf{X}_3 = (0, 0, -1)$.

8. For each of the following pairs of points in R^3, determine the vector that is associated with the directed line segment whose tail is the first point and whose head is the second point.
(a) $(2, 3, -1)$, $(0, 0, 2)$. (b) $(1, 1, 0)$, $(0, 1, 1)$.
(c) $(-1, -2, -3)$, $(3, 4, 5)$. (d) $(1, 1, 3)$, $(0, 0, 1)$.

9. Determine the head of the vector $(3, 4, -1)$ whose tail is $(1, -2, 3)$.

10. Find the length of the following vectors.
(a) $(1, 2, -3)$. (b) $(2, 3, -1, 4)$. (c) $(1, 0, 3)$. (d) $(0, 0, 3, 4)$.

11. Find the length of the following vectors.
(a) $(2, 3, 4)$. (b) $(0, -1, 2, 3)$.
(c) $(-1, -2, 0)$. (d) $(1, 2, -3, -4)$.

12. Find the distance between the following pairs of points.
(a) $(1, -1, 2)$, $(3, 0, 2)$. (b) $(4, 2, -1, 5)$, $(2, 3, -1, 4)$.
(c) $(0, 0, 2)$, $(-3, 0, 0)$. (d) $(1, 0, 0, 2)$, $(3, -1, 5, 2)$.

13. Find the distance between the following pairs of points.
(a) $(1, 1, 0)$, $(2, -3, 1)$. (b) $(4, 2, -1, 6)$, $(4, 3, 1, 5)$.
(c) $(0, 2, 3)$, $(1, 2, -4)$. (d) $(3, 4, 0, 1)$, $(2, 2, 1, -1)$.

14. Find $\mathbf{X} \cdot \mathbf{Y}$.
(a) $\mathbf{X} = (1, 2, 3)$, $\mathbf{Y} = (-4, 4, 5)$.
(b) $\mathbf{X} = (0, 2, 3, 1)$, $\mathbf{Y} = (-3, 1, -2, 0)$.
(c) $\mathbf{X} = (0, 0, 1)$, $\mathbf{Y} = (2, 2, 0)$.
(d) $\mathbf{X} = (2, 0, -1, 3)$, $\mathbf{Y} = (-3, -5, 2, -1)$.

15. Find $\mathbf{X} \cdot \mathbf{Y}$.
(a) $\mathbf{X} = (2, 3, 1)$, $\mathbf{Y} = (3, -2, 0)$.
(b) $\mathbf{X} = (1, 2, -1, 3)$, $\mathbf{Y} = (0, 0, -1, -2)$.
(c) $\mathbf{X} = (2, 0, 1)$, $\mathbf{Y} = (2, 2, -1)$.
(d) $\mathbf{X} = (0, 4, 2, 3)$, $\mathbf{Y} = (0, -1, 2, 0)$.

16. Verify Theorem 3.3 for $c = 3$ and $\mathbf{X} = (1, 2, 3)$, $\mathbf{Y} = (1, 2, -4)$, and $\mathbf{Z} = (1, 0, 2)$.

17. Verify Theorem 3.4 for $\mathbf{X}$ and $\mathbf{Y}$ as in Exercise 16.

18. Find the cosine of the angle between each pair of vectors $\mathbf{X}$ and $\mathbf{Y}$ in Exercise 14.

19. Find the cosine of the angle between each pair of vectors $\mathbf{X}$ and $\mathbf{Y}$ in Exercise 15.

20. Show that
(a) $\mathbf{i} \cdot \mathbf{i} = \mathbf{j} \cdot \mathbf{j} = \mathbf{k} \cdot \mathbf{k} = 1$.
(b) $\mathbf{i} \cdot \mathbf{j} = \mathbf{i} \cdot \mathbf{k} = \mathbf{j} \cdot \mathbf{k} = 0$.

21. Which of the vectors $\mathbf{X}_1 = (4, 2, 6, -8)$, $\mathbf{X}_2 = (-2, 3, -1, -1)$, $\mathbf{X}_3 = (-2, -1, -3, 4)$, $\mathbf{X}_4 = (1, 0, 0, 2)$, $\mathbf{X}_5 = (1, 2, 3, -4)$, $\mathbf{X}_6 = (0, -3, 1, 0)$ are
(a) Orthogonal? (b) Parallel? (c) In the same direction?

22. Verify the triangle inequality for $\mathbf{X} = (1, 2, 3, -1)$ and $\mathbf{Y} = (1, 0, -2, 3)$.

23. Find a unit vector in the direction of $\mathbf{X}$.
(a) $\mathbf{X} = (2, -1, 3)$. (b) $\mathbf{X} = (1, 2, 3, 4)$. (c) $\mathbf{X} = (0, 1, -1)$.
(d) $\mathbf{X} = (0, -1, 2, -1)$.

24. Find a unit vector in the direction of $\mathbf{X}$.
(a) $\mathbf{X} = (1, 2, -1)$. (b) $\mathbf{X} = (0, 0, 2, 0)$. (c) $\mathbf{X} = (-1, 0, -2)$.
(d) $\mathbf{X} = (0, 0, 3, 4)$.

25. Write each of the following vectors in R^3 in terms of $\mathbf{i}$, $\mathbf{j}$, and $\mathbf{k}$.
(a) $(1, 2, -3)$. (b) $(2, 3, -1)$. (c) $(0, 1, 2)$. (d) $(0, 0, -2)$.

26. Write each of the following vectors in R^3 as a 3×1 matrix.
(a) $2\mathbf{i} + 3\mathbf{j} - 4\mathbf{k}$. (b) $\mathbf{i} + 2\mathbf{j}$. (c) $-3\mathbf{i}$. (d) $3\mathbf{i} - 2\mathbf{k}$.

27. Verify that the triangle with vertices $P_1(2, 3, -4)$, $P_2(3, 1, 2)$, and $P_3(-3, 0, 4)$, is isosceles.

28. Verify that the triangle with vertices $P_1(2, 3, -4)$, $P_2(3, 1, 2)$, and $P_3(7, 0, 1)$ is a right triangle.

Theoretical Exercises

T.1. Prove the rest of Theorem 3.2.

T.2. Show that $-\mathbf{X} = (-1)\mathbf{X}$.

T.3. Establish Equations (1) and (2) in R^3, for the length of a vector and the distance between two points, by using the Pythagorean Theorem.

T.4. Prove Theorem 3.3.

T.5. Suppose that $\mathbf{X}$ is orthogonal to both $\mathbf{Y}$ and $\mathbf{Z}$. Show that $\mathbf{X}$ is orthogonal to any vector of the form $r\mathbf{Y} + s\mathbf{Z}$, where r and s are scalars.

T.6. Show that if $\mathbf{X} \cdot \mathbf{Y} = \mathbf{0}$ for all vectors $\mathbf{Y}$, then $\mathbf{X} = \mathbf{0}$.

T.7. Show that $\mathbf{X} \cdot (\mathbf{Y} + \mathbf{Z}) = \mathbf{X} \cdot \mathbf{Y} + \mathbf{X} \cdot \mathbf{Z}$.

T.8. Show that if $\mathbf{X} \cdot \mathbf{Y} = \mathbf{X} \cdot \mathbf{Z}$ for all $\mathbf{X}$, then $\mathbf{Y} = \mathbf{Z}$.

T.9. Show that if c is a scalar, then $\|c\mathbf{X}\| = |c|\,\|\mathbf{X}\|$, where $|c|$ is the absolute value of c.

T.10. Show that $\|\mathbf{X} + \mathbf{Y}\|^2 = \|\mathbf{X}\|^2 + \|\mathbf{Y}\|^2$ if and only if $\mathbf{X} \cdot \mathbf{Y} = 0$.

T.11. If $\mathbf{X}$ and $\mathbf{Y}$ are viewed as $n \times 1$ matrices, show that $\mathbf{X} \cdot \mathbf{Y} = \mathbf{X}^T\mathbf{Y}$.

T.12. Define the **distance** between two vectors $\mathbf{X}$ and $\mathbf{Y}$ in R^n as $d(\mathbf{X}, \mathbf{Y}) = \|\mathbf{X} - \mathbf{Y}\|$. Show that:
(a) $d(\mathbf{X}, \mathbf{Y}) \geqslant 0$.
(b) $d(\mathbf{X}, \mathbf{Y}) = 0$ if and only if $\mathbf{X} = \mathbf{Y}$.
(c) $d(\mathbf{X}, \mathbf{Y}) = d(\mathbf{Y}, \mathbf{X})$.
(d) $d(\mathbf{X}, \mathbf{Z}) \leqslant d(\mathbf{X}, \mathbf{Y}) + d(\mathbf{Y}, \mathbf{Z})$.

T.13. Prove the **parallelogram law**:

$$\|\mathbf{X} + \mathbf{Y}\|^2 + \|\mathbf{X} - \mathbf{Y}\|^2 = 2\|\mathbf{X}\|^2 + 2\|\mathbf{Y}\|^2.$$

3.3 Cross Product in R^3 (Optional—Needed in Section 7.1)

In this section we discuss an operation that is only meaningful in R^3. Despite this limitation, it has many applications in a number of different situations. We shall consider several of these applications in this section; others will occur in Section 7.1.

DEFINITION If

$$\mathbf{X} = x_1\mathbf{i} + x_2\mathbf{j} + x_3\mathbf{k} \qquad \text{and} \qquad \mathbf{Y} = y_1\mathbf{i} + y_2\mathbf{j} + y_3\mathbf{k}$$

are two vectors in R^3, then their **cross product** is the vector $\mathbf{X} \times \mathbf{Y}$ defined by

$$\mathbf{X} \times \mathbf{Y} = (x_2y_3 - x_3y_2)\mathbf{i} + (x_3y_1 - x_1y_3)\mathbf{j} + (x_1y_2 - x_2y_1)\mathbf{k}. \quad (1)$$

The cross product $\mathbf{X} \times \mathbf{Y}$ can also be written as a "determinant,"

$$\mathbf{X} \times \mathbf{Y} = \begin{vmatrix} \mathbf{i} & \mathbf{j} & \mathbf{k} \\ x_1 & x_2 & x_3 \\ y_1 & y_2 & y_3 \end{vmatrix}. \quad (2)$$

The right side of (2) is not really a determinant, but it is convenient to think of the computation in this manner. If we expand (2) about the first row, we obtain

$$\mathbf{X} \times \mathbf{Y} = \begin{vmatrix} x_2 & x_3 \\ y_2 & y_3 \end{vmatrix}\mathbf{i} - \begin{vmatrix} x_1 & x_3 \\ y_1 & y_3 \end{vmatrix}\mathbf{j} + \begin{vmatrix} x_1 & x_2 \\ y_1 & y_2 \end{vmatrix}\mathbf{k},$$

which is the right side of (1). Observe that the cross product $\mathbf{X} \times \mathbf{Y}$ is a vector while the dot product $\mathbf{X} \cdot \mathbf{Y}$ is a number.

Example 1. Let $\mathbf{X} = 2\mathbf{i} + \mathbf{j} + 2\mathbf{k}$ and $\mathbf{Y} = 3\mathbf{i} - \mathbf{j} - 3\mathbf{k}$. Then

$$\mathbf{X} \times \mathbf{Y} = \begin{vmatrix} \mathbf{i} & \mathbf{j} & \mathbf{k} \\ 2 & 1 & 2 \\ 3 & -1 & -3 \end{vmatrix} = -\mathbf{i} + 12\mathbf{j} - 5\mathbf{k}.$$

Some of the algebraic properties of cross product are described in the following theorem. The proof, which follows easily from the properties of determinants, is left to the reader (Exercise T.1).

THEOREM 3.6. *If* **X**, **Y**, *and* **Z** *are vectors and c is a scalar, then*:

(a) $\mathbf{X} \times \mathbf{Y} = -(\mathbf{Y} \times \mathbf{X})$.
(b) $\mathbf{X} \times (\mathbf{Y} + \mathbf{Z}) = \mathbf{X} \times \mathbf{Y} + \mathbf{X} \times \mathbf{Z}$.
(c) $(\mathbf{X} + \mathbf{Y}) \times \mathbf{Z} = \mathbf{X} \times \mathbf{Z} + \mathbf{Y} \times \mathbf{Z}$.
(d) $c(\mathbf{X} \times \mathbf{Y}) = (c\mathbf{X}) \times \mathbf{Y} = \mathbf{X} \times (c\mathbf{Y})$.
(e) $\mathbf{X} \times \mathbf{X} = \mathbf{0}$.
(f) $\mathbf{0} \times \mathbf{X} = \mathbf{X} \times \mathbf{0} = \mathbf{0}$.
(g) $\mathbf{X} \times (\mathbf{Y} \times \mathbf{Z}) = (\mathbf{X} \cdot \mathbf{Z})\mathbf{Y} - (\mathbf{X} \cdot \mathbf{Y})\mathbf{Z}$.
(h) $(\mathbf{X} \times \mathbf{Y}) \times \mathbf{Z} = (\mathbf{Z} \cdot \mathbf{X})\mathbf{Y} - (\mathbf{Z} \cdot \mathbf{Y})\mathbf{X}$.

Example 2. It follows from (1) or (2) that

$$\mathbf{i} \times \mathbf{i} = \mathbf{j} \times \mathbf{j} = \mathbf{k} \times \mathbf{k} = \mathbf{0}; \qquad \mathbf{i} \times \mathbf{j} = \mathbf{k}, \quad \mathbf{j} \times \mathbf{k} = \mathbf{i}, \quad \mathbf{k} \times \mathbf{i} = \mathbf{j}.$$

Also,

$$\mathbf{j} \times \mathbf{i} = -\mathbf{k}, \qquad \mathbf{k} \times \mathbf{j} = -\mathbf{i}, \qquad \mathbf{i} \times \mathbf{k} = -\mathbf{j}.$$

These vectors can be remembered by the method indicated in Figure 3.32. Moving around the circle in a clockwise direction, we see that the cross product of two vectors taken in the indicated order is the third vector; moving in a counterclockwise direction, we see that the cross product taken in the indicated order is the negative of the third vector. The cross product of a vector with itself is the zero vector.

Although many of the familiar properties of the real numbers hold for the cross product, it should be noted that two important properties do not hold. The commutative law does not hold, since $\mathbf{X} \times \mathbf{Y} = -(\mathbf{Y} \times \mathbf{X})$. Also, the associative law does not hold, since $\mathbf{i} \times (\mathbf{i} \times \mathbf{j}) = \mathbf{i} \times \mathbf{k} = -\mathbf{j}$, while $(\mathbf{i} \times \mathbf{i}) \times \mathbf{j} = \mathbf{0} \times \mathbf{j} = \mathbf{0}$.

We shall now take a closer look at the geometric properties of cross product. First, we observe the following two additional properties of cross

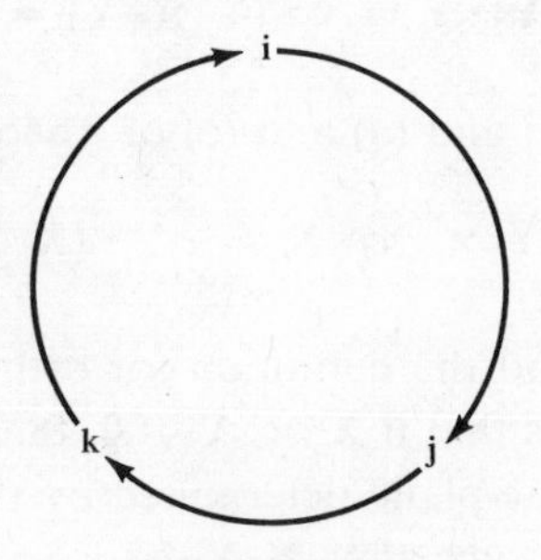

Figure 3.32

product, whose proofs we leave to the reader:

$$(\mathbf{X} \times \mathbf{Y}) \cdot \mathbf{Z} = \mathbf{X} \cdot (\mathbf{Y} \times \mathbf{Z}) \qquad \text{(Exercise T.2);} \qquad (3)$$

$$\mathbf{X} \times (\mathbf{Y} \times \mathbf{Z}) = (\mathbf{X} \cdot \mathbf{Z})\mathbf{Y} - (\mathbf{X} \cdot \mathbf{Y})\mathbf{Z} \qquad \text{(Exercise T.3).} \qquad (4)$$

It is also easy to show (Exercise T.4) that

$$(\mathbf{X} \times \mathbf{Y}) \cdot \mathbf{Z} = \begin{vmatrix} x_1 & x_2 & x_3 \\ y_1 & y_2 & y_3 \\ z_1 & z_2 & z_3 \end{vmatrix}.$$

Example 3. Let **X** and **Y** be as in Example 1, and let $\mathbf{Z} = \mathbf{i} + 2\mathbf{j} + 3\mathbf{k}$. Then

$$\mathbf{X} \times \mathbf{Y} = -\mathbf{i} + 12\mathbf{j} - 5\mathbf{k} \quad \text{and} \quad (\mathbf{X} \times \mathbf{Y}) \cdot \mathbf{Z} = 8$$

$$\mathbf{Y} \times \mathbf{Z} = 3\mathbf{i} - 12\mathbf{j} + 7\mathbf{k} \quad \text{and} \quad \mathbf{X} \cdot (\mathbf{Y} \times \mathbf{Z}) = 8,$$

which verifies Equation (3). Also,

$$\mathbf{X} \times (\mathbf{Y} \times \mathbf{Z}) = 31\mathbf{i} - 8\mathbf{j} - 27\mathbf{k},$$

$$\mathbf{X} \cdot \mathbf{Z} = 10, \qquad \mathbf{X} \cdot \mathbf{Y} = -1,$$

$$(\mathbf{X} \cdot \mathbf{Z})\mathbf{Y} = 30\mathbf{i} - 10\mathbf{j} - 30\mathbf{k}, \qquad (\mathbf{X} \cdot \mathbf{Y})\mathbf{Z} = -\mathbf{i} - 2\mathbf{j} - 3\mathbf{k}.$$

Hence

$$(\mathbf{X} \cdot \mathbf{Z})\mathbf{Y} - (\mathbf{X} \cdot \mathbf{Y})\mathbf{Z} = 31\mathbf{i} - 8\mathbf{j} - 27\mathbf{k}.$$

From Equation (3) and (e) of Theorem 3.6, we have

$$(\mathbf{X} \times \mathbf{Y}) \cdot \mathbf{Y} = \mathbf{X} \cdot (\mathbf{Y} \times \mathbf{Y}) = \mathbf{X} \cdot \mathbf{0} = 0. \qquad (5)$$

Also, from Equation (3) and (a) and (e) of Theorem 3.6, we have

$$(\mathbf{X} \times \mathbf{Y}) \cdot \mathbf{X} = -(\mathbf{Y} \times \mathbf{X}) \cdot \mathbf{X} = -\mathbf{Y} \cdot (\mathbf{X} \times \mathbf{X}) = -\mathbf{Y} \cdot \mathbf{0} = 0. \qquad (6)$$

From (5) and (6) and the definition for orthogonality of vectors given in Section 3.2, it follows that if $\mathbf{X} \times \mathbf{Y} \neq \mathbf{0}$, then $\mathbf{X} \times \mathbf{Y}$ is orthogonal to both **X** and **Y** and to the plane determined by them. Since $\mathbf{i} \times \mathbf{j} = \mathbf{k}$, the direction of $\mathbf{X} \times \mathbf{Y}$ is such that **X**, **Y**, and $\mathbf{X} \times \mathbf{Y}$ form a right-handed coordinate system. That is, the direction of $\mathbf{X} \times \mathbf{Y}$ is that in which a

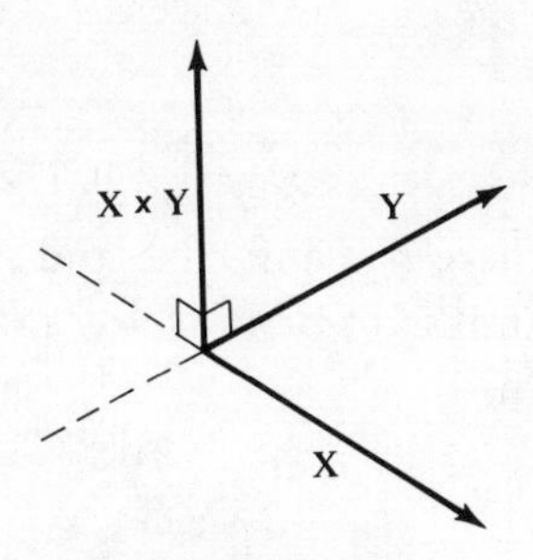

Figure 3.33

right-hand screw perpendicular to the plane of **X** and **Y** will move if we rotate it through an acute angle from **X** to **Y** (Figure 3.33).

The magnitude of $\mathbf{X} \times \mathbf{Y}$ can be determined as follows. From (a) of Theorem 3.3 (Section 3.2) we have

$$\begin{aligned}
\|\mathbf{X} \times \mathbf{Y}\|^2 &= (\mathbf{X} \times \mathbf{Y}) \cdot (\mathbf{X} \times \mathbf{Y}) \\
&= \mathbf{X} \cdot [\mathbf{Y} \times (\mathbf{X} \times \mathbf{Y})], \qquad \text{by (3)} \\
&= \mathbf{X} \cdot [(\mathbf{Y} \cdot \mathbf{Y})\mathbf{X} - (\mathbf{Y} \cdot \mathbf{X})\mathbf{Y}], \qquad \text{by (4)} \\
&= (\mathbf{X} \cdot \mathbf{X})(\mathbf{Y} \cdot \mathbf{Y}) - (\mathbf{Y} \cdot \mathbf{X})(\mathbf{Y} \cdot \mathbf{X}), \\
&\qquad \text{by (d) and (b) of Theorem 3.3} \\
&= \|\mathbf{X}\|^2\|\mathbf{Y}\|^2 - (\mathbf{X} \cdot \mathbf{Y})^2, \qquad \text{by (a) of Theorem 3.3}.
\end{aligned}$$

From Equation (4) of Section 3.2 it follows that

$$\mathbf{X} \cdot \mathbf{Y} = \|\mathbf{X}\| \, \|\mathbf{Y}\| \cos \theta,$$

where θ is the angle between **X** and **Y**. Hence

$$\begin{aligned}
\|\mathbf{X} \times \mathbf{Y}\|^2 &= \|\mathbf{X}\|^2\|\mathbf{Y}\|^2 - \|\mathbf{X}\|^2\|\mathbf{Y}\|^2 \cos^2 \theta \\
&= \|\mathbf{X}\|^2\|\mathbf{Y}\|^2(1 - \cos^2 \theta) \\
&= \|\mathbf{X}\|^2\|\mathbf{Y}\|^2 \sin^2 \theta.
\end{aligned}$$

Taking square roots, we obtain

$$\|\mathbf{X} \times \mathbf{Y}\| = \|\mathbf{X}\| \, \|\mathbf{Y}\| \sin \theta. \qquad (7)$$

It follows that the vectors **X** and **Y** are parallel if and only if $\mathbf{X} \times \mathbf{Y} = \mathbf{0}$ (Exercise T.5).

We now consider several applications of cross product.

Area of a Triangle

Consider the triangle with vertices $\mathbf{X}_1$, $\mathbf{X}_2$, and $\mathbf{X}_3$ (Figure 3.34). The area of this triangle is $\frac{1}{2}bh$, where b is the base and h is the height. If we take the segment between $\mathbf{X}_1$ and $\mathbf{X}_2$ to be the base, then

$$b = \|\mathbf{X}_2 - \mathbf{X}_1\|.$$

The height h is given by

$$h = \|\mathbf{X}_3 - \mathbf{X}_1\| \sin \theta.$$

Hence the area A_T of the triangle is

$$\begin{aligned} A_T &= \tfrac{1}{2}\|\mathbf{X}_2 - \mathbf{X}_1\|\,\|\mathbf{X}_3 - \mathbf{X}_1\| \sin \theta \\ &= \tfrac{1}{2}\|(\mathbf{X}_2 - \mathbf{X}_1) \times (\mathbf{X}_3 - \mathbf{X}_1)\|, \end{aligned}$$

by (7).

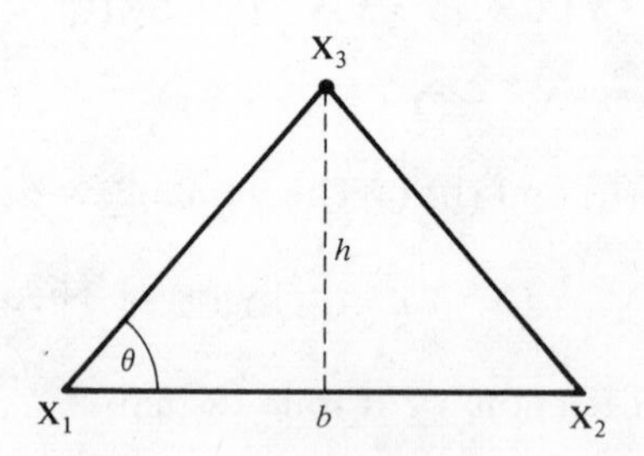

Figure 3.34

Example 4. To find the area of the triangle with vertices $\mathbf{X}_1 = (2, 2, 4)$, $\mathbf{X}_2 = (-1, 0, 5)$, and $\mathbf{X}_3 = (3, 4, 3)$, we write

$$\begin{aligned} \mathbf{X}_2 - \mathbf{X}_1 &= -3\mathbf{i} - 2\mathbf{j} + \mathbf{k}, \\ \mathbf{X}_3 - \mathbf{X}_1 &= \mathbf{i} + 2\mathbf{j} - \mathbf{k}. \end{aligned}$$

Then

$$\begin{aligned} A_T &= \tfrac{1}{2}\|(-3\mathbf{i} - 2\mathbf{j} + \mathbf{k}) \times (\mathbf{i} + 2\mathbf{j} - \mathbf{k})\| = \tfrac{1}{2}\|-2\mathbf{j} - 4\mathbf{k}\| \\ &= \|-\mathbf{j} - 2\mathbf{k}\| = \sqrt{5}\,. \end{aligned}$$

Area of a Parallelogram

The area A_P of the parallelogram with adjacent sides $\mathbf{X}_2 - \mathbf{X}_1$ and $\mathbf{X}_3 - \mathbf{X}_1$ (Figure 3.35) is $2A_T$, so

$$A_P = \|(\mathbf{X}_2 - \mathbf{X}_1) \times (\mathbf{X}_3 - \mathbf{X}_1)\|.$$

Example 5. If $\mathbf{X}_1$, $\mathbf{X}_2$, and $\mathbf{X}_3$ are as in Example 4, then the area of the parallelogram with adjacent sides $\mathbf{X}_2 - \mathbf{X}_1$ and $\mathbf{X}_3 - \mathbf{X}_1$ is $2\sqrt{5}$.

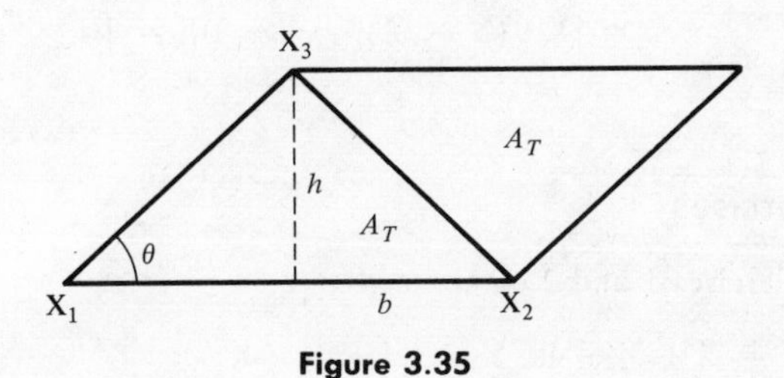

Figure 3.35

Volume of a Parallelepiped

Consider the parallelepiped with a vertex at the origin and edges $\mathbf{X}$, $\mathbf{Y}$, and $\mathbf{Z}$ (Figure 3.36). The volume v of the parallelepiped is the product of the area of the face containing $\mathbf{X}$ and $\mathbf{Y}$ and the distance h from this face to the face parallel to it. Now

$$h = \|\mathbf{X}\| \, |\cos \theta|,$$

where θ is the angle between $\mathbf{X}$ and $\mathbf{Y} \times \mathbf{Z}$, and the area of the face

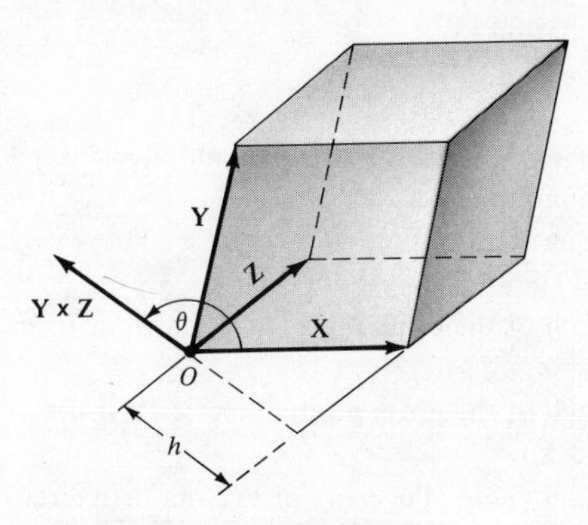

Figure 3.36

determined by $\mathbf{Y}$ and $\mathbf{Z}$ is $\|\mathbf{Y} \times \mathbf{Z}\|$. Hence

$$v = \|\mathbf{Y} \times \mathbf{Z}\| \, \|\mathbf{X}\| \, |\cos \theta| = |\mathbf{X} \cdot (\mathbf{Y} \times \mathbf{Z})|.$$

Example 6. Consider the parallelepiped with a vertex at the origin and edges $\mathbf{X} = \mathbf{i} - 2\mathbf{j} + 3\mathbf{k}$, $\mathbf{Y} = \mathbf{i} + 3\mathbf{j} + \mathbf{k}$, and $\mathbf{Z} = 2\mathbf{i} + \mathbf{j} + 2\mathbf{k}$. Then

$$\mathbf{Y} \times \mathbf{Z} = 5\mathbf{i} - 5\mathbf{k}.$$

Hence $\mathbf{X} \cdot (\mathbf{Y} \times \mathbf{Z}) = -10$. Thus

$$v = |\mathbf{X} \cdot (\mathbf{Y} \times \mathbf{Z})| = |-10| = 10.$$

3.3 Exercises

In Exercises 1 and 2 compute $\mathbf{X} \times \mathbf{Y}$.

1. (a) $\mathbf{X} = 2\mathbf{i} + 3\mathbf{j} + 4\mathbf{k}$, $\mathbf{Y} = -\mathbf{i} + 3\mathbf{j} - \mathbf{k}$.
(b) $\mathbf{X} = (1, 0, 1)$, $\mathbf{Y} = (2, 3, -1)$.
(c) $\mathbf{X} = \mathbf{i} - \mathbf{j} + 2\mathbf{k}$, $\mathbf{Y} = 3\mathbf{i} - 4\mathbf{j} + \mathbf{k}$.
(d) $\mathbf{X} = (2, -1, 1)$, $\mathbf{Y} = -2\mathbf{X}$.

2. (a) $\mathbf{X} = (1, -1, 2)$, $\mathbf{Y} = (3, 1, 2)$.
(b) $\mathbf{X} = 2\mathbf{i} + \mathbf{j} - 2\mathbf{k}$, $\mathbf{Y} = \mathbf{i} + 3\mathbf{k}$.
(c) $\mathbf{X} = 2\mathbf{j} + \mathbf{k}$, $\mathbf{Y} = 3\mathbf{X}$.
(d) $\mathbf{X} = (4, 0, -2)$, $\mathbf{Y} = (0, 2, -1)$.

3. Let $\mathbf{X} = \mathbf{i} + 2\mathbf{j} - 3\mathbf{k}$, $\mathbf{Y} = 2\mathbf{i} + 3\mathbf{j} + \mathbf{k}$, $\mathbf{Z} = 2\mathbf{i} - \mathbf{j} + 2\mathbf{k}$, and $c = -3$. Verify properties (a) through (d) of Theorem 3.6.

4. Let $\mathbf{X} = 2\mathbf{i} - \mathbf{j} + 3\mathbf{k}$, $\mathbf{Y} = 3\mathbf{i} + \mathbf{j} - \mathbf{k}$, and $\mathbf{Z} = 3\mathbf{i} + \mathbf{j} + 2\mathbf{k}$.
(a) Verify Equation (3).
(b) Verify Equation (4).

5. Let $\mathbf{X} = \mathbf{i} - \mathbf{j} + 2\mathbf{k}$, $\mathbf{Y} = 2\mathbf{i} + 2\mathbf{j} - \mathbf{k}$, and $\mathbf{Z} = \mathbf{i} + \mathbf{j} - \mathbf{k}$.
(a) Verify Equation (3).
(b) Verify Equation (4).

6. Verify that each of the cross products $\mathbf{X} \times \mathbf{Y}$ in Exercise 1 is orthogonal to both $\mathbf{X}$ and $\mathbf{Y}$.

7. Verify that each of the cross products $\mathbf{X} \times \mathbf{Y}$ in Exercise 2 is orthogonal to both $\mathbf{X}$ and $\mathbf{Y}$.

8. Verify Equation (7) for the pairs of vectors in Exercise 1.

9. Find the area of the triangle with vertices $\mathbf{X}_1 = (1, -2, 3)$, $\mathbf{X}_2 = (-3, 1, 4)$, and $\mathbf{X}_3 = (0, 4, 3)$.

10. Find the area of the triangle with vertices $\mathbf{X}_1$, $\mathbf{X}_2$, and $\mathbf{X}_3$, where $\mathbf{X}_2 - \mathbf{X}_1 = 2\mathbf{i} + 3\mathbf{j} - \mathbf{k}$ and $\mathbf{X}_3 - \mathbf{X}_1 = \mathbf{i} + 2\mathbf{j} + 2\mathbf{k}$.

11. Find the area of the parallelogram with adjacent sides $\mathbf{X}_2 - \mathbf{X}_1 = \mathbf{i} + 3\mathbf{j} - 2\mathbf{k}$ and $\mathbf{X}_3 - \mathbf{X}_1 = 3\mathbf{i} - \mathbf{j} - \mathbf{k}$.

12. Find the volume of the parallelepiped with a vertex at the origin and edges $\mathbf{X} = 2\mathbf{i} - \mathbf{j}$, $\mathbf{Y} = \mathbf{i} - 2\mathbf{j} - 2\mathbf{k}$, and $\mathbf{Z} = 3\mathbf{i} - \mathbf{j} + \mathbf{k}$.

13. Repeat Exercise 12 for $\mathbf{X} = \mathbf{i} - 2\mathbf{j} + 4\mathbf{k}$, $\mathbf{Y} = 3\mathbf{i} + 4\mathbf{j} + \mathbf{k}$, and $\mathbf{Z} = -\mathbf{i} + \mathbf{j} + \mathbf{k}$.

Theoretical Exercises

T.1. Prove Theorem 3.6.

T.2. Show that $(\mathbf{X} \times \mathbf{Y}) \cdot \mathbf{Z} = \mathbf{X} \cdot (\mathbf{Y} \times \mathbf{Z})$.

T.3. Show that $\mathbf{X} \times (\mathbf{Y} \times \mathbf{Z}) = (\mathbf{X} \cdot \mathbf{Z})\mathbf{Y} - (\mathbf{X} \cdot \mathbf{Y})\mathbf{Z}$.

T.4. Show that

$$(\mathbf{X} \times \mathbf{Y}) \cdot \mathbf{Z} = \begin{vmatrix} x_1 & x_2 & x_3 \\ y_1 & y_2 & y_3 \\ z_1 & z_2 & z_3 \end{vmatrix}.$$

T.5. Show that $\mathbf{X}$ and $\mathbf{Y}$ are parallel if and only if $\mathbf{X} \times \mathbf{Y} = \mathbf{0}$.

T.6. Show that $\|\mathbf{X} \times \mathbf{Y}\|^2 + (\mathbf{X} \cdot \mathbf{Y})^2 = \|\mathbf{X}\|^2\|\mathbf{Y}\|^2$.

T.7. Prove the *Jacobi identity* $(\mathbf{X} \times \mathbf{Y}) \times \mathbf{Z} + (\mathbf{Y} \times \mathbf{Z}) \times \mathbf{X} + (\mathbf{Z} \times \mathbf{X}) \times \mathbf{Y} = \mathbf{0}$.

3.4 Vector Spaces and Subspaces

We have already defined R^n and examined some of its basic properties. We must now study the fundamental structure of R^n. In many applications in mathematics, the sciences, and engineering, the notion of a vector space arises. This idea is merely a carefully constructed generalization of R^n. In studying the properties and structure of a vector space, we can study not only R^n, in particular, but many other important vector spaces. In this section we define the notion of a vector space and a subspace. In the following section we study their structure.

DEFINITION 1. A **real vector space** is a set V of elements, together with two operations $\oplus$ and $\odot$ satisfying the following properties:

(α) If $\mathbf{X}$ and $\mathbf{Y}$ are any elements of V, then $\mathbf{X} \oplus \mathbf{Y}$ is in V (that is, V is closed under the operation $\oplus$).

(a) $\mathbf{X} \oplus \mathbf{Y} = \mathbf{Y} \oplus \mathbf{X}$, for $\mathbf{X}$ and $\mathbf{Y}$ in V.

(b) $\mathbf{X} \oplus (\mathbf{Y} \oplus \mathbf{Z}) = (\mathbf{X} \oplus \mathbf{Y}) \oplus \mathbf{Z}$, for $\mathbf{X}$, $\mathbf{Y}$, and $\mathbf{Z}$ in V.

(c) There is a unique element $\mathbf{0}$ in V such that

$$\mathbf{X} \oplus \mathbf{0} = \mathbf{0} \oplus \mathbf{X} = \mathbf{X}, \qquad \text{for all } \mathbf{X} \text{ in } V.$$

(d) For each $\mathbf{X}$ in V, there is a unique element $-\mathbf{X}$ in V such that

$$\mathbf{X} \oplus -\mathbf{X} = \mathbf{0}.$$

(β) If $\mathbf{X}$ is any element of V and c is any real number, then $c \odot \mathbf{X}$ is in V.

(e) $c \odot (\mathbf{X} \oplus \mathbf{Y}) = c \odot \mathbf{X} \oplus c \odot \mathbf{Y}$, for all real numbers c and all $\mathbf{X}$ and $\mathbf{Y}$ in V.

(f) $(c + d) \odot \mathbf{X} = c \odot \mathbf{X} \oplus d \odot \mathbf{X}$, for all real numbers c and d, and all $\mathbf{X}$ in V.

(g) $c \odot (d \odot \mathbf{X}) = (cd) \odot \mathbf{X}$, for all real numbers c and d and all $\mathbf{X}$ on V.

(h) $1 \odot \mathbf{X} = \mathbf{X}$, for all $\mathbf{X}$ in V.

The elements of V are called **vectors**; the elements of R are called **scalars**. The operation $\oplus$ is called **vector addition**; the operation $\odot$ is called **scalar multiplication**. The vector $\mathbf{0}$ is called the **zero vector**. The vector $-\mathbf{X}$ is called the **negative** of $\mathbf{X}$.

Example 1. Consider the set R^n together with the operations of vector addition and scalar multiplication as defined in Section 3.2. Theorem 3.2 in Section 3.2 established the fact that R^n is a vector space under the above-mentioned operations.

Example 2. Consider the set V of all ordered triples of real numbers of the form $(x, y, 0)$ and define the operations $\oplus$ and $\odot$ by

$$(x, y, 0) \oplus (x', y', 0) = (x + x', y + y', 0)$$

$$c \odot (x, y, 0) = (cx, cy, 0).$$

It is then easy to show (Exercise 3) that V is a vector space, since it satisfies all the properties of Definition 1.

Example 3. Consider the set V of all ordered triples of real numbers (x, y, z) and define the operations $\oplus$ and $\odot$ by

$$(x, y, z) \oplus (x', y', z') = (x + x', y + y', z + z')$$

$$c \odot (x, y, z) = (cx, y, z).$$

It is then easy to verify (Exercise 4) that properties (α), (β), (a), (b), (c), (d), and (e) of Definition 1 hold. Here $\mathbf{0} = (0, 0, 0)$ and the negative of the vector (x, y, z) is the vector $(-x, -y, -z)$. For example, to verify property (e) we proceed as follows. First,

$$c \odot [(x, y, z) \oplus (x', y', z')] = c \odot (x + x', y + y', z + z')$$
$$= (c(x + x'), y + y', z + z').$$

Also,

$$c \odot (x, y, z) \oplus c \odot (x', y', z')$$
$$= (cx, y, z) \oplus (cx', y', z')$$
$$= (cx + cx', y + y', z + z') = (c(x + x'), y + y', z + z').$$

However, property (f) fails to hold. Thus

$$(c + d) \odot (x, y, z) = ((c + d)x, y, z).$$

On the other hand,

$$c \odot (x, y, z) \oplus d \odot (x, y, z) = (cx, y, z) \oplus (dx, y, z)$$
$$= (cx + dx, y + y, z + z)$$
$$= ((c + d)x, 2y, 2z).$$

Thus V is not a vector space under the prescribed operations. Incidentally, properties (g) and (h) *do* hold for this example.

Example 4. Consider the set V of all 2×3 matrices under the usual operations of matrix addition and scalar multiplication. In Section 1.2 (Theorems 1.1 and 1.3) we have established that the properties in Definition 1 hold, thereby making V into a vector space.

To specify a vector space we must be given a set V and two operations $\oplus$ and $\odot$ satisfying the properties of Definition 1. The first thing to check is whether (α) and (β) hold, for if either of these fails, we do not have a vector space. We frequently refer to a real vector space simply as a **vector space**. One also writes $\mathbf{X} \oplus \mathbf{Y}$ simply as $\mathbf{X} + \mathbf{Y}$ and $c \odot \mathbf{X}$ simply as $c\mathbf{X}$, being careful to keep the particular operation in mind.

Many other important examples of vector spaces occur in numerous areas of mathematics. However, our purposes will be adequately served by concentrating on R^n. The advantage of Definition 1 is that it is not concerned with the question of what is a vector. For example, is a vector in R^3 a point, a directed line segment, or a 3×1 matrix? Definition 1 only deals with the algebraic behavior of the elements in a vector space. In the case of R^3, whichever point of view we take, the algebraic behavior is the same. The mathematician abstracts those features that all such objects have in common (that is, those properties that make them all behave alike) and defines a new structure, called a real vector space. We can now talk about properties of all vector spaces without having to refer to any one vector space in particular. Thus a "vector" is now merely an element of a vector space, and it no longer needs to be associated with a directed line segment. In the interest of completeness and since it is no more difficult to do so, some of the results of this chapter will be stated for vector spaces. The reader can, however, replace the word "vector space" by R^n for a more concrete point of view. The following theorem presents several useful properties common to all vector spaces.

THEOREM 3.7. *If V is a vector space, then*:

(a) $0\mathbf{X} = \mathbf{0}$, *for every* $\mathbf{X}$ *in* V.
(b) $c\mathbf{0} = \mathbf{0}$, *for every scalar* c.
(c) *If* $c\mathbf{X} = \mathbf{0}$, *then* $c = 0$ *or* $\mathbf{X} = \mathbf{0}$.
(d) $(-1)\mathbf{X} = -\mathbf{X}$, *for every* $\mathbf{X}$ *in* V.

proof

(a) We have

$$0\mathbf{X} = (0 + 0)\mathbf{X} = 0\mathbf{X} + 0\mathbf{X}, \tag{1}$$

by (f) of Definition 1. Adding $-0\mathbf{X}$ to both sides of (1), we obtain by (b), (c), and (d) of the above definition,

$$\begin{aligned} \mathbf{0} &= 0\mathbf{X} + (-0\mathbf{X}) = (0\mathbf{X} + 0\mathbf{X}) + (-0\mathbf{X}) \\ &= 0\mathbf{X} + [0\mathbf{X} + (-0\mathbf{X})] \\ &= 0\mathbf{X} + \mathbf{0} = 0\mathbf{X}. \end{aligned}$$

(b) Exercise T.1.

(c) Suppose that $c\mathbf{X} = \mathbf{0}$ and $c \neq 0$. Then $\mathbf{0} = (1/c)\mathbf{0} = (1/c)(c\mathbf{X}) = [(1/c)c]\mathbf{X} = 1\mathbf{X} = \mathbf{X}$, by (b) of this theorem and (g) and (h) of the above definition.

(d) $(-1)\mathbf{X} + \mathbf{X} = (-1)\mathbf{X} + (1)\mathbf{X} = (-1 + 1)\mathbf{X} = 0\mathbf{X} = \mathbf{0}$ so that

$$(-1)\mathbf{X} = -\mathbf{X}.$$

Subspaces

DEFINITION Let V be a vector space and W a subset of V. If W is a vector space with respect to the operations in V, then W is called a **subspace of** V.

Example 5. If V is a vector space, then V is a subspace of itself. Also $\{\mathbf{0}\}$ is a subspace (Exercise 12) of V. Thus every vector space has at least two subspaces.

Example 6. Let W be the subset of R^3 consisting of all vectors of the form $(a, b, 0)$, where a and b are any real numbers. To check if W is a subspace of R^3, we first see whether properties (α) and (β) of Definition 1 hold. Thus let $\mathbf{X} = (a_1, b_1, 0)$ and $\mathbf{Y} = (a_2, b_2, 0)$ be vectors in W. Then $\mathbf{X} + \mathbf{Y} = (a_1, b_1, 0) + (a_2, b_2, 0) = (a_1 + a_2, b_1 + b_2, 0)$ is in W, since the third component is zero. Also, if c is a scalar, then $c\mathbf{X} = c(a_1, b_1, 0) = (ca_1, cb_1, 0)$ is in W. Thus properties (α) and (β) of Definition 1 hold. We can also easily verify that properties (a)–(h) hold. Hence W is a subspace of R^3.

We now pause in our listing of subspaces to develop a labor-saving result. We just noted that to verify that a subset of W of a vector space V is a subspace, one must check that (α), (β), and (a)–(h) of Definition 1 hold. However, the following theorem says that it is enough to merely check that (α) and (β) hold.

THEOREM 3.8. *Let V be a vector space with operations $\oplus$ and $\odot$ and let W be a nonempty subset of V. Then W is a subspace of V if and only if the following conditions hold:*

(α) *If* $\mathbf{X}$ *and* $\mathbf{Y}$ *are any vectors in W, then* $\mathbf{X} \oplus \mathbf{Y}$ *is in W.*

(β) *If c is any real number and* $\mathbf{X}$ *is any vector in W, then $c \odot \mathbf{X}$ is in W.*

proof. Exercise T.7.

Example 7. Consider the set W consisting of all 2×3 matrices of the form

$$\begin{bmatrix} a & b & 0 \\ 0 & c & d \end{bmatrix},$$

where a, b, c, and d are arbitrary real numbers. Then W is a subset of the vector space V of Example 4. Show that W is a subspace of V.

solution. Consider

$$\mathbf{X} = \begin{bmatrix} a_1 & b_1 & 0 \\ 0 & c_1 & d_1 \end{bmatrix} \quad \text{and} \quad \mathbf{Y} = \begin{bmatrix} a_2 & b_2 & 0 \\ 0 & c_2 & d_2 \end{bmatrix} \text{ in } W.$$

Then

$$\mathbf{X} + \mathbf{Y} = \begin{bmatrix} a_1 + a_2 & b_1 + b_2 & 0 \\ 0 & c_1 + c_2 & d_1 + d_2 \end{bmatrix} \quad \text{is in } W,$$

so that (α) of Theorem 3.8 is satisfied. Also, if k is a scalar, then

$$k\mathbf{X} = \begin{bmatrix} ka_1 & kb_1 & 0 \\ 0 & kc_1 & kd_1 \end{bmatrix} \quad \text{is in } W,$$

so that (β) of Theorem 3.8 is satisfied. Hence W is a subspace of V.

We shall now confine our attention to subspaces of R^n.

Example 8. Let W be the subset of R^3 consisting of all vectors of the form $(a, b, 1)$, where a and b are any real numbers. To check whether properties (α) and (β) of Theorem 3.8 hold, we let $\mathbf{X} = (a_1, b_1, 1)$ and $\mathbf{Y} = (a_2, b_2, 1)$ be vectors in W. Then $\mathbf{X} + \mathbf{Y} = (a_1, b_1, 1) + (a_2, b_2, 1) = (a_1 + a_2, b_1 + b_2, 2)$, which is not in W, since the third component is 2 and not 1. Since (α) of Theorem 3.8 does not hold, W is not a subspace of R^3.

Example 9. Let W be the subset of R^4 consisting of all vectors of the form $(a, b, a - b, a + 2b)$, where a and b are any real numbers. To check whether (α) and (β) of Theorem 3.8 hold, we let $\mathbf{X} = (a_1, b_1, a_1 - b_1, a_1 + 2b_1)$ and $\mathbf{Y} = (a_2, b_2, a_2 - b_2, a_2 + 2b_2)$ be vectors in W. Then $\mathbf{X} + \mathbf{Y} = (a_1, b_1, a_1 - b_1, a_1 + 2b_1) + (a_2, b_2, a_2 - b_2, a_2 + 2b_2) = [a_1 + a_2, b_1 + b_2, (a_1 - b_1) + (a_2 - b_2), (a_1 + 2b_1) + (a_2 + 2b_2)] = [a_1 + a_2, b_1 + b_2, (a_1 + a_2) - (b_1 + b_2), (a_1 + a_2) + 2(b_1 + b_2)]$, which is in W, since the third component is the first component minus its second component and the fourth component is the first component plus twice its second component. Also, if c is a scalar, then $c\mathbf{X} = c(a_1, b_1, a_1 - b_1, a_1 + 2b_1) = [ca_1, cb_1, ca_1 - cb_1, ca_1 + 2(cb_1)]$, which is in W. Since (α) and (β) hold, W is a subspace of R^4.

Example 10. Consider the homogeneous system $\mathbf{AX} = \mathbf{0}$, where $\mathbf{A}$ is an $m \times n$ matrix. A solution consists of a vector $\mathbf{X}$ in R^n. Let W be the subset of R^n consisting of all solutions to the homogeneous system. To check that W is a subspace of R^n, we verify properties (α) and (β) of Theorem 3.8. Thus let $\mathbf{X}$ and $\mathbf{Y}$ be solutions. Then

$$\mathbf{AX} = \mathbf{0} \quad \text{and} \quad \mathbf{AY} = \mathbf{0}.$$

Now

$$\mathbf{A(X + Y) = AX + AY = 0 + 0 = 0},$$

so $\mathbf{X} + \mathbf{Y}$ is a solution. Also, if c is a scalar, then

$$\mathbf{A}(c\mathbf{X}) = c(\mathbf{AX}) = c\mathbf{0} = \mathbf{0},$$

so $c\mathbf{X}$ is also a solution. Hence W is a subspace of R^n.

It should be noted that the set of all solutions to the linear system $\mathbf{AX} = \mathbf{B}$, where $\mathbf{A}$ is $m \times n$, is not a subspace of R^n if $\mathbf{B} \neq \mathbf{0}$ (Exercise T.2).

Example 11. A simple way of constructing subspaces in a vector space is as follows. Let $\mathbf{U}_1$ and $\mathbf{U}_2$ be fixed vectors in a vector space V and let W be the set of all vectors in V of the form

$$a_1\mathbf{U}_1 + \mathbf{a}_2\mathbf{U}_2,$$

where a_1 and a_2 are any real numbers. To show that W is a subspace of V, we verify properties (α) and (β) of Theorem 3.8. Thus let

$$\mathbf{X} = a_1\mathbf{U}_1 + a_2\mathbf{U}_2 \quad \text{and} \quad \mathbf{Y} = b_1\mathbf{U}_1 + b_2\mathbf{U}_2$$

be vectors in W. Then

$$\mathbf{X} + \mathbf{Y} = (a_1\mathbf{U}_1 + a_2\mathbf{U}_2) + (b_1\mathbf{U}_1 + b_2\mathbf{U}_2) = (a_1 + b_1)\mathbf{U}_1 + (a_2 + b_2)\mathbf{U}_2,$$

which is in W. Also, if c is a scalar, then

$$c\mathbf{X} = (ca_1)\mathbf{U}_1 + (ca_2)\mathbf{U}_2$$

is in W. Hence W is a subspace of V.

3.4 Exercises

1. Verify in detail that R^2 is a vector space.
2. Verify in detail that R^3 is a vector space.
3. Verify that the set in Example 2 is a vector space.

4. Verify that all the properties of Definition 1, except property (f), hold for the set in Example 3.

In Exercises 5 through 11 determine whether the given set together with the given operations is a vector space. If it is not a vector space, list the properties of Definition 1 that fail to hold.

5. The set of all ordered triples of real numbers (x, y, z) with the operations $(x, y, z) \oplus (x', y', z') = (x', y + y', z')$ and $c \odot (x, y, z) = (cx, cy, cz)$.

6. The set of all ordered triples of real numbers (x, y, z) with the operations $(x, y, z) \oplus (x', y', z') = (x + x', y + y', z + z')$ and $c \odot (x, y, z) = (x, 1, z)$.

7. The set of all ordered triples of real numbers of the form $(0, 0, z)$ with the operations $(0, 0, z) \oplus (0, 0, z') = (0, 0, z + z')$ and $c \odot (0, 0, z) = (0, 0, cz)$.

8. The set of all real numbers with the usual operations of addition and multiplication.

9. The set of all ordered pairs of real numbers (x, y), where $x \leqslant 0$, with the usual operations in R^2.

10. The set of all ordered pairs of real numbers (x, y) with the operations $(x, y) \oplus (x', y') = (x + x', y + y')$ and $c \odot (x, y) = (0, 0)$.

11. The set of all positive real numbers with the operations $x \oplus y = xy$ and $c \odot x = x^c$.

12. If V is a vector space, show that the subset $\{\mathbf{0}\}$ is a subspace.

13. Which of the following subsets of R^3 are subspaces of R^3? The set of all vectors of the form
(a) $(a, b, 2)$. (b) (a, b, c), where $c = a + b$.
(c) (a, b, c), where $c > 0$.

14. Which of the following subsets of R^3 are subspaces of R^3? The set of all vectors of the form
(a) (a, b, c), where $a = c = 0$. (b) (a, b, c), where $a = -c$.
(c) (a, b, c), where $b = 2a + 1$.

15. Which of the following subsets of R^4 are subspaces of R^4? The set of all vectors of the form
(a) (a, b, c, d), where $a - b = 2$. (b) (a, b, c, d), where $c = a + 2b$ and $d = a - 3b$. (c) (a, b, c, d), where $a = 0$ and $b = -d$.

16. Which of the following subsets of R^4 are subspaces of R^4? The set of all vectors of the form
(a) (a, b, c, d), where $a = b = 0$. (b) (a, b, c, d), where $a = 1, b = 0$, and $a + d = 1$. (c) (a, b, c, d), where $a > 0$ and $b < 0$.

17. Let $\mathbf{X} = (1, 2, -3)$ and $\mathbf{Y} = (-2, 3, 0)$ be two vectors in R^3 and let W be the subset of R^3 consisting of all vectors of the form $a\mathbf{X} + b\mathbf{Y}$, where a and b are any real numbers. Verify that W is a subspace of R^3.

18. Let $\mathbf{X} = (2, 0, 3, -4)$ and $\mathbf{Y} = (4, 2, -5, 1)$ be two vectors in R^4 and let W be the subset of R^4 consisting of all vectors of the form $a\mathbf{X} + b\mathbf{Y}$, where a and b are any real numbers. Verify that W is a subspace of R^4.

19. Which of the following subsets of the vector space of all 2×3 matrices defined in Example 4 are subspaces? The set of all matrices of the form

(a) $\begin{bmatrix} a & b & c \\ d & 0 & 0 \end{bmatrix}$, where $b = a + c$. (b) $\begin{bmatrix} a & b & c \\ d & 0 & 0 \end{bmatrix}$, where $c > 0$.

(c) $\begin{bmatrix} a & b & c \\ d & e & f \end{bmatrix}$, where $a = -2c$ and $f = 2e + d$.

20. Which of the following subsets of the vector space of all 2×3 matrices defined in Example 4 are subspaces? The set of all matrices of the form

(a) $\begin{bmatrix} a & b & c \\ d & e & f \end{bmatrix}$, where $a = 2c + 1$. (b) $\begin{bmatrix} 0 & 1 & a \\ b & c & 0 \end{bmatrix}$.

(c) $\begin{bmatrix} a & b & c \\ d & e & f \end{bmatrix}$, where $a + c = 0$ and $b + d + f = 0$.

Theoretical Exercises

T.1. Show that $c\mathbf{0} = \mathbf{0}$ for every scalar c.

T.2. Show that the set of all solutions to $\mathbf{AX} = \mathbf{B}$, where $\mathbf{A}$ is $m \times n$, is not a subspace of R^n if $\mathbf{B} \neq \mathbf{0}$.

T.3. Prove that $-(-\mathbf{X}) = \mathbf{X}$.

T.4. Show that a subset W of a vector space V is a subspace of V if and only if the following condition holds: If $\mathbf{X}$ and $\mathbf{Y}$ are any vectors in W and a and b are any scalars, then $a\mathbf{X} + b\mathbf{Y}$ is in W.

T.5. Prove that if $\mathbf{X} + \mathbf{Y} = \mathbf{X} + \mathbf{Z}$, then $\mathbf{Y} = \mathbf{Z}$.

T.6. Show that if $\mathbf{X} \neq \mathbf{0}$ and $a\mathbf{X} = b\mathbf{X}$, then $a = b$.

T.7. Prove Theorem 3.8.

3.5 Linear Independence and Bases

In this section we turn to a more thorough study of the structure of a vector space and its subspaces. The only vector space having a finite number of vectors in it is the vector space whose only member is $\mathbf{0}$. For if $\mathbf{X} \neq \mathbf{0}$ is in a vector space V, then $c\mathbf{X}$ is also in V, where c is any real number, and so V has infinitely many members. However, in this section we shall show that each vector space V studied here has a finite number of vectors that completely describe V. Again, recall that you may substitute "R^n" or "a subspace of R^n" for "vector space."

DEFINITION A vector $\mathbf{X}$ is defined to be a **linear combination** of the vectors $\mathbf{X}_1, \mathbf{X}_2, \ldots, \mathbf{X}_k$ if it can be written as

$$\mathbf{X} = c_1\mathbf{X}_1 + c_2\mathbf{X}_2 + \cdots + c_k\mathbf{X}_k,$$

where $c_1, c_2, \ldots, c_k$ are scalars.

Example 1. In R^4, let $\mathbf{X}_1 = (1, 2, 1, -1)$, $\mathbf{X}_2 = (1, 0, 2, -3)$, and $\mathbf{X}_3 = (1, 1, 0, -2)$. The vector

$$\mathbf{X} = (2, 1, 5, -5)$$

is a linear combination of $\mathbf{X}_1$, $\mathbf{X}_2$, and $\mathbf{X}_3$ if we can find c_1, c_2, and c_3 such that

$$c_1\mathbf{X}_1 + c_2\mathbf{X}_2 + c_3\mathbf{X}_3 = \mathbf{X}.$$

Substituting for $\mathbf{X}_1$, $\mathbf{X}_2$, $\mathbf{X}_3$, and $\mathbf{X}$, we obtain

$$c_1(1, 2, 1, -1) + c_2(1, 0, 2, -3) + c_3(1, 1, 0, -2) = (2, 1, 5, -5)$$

or

$$(c_1 + c_2 + c_3, 2c_1 + 0c_2 + c_3, c_1 + 2c_2 + 0c_3, -c_1 - 3c_2 - 2c_3) = (2, 1, 5, -5).$$

Equating corresponding components leads to the linear system

$$\begin{aligned} c_1 + c_2 + c_3 &= 2 \\ 2c_1 \qquad + c_3 &= 1 \\ c_1 + 2c_2 \qquad &= 5 \\ -c_1 - 3c_2 - 2c_3 &= -5. \end{aligned}$$

Solving this linear system by Gauss–Jordan reduction (the method discussed in Section 1.3), we obtain $c_1 = 1$, $c_2 = 2$, and $c_3 = -1$, which means that $\mathbf{X}$ is a linear combination of $\mathbf{X}_1$, $\mathbf{X}_2$, and $\mathbf{X}_3$. Thus

$$1(1, 2, 1, -1) + 2(1, 0, 2, -3) - 1(1, 1, 0, -2) = (2, 1, 5, -5).$$

DEFINITION Let $S = \{\mathbf{X}_1, \mathbf{X}_2, \ldots, \mathbf{X}_m\}$ be a set of vectors in a vector space V. The set S **spans** V, or V **is spanned by** S, if every vector in V is a linear combination of the vectors in S.

Example 2. Let V be the vector space R^3 and let $S = \{\mathbf{X}_1, \mathbf{X}_2, \mathbf{X}_3\}$, where $\mathbf{X}_1 = (1, 2, 1)$, $\mathbf{X}_2 = (1, 0, 2)$, and $\mathbf{X}_3 = (1, 1, 0)$. To find out whether S spans R^3, we pick any vector $\mathbf{X} = (a, b, c)$ in R^3 (a, b, and c are arbitrary real numbers) and must find out whether there are constants c_1, c_2, and c_3 such that

$$c_1\mathbf{X}_1 + c_2\mathbf{X}_2 + c_3\mathbf{X}_3 = \mathbf{X}.$$

This leads to the linear system

$$\begin{aligned} c_1 + c_2 + c_3 &= a \\ 2c_1 + c_3 &= b \\ c_1 + 2c_2 &= c. \end{aligned}$$

A solution is (verify)

$$c_1 = \frac{-2a + 2b + c}{3}, \qquad c_2 = \frac{a - b + c}{3}, \qquad c_3 = \frac{4a - b - 2c}{3}.$$

Thus $\{\mathbf{X}_1, \mathbf{X}_2, \mathbf{X}_3\}$ spans R^3.

Example 3. Let V be the vector space R^3 and let $S = \{\mathbf{X}_1, \mathbf{X}_2\}$, where $\mathbf{X}_1 = (1, 2, 1)$ and $\mathbf{X}_2 = (1, 0, 2)$. To find out whether S spans R^3, we again pick any vector $\mathbf{X} = (a, b, c)$ in R^3 (a, b, and c are arbitrary real numbers) and must find out whether there are constants c_1 and c_2 such that

$$c_1\mathbf{X} + c_2\mathbf{X}_2 = \mathbf{X}.$$

Transforming the augmented matrix of the resulting linear system, we obtain

$$\left[\begin{array}{cc|c} 1 & 0 & 2a - c \\ 0 & 1 & c - a \\ 0 & 0 & b - 4a + c \end{array}\right].$$

Thus a solution exists only when $b - 4a + c = 0$. Since we need a solution for arbitrary a, b, and c, we conclude that $\{\mathbf{X}_1, \mathbf{X}_2\}$ does not span R^3.

Example 4. The vectors $\mathbf{E}_1 = \mathbf{i} = (1, 0)$ and $\mathbf{E}_2 = \mathbf{j} = (0, 1)$, span R^2, for as was observed in Section 3.1, if $\mathbf{X} = (x_1, x_2)$ is any vector in R^2, then $\mathbf{X} = x_1\mathbf{E}_1 + x_2\mathbf{E}_2$. As was noted in Section 3.2, every vector $\mathbf{X}$ in R^3 can be written as a linear combination of the vectors $\mathbf{E}_1 = \mathbf{i} = (1, 0, 0)$, $\mathbf{E}_2 = \mathbf{j} = (0, 1, 0)$, and $\mathbf{E}_3 = \mathbf{k} = (0, 0, 1)$. Thus $\mathbf{E}_1$, $\mathbf{E}_2$, and $\mathbf{E}_3$ span R^3. Similarly, the vectors $\mathbf{E}_1 = (1, 0, \ldots, 0)$, $\mathbf{E}_2 = (0, 1, 0, \ldots, 0), \ldots, \mathbf{E}_n = (0, 0, \ldots, 0, 1)$ span R^n, since any vector $\mathbf{X} = (x_1, x_2, \ldots, x_n)$ in R^n can be written as

$$\mathbf{X} = x_1\mathbf{E}_1 + x_2\mathbf{E}_2 + \cdots + x_n\mathbf{E}_n.$$

Linear Dependence

DEFINITION Let $S = \{\mathbf{X}_1, \mathbf{X}_2, \ldots, \mathbf{X}_k\}$ be a set of distinct vectors in a vector space V. Then S is said to be **linearly dependent** if there exist constants $c_1, c_2, \ldots, c_k$ not all zero, such that

$$c_1\mathbf{X}_1 + c_2\mathbf{X}_2 + \cdots + c_k\mathbf{X}_k = \mathbf{0}. \tag{1}$$

Otherwise, S is called **linearly independent**. That is, S is linearly independent if (1) holds only for

$$c_1 = c_2 = \cdots = c_k = 0.$$

It should be emphasized that for any set of distinct vectors $S = \{\mathbf{X}_1, \mathbf{X}_2, \ldots, \mathbf{X}_k\}$, Equation (1) always holds if we choose all the scalars $c_1, c_2, \ldots, c_k$ equal to zero. The important point in this definition is whether or not it is possible to satisfy (1) with at least one of the scalars not equal to zero.

Example 5. In R^4 consider the vectors $\mathbf{X}_1 = (1, 0, 1, 2)$, $\mathbf{X}_2 = (0, 1, 1, 2)$, and $\mathbf{X}_3 = (1, 1, 1, 3)$. To find out whether $S = \{\mathbf{X}_1, \mathbf{X}_2, \mathbf{X}_3\}$ is linearly independent or linearly dependent, we form Equation (1):

$$c_1\mathbf{X}_1 + c_2\mathbf{X}_2 + c_3\mathbf{X}_3 = \mathbf{0},$$

and solve for c_1, c_2, and c_3. The resulting linear system is

$$\begin{aligned} c_1 \qquad\quad + \; c_3 &= 0 \\ c_2 + \; c_3 &= 0 \\ c_1 + \; c_2 + \; c_3 &= 0 \\ 2c_1 + 2c_2 + 3c_3 &= 0, \end{aligned}$$

which has as its only solution $c_1 = c_2 = c_3 = 0$ (verify), showing that S is linearly independent.

Example 6. Consider the vectors

$$\mathbf{X}_1 = (1, 2, -1), \qquad \mathbf{X}_2 = (1, -2, 1), \qquad \mathbf{X}_3 = (-3, 2, -1),$$

and

$$\mathbf{X}_4 = (2, 0, 0) \text{ in } R^3.$$

Setting up Equation (1) we have

$$\begin{aligned} c_1 + \ c_2 - 3c_3 + 2c_4 &= 0 \\ 2c_1 - 2c_2 + 2c_3 \qquad &= 0 \\ -c_1 + \ c_2 - \ c_3 \qquad &= 0, \end{aligned}$$

a homogeneous system of three equations in four unknowns. By Theorem 1.7, Section 1.3, we are assured of the existence of a nontrivial solution. Hence $\mathbf{X}_1$, $\mathbf{X}_2$, $\mathbf{X}_3$, and $\mathbf{X}_4$ are linearly dependent. In fact, two of the infinitely many solutions are

$$c_1 = 1, \quad c_2 = 2, \quad c_3 = 1, \quad \text{and} \quad c_4 = 0;$$

$$c_1 = 1, \quad c_2 = 1, \quad c_3 = 0, \quad \text{and} \quad c_4 = -1.$$

Example 7. The vectors $\mathbf{E}_1$ and $\mathbf{E}_2$ in R^2, defined in Example 4, are linearly independent, since

$$c_1(1, 0) + c_2(0, 1) = (0, 0)$$

can hold only if $c_1 = c_2 = 0$. Similarly, the vectors $\mathbf{E}_1$, $\mathbf{E}_2$, and $\mathbf{E}_3$ in R^3, and more generally, the vectors $\mathbf{E}_1, \mathbf{E}_2, \ldots, \mathbf{E}_n$ in R^n are linearly independent (Exercise T.1).

Example 8. If $S = \{\mathbf{X}_1, \mathbf{X}_2, \ldots, \mathbf{X}_k\}$ and $\mathbf{X}_i$ is the zero vector, Equation (1) holds by letting $c_i = 1$ and $c_j = 0$ for $j \neq i$.

Let S_1 and S_2 be finite subsets of a vector space and let S_1 be a subset of S_2. Then (a) if S_1 is linearly dependent, so is S_2; and (b) if S_2 is linearly independent, so is S_1 (Exercise T.2).

We now consider the meaning of linear dependence in R^2 and R^3. Suppose that $\{\mathbf{X}_1, \mathbf{X}_2\}$ is linearly dependent in R^2. Then there exist scalars c_1 and c_2 not both zero such that $c_1\mathbf{X}_1 + c_2\mathbf{X}_2 = \mathbf{0}$. If $c_1 \neq 0$, then $\mathbf{X}_1 =$

$-(c_2/c_1)\mathbf{X}_2$. If $c_2 \neq 0$, then $\mathbf{X}_2 = -(c_1/c_2)\mathbf{X}_1$. Thus one of the vectors is a multiple of the other. Conversely, suppose that $\mathbf{X}_1 = c\mathbf{X}_2$. Then $1\mathbf{X}_1 - c\mathbf{X}_2 = \mathbf{0}$, and since the coefficients of $\mathbf{X}_1$ and $\mathbf{X}_2$ are not both zero, it follows that $\mathbf{X}_1$ and $\mathbf{X}_2$ are linearly dependent. Thus $\{\mathbf{X}_1, \mathbf{X}_2\}$ is linearly dependent in R^2 if and only if one of the vectors is a multiple of the other [Figure 3.37(a)]. Hence two vectors in R^2 are linearly dependent if and only if they both lie on the same line passing through the origin [Figure 3.37(a) and (b)].

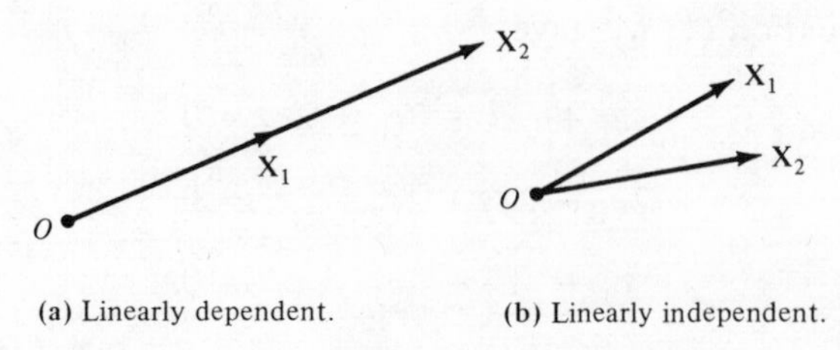

(a) Linearly dependent. (b) Linearly independent.

Figure 3.37

Suppose now that $\{\mathbf{X}_1, \mathbf{X}_2, \mathbf{X}_3\}$ is linearly dependent in R^3. Then we can write

$$c_1\mathbf{X}_1 + c_2\mathbf{X}_2 + c_3\mathbf{X}_3 = \mathbf{0},$$

where c_1, c_2, and c_3 are not all zero. Say that $c_2 \neq 0$. Then

$$\mathbf{X}_2 = -\frac{c_1}{c_2}\mathbf{X}_1 - \frac{c_3}{c_2}\mathbf{X}_3,$$

which means that $\mathbf{X}_2$ is in the subspace of R^3 spanned by the vectors $\mathbf{X}_1$ and $\mathbf{X}_3$. The subspace of R^3 spanned by $\mathbf{X}_1$ and $\mathbf{X}_3$ is the plane through the origin determined by $\mathbf{X}_1$ and $\mathbf{X}_3$. Conversely, if $\mathbf{X}_2$ is in the plane through the origin determined by $\mathbf{X}_1$ and $\mathbf{X}_3$, then $\mathbf{X}_2$ is a linear combination of $\mathbf{X}_1$ and $\mathbf{X}_3$:

$$\mathbf{X}_2 = c_1\mathbf{X}_1 + c_3\mathbf{X}_3.$$

Then

$$c_1\mathbf{X}_1 - 1\mathbf{X}_2 + c_3\mathbf{X}_3 = \mathbf{0},$$

which means that $\{\mathbf{X}_1, \mathbf{X}_2, \mathbf{X}_3\}$ is linearly dependent. Thus three vectors in R^3 are linearly dependent if and only if they all lie in the same plane passing through the origin [Figure 3.38(a) and (b)].

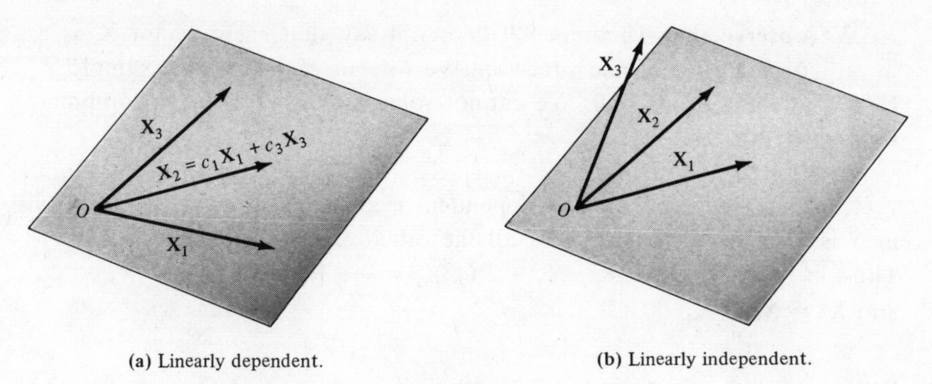

(a) Linearly dependent. (b) Linearly independent.

Figure 3.38

THEOREM 3.9. *Let $S = \{\mathbf{X}_1, \mathbf{X}_2, \ldots, \mathbf{X}_k\}$ be a set of nonzero vectors in a vector space V. Then S is linearly dependent if and only if one of the vectors $\mathbf{X}_j$ is a linear combination of the preceding vectors in S.*

proof. Suppose that $\mathbf{X}_j$ is a linear combination of the preceding vectors,

$$\mathbf{X}_j = c_1\mathbf{X}_1 + c_2\mathbf{X}_2 + \cdots + c_{j-1}\mathbf{X}_{j-1}.$$

Then

$$c_1\mathbf{X}_1 + c_2\mathbf{X}_2 + \cdots + c_{j-1}\mathbf{X}_{j-1} + (-1)\mathbf{X}_j + 0\mathbf{X}_{j+1} + \cdots + 0\mathbf{X}_k = \mathbf{0}. \tag{2}$$

Since at least one coefficient, -1, is not zero in (2), we conclude that S is linearly dependent.

Conversely, suppose that S is linearly dependent. Then there exist scalars $c_1, c_2, \ldots, c_k$, not all zero such that (1) holds. Now let j be the largest subscript for which $c_j \neq 0$. If $j > 1$, then

$$\mathbf{X}_j = -\left(\frac{c_1}{c_j}\right)\mathbf{X}_1 - \left(\frac{c_2}{c_j}\right)\mathbf{X}_2 - \cdots - \left(\frac{c_{j-1}}{c_j}\right)\mathbf{X}_{j-1}.$$

If $j = 1$, then $c_j\mathbf{X}_j = 0$, which implies that $\mathbf{X}_j = 0$, a contradiction to the hypothesis that none of the vectors in S is the zero vector.

Example 9. If $\mathbf{X}_1$, $\mathbf{X}_2$, $\mathbf{X}_3$, and $\mathbf{X}_4$ are as in Example 6, then $\mathbf{X}_1 + 2\mathbf{X}_2 + \mathbf{X}_3 + 0\mathbf{X}_4 = \mathbf{0}$, which implies that $\mathbf{X}_3 = -\mathbf{X}_1 - 2\mathbf{X}_2$. Also, $\mathbf{X}_1 + \mathbf{X}_2 + 0\mathbf{X}_3 - \mathbf{X}_4 = \mathbf{0}$, which implies that $\mathbf{X}_4 = \mathbf{X}_1 + \mathbf{X}_2$.

We observe that Theorem 3.9 does not say that *every* vector $\mathbf{X}_j$ is a linear combination of the preceding vectors in S. Thus, in Example 9, $\mathbf{X}_1 + 2\mathbf{X}_2 + \mathbf{X}_3 + 0\mathbf{X}_4 = \mathbf{0}$. We cannot solve for $\mathbf{X}_4$ as a linear combination of $\mathbf{X}_1$, $\mathbf{X}_2$, and $\mathbf{X}_3$.

We can also show that if $S = \{\mathbf{X}_1, \mathbf{X}_2, \ldots, \mathbf{X}_k\}$ is a set of vectors in a vector space, then S is linearly dependent if and only if one of the vectors in S is a linear combination of all the other vectors in S (Exercise T.3). Thus, in Example 9, $\mathbf{X}_1 = -\mathbf{X}_3 - 2\mathbf{X}_2$, $\mathbf{X}_2 = -\frac{1}{2}\mathbf{X}_1 - \frac{1}{2}\mathbf{X}_3$, $\mathbf{X}_1 = \mathbf{X}_4 - \mathbf{X}_2$, and $\mathbf{X}_2 = \mathbf{X}_4 - \mathbf{X}_1$.

Basis

DEFINITION A set of vectors $S = \{\mathbf{X}_1, \mathbf{X}_2, \ldots, \mathbf{X}_n\}$ in a vector space V is called a **basis** for V if S spans V and S is linearly independent.

Example 10. The vectors $\mathbf{E}_1$ and $\mathbf{E}_2$ form a basis for R^2, the vectors $\mathbf{E}_1$, $\mathbf{E}_2$, and $\mathbf{E}_3$ form a basis for R^3 and, in general, the vectors $\mathbf{E}_1$, $\mathbf{E}_2, \ldots, \mathbf{E}_n$ form a basis for R^n. Each of these is called the **natural basis** for R^2, R^3, and R^n, respectively.

Example 11. Show that the set $S = \{\mathbf{X}_1, \mathbf{X}_2, \mathbf{X}_3, \mathbf{X}_4\}$, where $\mathbf{X}_1 = (1, 0, 1, 0)$, $\mathbf{X}_2 = (0, 1, -1, 2)$, $\mathbf{X}_3 = (0, 2, 2, 1)$, and $\mathbf{X}_4 = (1, 0, 0, 1)$ is a basis for R^4.

solution. To show that S is linearly independent, we form the equation

$$c_1\mathbf{X}_1 + c_2\mathbf{X}_2 + c_3\mathbf{X}_3 + c_4\mathbf{X}_4 = \mathbf{0}$$

and solve for c_1, c_2, c_3, and c_4. Substituting for $\mathbf{X}_1$, $\mathbf{X}_2$, $\mathbf{X}_3$, and $\mathbf{X}_4$, we obtain the linear system

$$\begin{aligned} c_1 \qquad\qquad\qquad + c_4 &= 0 \\ c_2 + 2c_3 \qquad &= 0 \\ c_1 - c_2 + 2c_3 \qquad &= 0 \\ 2c_2 + c_3 + c_4 &= 0, \end{aligned}$$

which has as its only solution $c_1 = c_2 = c_3 = c_4 = 0$ (verify), showing that S is linearly independent.

To show that S spans R^4, we let $\mathbf{X} = (a, b, c, d)$ be any vector in R^4.

We now seek constants k_1, k_2, k_3, and k_4 such that

$$k_1\mathbf{X}_1 + k_2\mathbf{X}_2 + k_3\mathbf{X}_3 + k_4\mathbf{X}_4 = \mathbf{X}.$$

Substituting for $\mathbf{X}_1$, $\mathbf{X}_2$, $\mathbf{X}_3$, $\mathbf{X}_4$, and $\mathbf{X}$, we find a solution for k_1, k_2, k_3, and k_4 to the resulting linear system. Hence S spans R^4 and is a basis for R^4.

THEOREM 3.10. *If $S = \{\mathbf{X}_1, \mathbf{X}_2, \ldots, \mathbf{X}_n\}$ is a basis for a vector space V, then every vector in V can be written in one and only one way as a linear combination of the vectors in S.*

proof. First, every vector $\mathbf{X}$ in V can be written as a linear combination of the vectors in S, because S spans V. Now let

$$\mathbf{X} = c_1\mathbf{X}_1 + c_2\mathbf{X}_2 + \cdots + c_n\mathbf{X}_n \tag{3}$$

and

$$\mathbf{X} = d_1\mathbf{X}_1 + d_2\mathbf{X}_2 + \cdots + d_n\mathbf{X}_n. \tag{4}$$

Subtracting (4) from (3), we obtain

$$0 = (c_1 - d_1)\mathbf{X}_1 + (c_2 - d_2)\mathbf{X}_2 + \cdots + (c_n - d_n)\mathbf{X}_n.$$

Since S is linearly independent, it follows that $c_i - d_i = 0$, $1 \leqslant i \leqslant n$, so $c_i = d_i$, $1 \leqslant i \leqslant n$.

We shall leave the proof of the following important theorem as an exercise. The example following the theorem completely imitates the proof.

THEOREM 3.11. *If $S = \{\mathbf{X}_1, \mathbf{X}_2, \ldots, \mathbf{X}_m\}$ is a set of nonzero vectors spanning a subspace W of a vector space V, then some subset of S is a basis for W.*

proof. Exercise T.4.

Example 12. Consider the subspace W of R^4 spanned by the vectors $\mathbf{X}_1 = (1, 2, -2, 1)$, $\mathbf{X}_2 = (-3, 0, -4, 3)$, $\mathbf{X}_3 = (2, 1, 1, -1)$, and $\mathbf{X}_4 = (-3, 3, -9, 6)$. Every vector in W is of the form

$$a\mathbf{X}_1 + b\mathbf{X}_2 + c\mathbf{X}_3 + d\mathbf{X}_4. \tag{5}$$

To find a basis for W we first determine whether the set $S = \{\mathbf{X}_1, \mathbf{X}_2, \mathbf{X}_3, \mathbf{X}_4\}$ is linearly dependent or linearly independent. If S were linearly independent, then since S already spans W, we would conclude that S is a basis for W. In this example we find that S is linearly dependent (verify) and that

$$\mathbf{X}_1 - \mathbf{X}_2 - 2\mathbf{X}_3 = \mathbf{0},$$

which implies that

$$\mathbf{X}_2 = \mathbf{X}_1 - 2\mathbf{X}_3. \tag{6}$$

Substituting (6) into (5), we conclude that every vector in W can be written as

$$(a + b)\mathbf{X}_1 + (c - 2b)\mathbf{X}_3 + d\mathbf{X}_4. \tag{7}$$

Thus W is spanned by $\mathbf{X}_1$, $\mathbf{X}_3$, and $\mathbf{X}_4$. We now check whether the set $\{\mathbf{X}_1, \mathbf{X}_3, \mathbf{X}_4\}$ is linearly dependent or linearly independent. We find that $\mathbf{X}_1$, $\mathbf{X}_3$, and $\mathbf{X}_4$ are linearly dependent and that

$$-3\mathbf{X}_1 + 3\mathbf{X}_3 + \mathbf{X}_4 = \mathbf{0},$$

which implies that

$$\mathbf{X}_4 = 3\mathbf{X}_1 - 3\mathbf{X}_3. \tag{8}$$

Substituting (8) into (7), we conclude that every vector in W can be written as a linear combination of $\mathbf{X}_1$ and $\mathbf{X}_3$ so that W is spanned by $\mathbf{X}_1$ and $\mathbf{X}_3$. We now check whether the set $\{\mathbf{X}_1, \mathbf{X}_3\}$ is linearly dependent or linearly independent. It turns out to be linearly independent. Since $\{\mathbf{X}_1, \mathbf{X}_3\}$ spans W and is linearly independent, it is a basis for W.

We are now about to establish a main result (Corollary 3.1) of this section, which will tell us about the number of vectors in two different bases. First, observe that if $\{\mathbf{X}_1, \mathbf{X}_2, \ldots, \mathbf{X}_n\}$ is a basis for a vector space V, then $\{c\mathbf{X}_1, \mathbf{X}_2, \ldots, \mathbf{X}_n\}$ is also a basis if $c \neq 0$. Thus a vector space always has infinitely many bases.

THEOREM 3.12. *If* $S = \{\mathbf{X}_1, \mathbf{X}_2, \ldots, \mathbf{X}_n\}$ *is a basis for a vector space* V *and* $T = \{\mathbf{Y}_1, \mathbf{Y}_2, \ldots, \mathbf{Y}_r\}$ *is a linearly independent set of vectors in* V, *then* $n \geqslant r$.

proof. Let $T_1 = \{\mathbf{Y}_1, \mathbf{X}_1, \ldots, \mathbf{X}_n\}$. Since S spans V, so does T_1. Since $\mathbf{Y}_1$ is a linear combination of the vectors in S, we find that T_1 is linearly dependent. Then, by Theorem 3.9, some $\mathbf{X}_j$ is a linear combination of the preceding vectors in T_1. Delete that particular vector $\mathbf{X}_j$.

Let $S_1 = \{\mathbf{Y}_1, \mathbf{X}_1, \ldots, \mathbf{X}_{j-1}, \mathbf{X}_{j+1}, \ldots, \mathbf{X}_n\}$. Note that S_1 spans V. Next, let $T_2 = \{\mathbf{Y}_2, \mathbf{Y}_1, \mathbf{X}_1, \ldots, \mathbf{X}_{j-1}, \mathbf{X}_{j+1}, \ldots, \mathbf{X}_n\}$. Then T_2 is linearly dependent and some vector in T_2 is a linear combination of the preceding vectors in T_2. Since T is linearly independent, this vector cannot be $\mathbf{Y}_1$, so it is $\mathbf{X}_i$, $i \neq j$. Repeat this process n times until either we have used up all the $\mathbf{Y}$'s, in which case $n \geqslant r$, or we have not used up all $\mathbf{Y}$'s but have removed all the $\mathbf{X}$'s so that $r > n$. In the latter case we get a set $T_k = \{\mathbf{Y}_k, \mathbf{Y}_{k-1}, \ldots, \mathbf{Y}_2, \mathbf{Y}_1\}$, which is linearly dependent. Since T_k is a subset of T, this implies that T is linearly dependent, a contradiction. Thus we can only have $n \geqslant r$.

COROLLARY 3.1. *If* $S = \{\mathbf{X}_1, \mathbf{X}_2, \ldots, \mathbf{X}_n\}$ *and* $T = \{\mathbf{Y}_1, \mathbf{Y}_2, \ldots, \mathbf{Y}_m\}$ *are bases for a vector space, then* $n = m$.

proof. Since T is a linearly independent set of vectors, Theorem 3.12 implies that $n \geqslant m$. Similarly, $m \geqslant n$ because S is linearly independent. Hence $n = m$.

Thus, although a vector space has many bases, all bases have the same number of vectors. We can then make the following definition.

DEFINITION The **dimension** of a nonzero vector space V is the number of vectors in a basis for V. We often write **dim** V for the dimension of V.

Example 13. The dimension of R^2 is 2; the dimension of R^3 is 3; and in general, the dimension of R^n is n. Since the set $\{\mathbf{0}\}$ is linearly dependent, then it is natural to say that the vector space $\{\mathbf{0}\}$ has dimension **zero**.

All vector spaces considered in this book are **finite dimensional**; that is, their dimension is a finite number. However, we point out that there are many vector spaces of infinite dimension that are extremely important in mathematics and physics; their study lies beyond the scope of this book.

It is easy to show that if V is a finite-dimensional vector space, then every nonzero subspace W of V has a finite basis and $\dim W \leqslant \dim V$ (Exercise T.6).

Example 14. The subspace W of R^4 considered in Example 12 has dimension 2.

We might also consider the subspaces of R^2 [recall that R^2 can be viewed as the (x, y)-plane]. First, we have $\{\mathbf{0}\}$ and R^2, the trivial subspaces of dimension 0 and 2, respectively. Now the subspace V of R^2 spanned by a vector $\mathbf{X} \neq \mathbf{0}$ is a one-dimensional subspace of R^2; V is represented by a line through the origin. Thus the subspaces of R^2 are $\{\mathbf{0}\}$, R^2, and all the lines through the origin.

It follows that if a vector space V has dimension n, then any set of $n + 1$ vectors in V is necessarily linearly dependent (Exercise T.7). Any set of more than n vectors in R^n is linearly dependent. Thus the four vectors in R^3 considered in Example 6 are linearly dependent. Also, if a vector space V is of dimension n, then no set of $n - 1$ vectors in V can span V (Exercise T.8). Thus in Example 3 the vectors $\mathbf{X}_1$ and $\mathbf{X}_2$ do not span R^3.

We now come to a theorem that we shall have occasion to use several times in constructing a basis containing a given set of linearly independent vectors. We shall leave the proof as an exercise (Exercise T.9). The example following the theorem completely imitates the proof.

THEOREM 3.13. *If S is a linearly independent set in a finite dimensional vector space V, then there is a basis T for V, which contains S.*

Example 15. Suppose that we wish to find a basis for R^4 which contains the vectors $\mathbf{X}_1 = (1, 0, 1, 0)$ and $\mathbf{X}_2 = (-1, 1, -1, 0)$.

We use Theorem 3.13 as follows. First, let $\{\mathbf{E}_1, \mathbf{E}_2, \mathbf{E}_3, \mathbf{E}_4\}$ be the natural basis for R^4, where

$$\mathbf{E}_1 = (1, 0, 0, 0), \quad \mathbf{E}_2 = (0, 1, 0, 0), \quad \mathbf{E}_3 = (0, 0, 1, 0),$$

and

$$\mathbf{E}_4 = (0, 0, 0, 1).$$

Form the set $S_1 = \{\mathbf{X}_1, \mathbf{X}_2, \mathbf{E}_1, \mathbf{E}_2, \mathbf{E}_3, \mathbf{E}_4\}$. Since S_1 spans R^4, it contains, by Theorem 3.11, a basis for R^4. The basis is obtained by deleting from S_1 every vector that is a linear combination of its preceding vectors. We check whether $\mathbf{E}_1$ is a linear combination of $\mathbf{X}_1$ and $\mathbf{X}_2$. Since the answer is no, we retain $\mathbf{E}_1$. Now we check whether $\mathbf{E}_2$ is a linear combination of $\mathbf{X}_1$, $\mathbf{X}_2$, and $\mathbf{E}_1$. Since the answer is yes, we delete $\mathbf{E}_2$. Next, we check whether $\mathbf{E}_3$ is a linear combination of $\mathbf{X}_1$, $\mathbf{X}_2$, and $\mathbf{E}_1$. Since the answer is yes, we delete $\mathbf{E}_3$. We now check whether $\mathbf{E}_4$ is a linear combination of $\mathbf{X}_1$, $\mathbf{X}_2$, and $\mathbf{E}_1$. The answer is no, so our basis consists of the vectors $\mathbf{X}_1$, $\mathbf{X}_2$, $\mathbf{E}_1$, and $\mathbf{E}_4$.

From the definition of a basis, a set of vectors in a vector space S is a basis for V if it spans V and is linearly independent. However, if we are given the additional information that the dimension of V is n, we need only verify one of the two conditions. This is the content of the following theorem.

THEOREM 3.14. *Let V be an n-dimensional vector space, and let $S = \{\mathbf{X}_1, \mathbf{X}_2, \ldots, \mathbf{X}_n\}$ be a set of vectors.*

(a) *If S is linearly independent, then it is a basis for V.*

(b) *If S spans V, then it is a basis for V.*

proof. Exercise T.10.

We conclude this section with one more important property of the solution space of the homogeneous system $\mathbf{AX} = \mathbf{0}$.

Example 16. Consider the homogeneous system

$$\begin{bmatrix} 1 & 2 & 0 & 3 & 1 \\ 2 & 3 & 0 & 3 & 1 \\ 1 & 1 & 2 & 2 & 1 \\ 3 & 5 & 0 & 6 & 2 \\ 2 & 3 & 2 & 5 & 2 \end{bmatrix} \begin{bmatrix} x_1 \\ x_2 \\ x_3 \\ x_4 \\ x_5 \end{bmatrix} = \begin{bmatrix} 0 \\ 0 \\ 0 \\ 0 \\ 0 \end{bmatrix}.$$

We wish to find a basis for the solution space V. Using Gauss–Jordan reduction, we find that we must solve the homogeneous system

$$\begin{bmatrix} 1 & 0 & 0 & -3 & -1 \\ 0 & 1 & 0 & 3 & 1 \\ 0 & 0 & 1 & 1 & \frac{1}{2} \\ 0 & 0 & 0 & 0 & 0 \\ 0 & 0 & 0 & 0 & 0 \end{bmatrix} \begin{bmatrix} x_1 \\ x_2 \\ x_3 \\ x_4 \\ x_5 \end{bmatrix} = \begin{bmatrix} 0 \\ 0 \\ 0 \\ 0 \\ 0 \end{bmatrix}.$$

The general solution is (verify)

$$\mathbf{X} = \begin{bmatrix} 3s + t \\ -3s - t \\ -s - \frac{1}{2}t \\ s \\ t \end{bmatrix},$$

where s and t are any real numbers, and we can then write

$$\mathbf{X} = s\begin{bmatrix} 3 \\ -3 \\ -1 \\ 1 \\ 0 \end{bmatrix} + t\begin{bmatrix} 1 \\ -1 \\ -\frac{1}{2} \\ 0 \\ 1 \end{bmatrix}.$$

Since s and t can take on any values, letting them first be 1 and 0, and then 0 and 1, we get as solutions

$$\mathbf{X}_1 = \begin{bmatrix} 3 \\ -3 \\ -1 \\ 1 \\ 0 \end{bmatrix} \quad \text{and} \quad \mathbf{X}_2 = \begin{bmatrix} 1 \\ -1 \\ -\frac{1}{2} \\ 0 \\ 1 \end{bmatrix}.$$

Clearly, $\{\mathbf{X}_1, \mathbf{X}_2\}$ spans V, the solution space, and since $\{\mathbf{X}_1, \mathbf{X}_2\}$ is linearly independent (verify), it is a basis for V. Thus the dimension of V is 2.

3.5 Exercises

1. Which of the following vectors are linear combinations of $\mathbf{X}_1 = (4, 2, -3)$, $\mathbf{X}_2 = (2, 1, -2)$, and $\mathbf{X}_3 = (-2, -1, 0)$?
(a) $(1, 1, 1)$. (b) $(4, 2, -6)$. (c) $(-2, -1, 1)$. (d) $(-1, 2, 3)$.

2. Which of the following vectors are linear combinations of $\mathbf{X}_1 = (1, 2, 1, 0)$, $\mathbf{X}_2 = (4, 1, -2, 3)$, $\mathbf{X}_3 = (1, 2, 6, -5)$, and $\mathbf{X}_4 = (-2, 3, -1, 2)$?
(a) $(3, 6, 3, 0)$. (b) $(1, 0, 0, 0)$. (c) $(3, 6, -2, 5)$. (d) $(0, 0, 0, 1)$.

3. Which of the following sets of vectors span R^2?
(a) $(1, 2), (-1, 1)$. (b) $(0, 0), (1, 1), (-2, -2)$.
(c) $(1, 3), (2, -3), (0, 2)$. (d) $(2, 4), (-1, 2)$.

4. Which of the following sets of vectors span R^3?
(a) $(1, -1, 2), (0, 1, 1)$.
(b) $(1, 2, -1), (6, 3, 0), (4, -1, 2), (-3, -2, 1)$.
(c) $(2, 2, 3), (-1, -2, 1), (0, 1, 0)$.
(d) $(1, 0, 0), (0, 1, 0), (0, 0, 1), (1, 1, 1)$.

5. Which of the following sets of vectors span R^4?

(a) (1, 0, 0, 1), (0, 1, 0, 0), (1, 1, 1, 1), (1, 1, 1, 0).

(b) (1, 2, 1, 0), (1, 1, − 1, 0), (0, 0, 0, 1).

(c) (6, 4, − 2, 4), (2, 0, 0, 1), (3, 2, − 1, 2), (5, 6, − 3, 2), (0, 4, − 2, − 1).

(d) (1, 1, 0, 0), (1, 2, − 1, 1), (0, 0, 1, 1), (2, 1, 2, 1).

6. Which of the following sets of vectors in R^3 are linearly dependent? For those that are, express one vector as a linear combination of the rest.

(a) $\mathbf{X}_1 = (4, 2, 1)$, $\mathbf{X}_2 = (2, 6, -5)$, $\mathbf{X}_3 = (1, -2, 3)$.

(b) $\mathbf{X}_1 = (1, 2, -1)$, $\mathbf{X}_2 = (3, 2, 5)$.

(c) $\mathbf{X}_1 = (1, 1, 0)$, $\mathbf{X}_2 = (0, 2, 3)$, $\mathbf{X}_3 = (1, 2, 3)$, $\mathbf{X}_4 = (3, 6, 6)$.

(d) $\mathbf{X}_1 = (1, 2, 3)$, $\mathbf{X}_2 = (1, 1, 1)$, $\mathbf{X}_3 = (1, 0, 1)$.

7. Consider the vector space R^4. Follow the directions of Exercise 6.

(a) $\mathbf{X}_1 = (1, 1, 2, 1)$, $\mathbf{X}_2 = (1, 0, 0, 2)$, $\mathbf{X}_3 = (4, 6, 8, 6)$, $\mathbf{X}_4 = (0, 3, 2, 1)$.

(b) $\mathbf{X}_1 = (1, 1, 1, 1)$, $\mathbf{X}_2 = (1, 0, 0, 2)$, $\mathbf{X}_3 = (0, 1, 0, 2)$.

(c) $\mathbf{X}_1 = (1, 1, 1, 1)$, $\mathbf{X}_2 = (2, 3, 1, 2)$, $\mathbf{X}_3 = (3, 1, 2, 1)$, $\mathbf{X}_4 = (2, 2, 1, 1)$.

(d) $\mathbf{X}_1 = (4, 2, -1, 3)$, $\mathbf{X}_2 = (6, 5, -5, 1)$, $\mathbf{X}_3 = (2, -1, 3, 5)$.

8. Consider the vector space R^3. Follow the directions of Exercise 6.

(a) $\mathbf{X}_1 = (1, 0, 1)$, $\mathbf{X}_2 = (0, 1, -2)$, $\mathbf{X}_3 = (0, 1, 3)$.

(b) $\mathbf{X}_1 = (2, 1, 0)$, $\mathbf{X}_2 = (1, 0, 3)$, $\mathbf{X}_3 = (0, 1, 0)$.

(c) $\mathbf{X}_1 = (0, 1, 3)$, $\mathbf{X}_2 = (3, 1, 0)$, $\mathbf{X}_3 = (2, 1, 1)$.

(d) $\mathbf{X}_1 = (1, 0, -4)$, $\mathbf{X}_2 = (5, -5, 6)$, $\mathbf{X}_3 = (3, -5, 2)$.

9. Which of the sets in Exercise 3 are bases for R^2?

10. Which of the sets in Exercise 4 are bases for R^3?

11. Which of the sets in Exercise 5 are bases for R^4?

12. Which of the sets in Exercise 6 are bases for R^3?

13. Which of the sets in Exercise 7 are bases for R^4?

14. Which of the sets in Exercise 8 are bases for R^3?

15. Which of the following sets form a basis for R^3? Express the vector $\mathbf{X} = (2, 1, 3)$ as a linear combination of the vectors in each set that is a basis.

(a) $\mathbf{X}_1 = (1, 1, 1)$, $\mathbf{X}_2 = (1, 2, 3)$, $\mathbf{X}_3 = (0, 1, 0)$.

(b) $\mathbf{X}_1 = (1, 2, 2)$, $\mathbf{X}_2 = (2, 1, 3)$, $\mathbf{X}_3 = (0, 0, 0)$.

(c) $\mathbf{X}_1 = (-2, 1, 3)$, $\mathbf{X}_2 = (-1, 2, 3)$, $\mathbf{X}_3 = (-1, -4, -3)$.

16. Which of the following sets form a basis for R^4? Express the vector $\mathbf{X} = (1, -5, 6, 9)$ as a linear combination of the vectors in each set that is a basis.

(a) $\mathbf{X}_1 = (1, -1, 2, 3)$, $\mathbf{X}_2 = (1, 1, 0, 1)$, $\mathbf{X}_3 = (0, 0, 1, 0)$, $\mathbf{X}_4 = (0, 1, 0, 0)$.

(b) $\mathbf{X}_1 = (1, 1, 0, 2)$, $\mathbf{X}_2 = (0, 1, -2, -1)$, $\mathbf{X}_3 = (1, 1, -3, -3)$, $\mathbf{X}_4 = (3, 2, -1, 2)$.

(c) $\mathbf{X}_1 = (2, -4, 5, -3)$, $\mathbf{X}_2 = (3, -1, 2, 0)$, $\mathbf{X}_3 = (0, -2, 3, -5)$, $\mathbf{X}_4 = (-1, 1, -2, 3)$.

17. Find a basis for the subspace W of R^3 spanned by $\mathbf{X}_1 = (1, 2, 2)$, $\mathbf{X}_2 = (3, 2, 1)$, $\mathbf{X}_3 = (11, 10, 7)$, and $\mathbf{X}_4 = (7, 6, 4)$. What is dim W?

18. Find a basis for the subspace W of R^4 spanned by $\mathbf{X}_1 = (1, 1, 0, -1)$, $\mathbf{X}_2 = (0, 1, 2, 1)$, $\mathbf{X}_3 = (1, 0, 1, -1)$, $\mathbf{X}_4 = (1, 1, -6, -3)$, and $\mathbf{X}_5 = (-1, -5, 1, 0)$. What is dim W?

19. Find the dimensions of the subspaces of R^2 spanned by the vectors in Exercise 3.

20. Find the dimensions of the subspaces of R^3 spanned by the vectors in Exercise 4.

21. Find the dimensions of the subspaces of R^4 spanned by the vectors in Exercise 5.

22. Find a basis for R^3 that includes the vectors
(a) $(1, 0, 2)$. (b) $(1, 0, 2)$ and $(0, 1, 3)$.

23. Find a basis for R^4 that includes the vectors $(1, 0, 1, 0)$ and $(0, 1, -1, 0)$.

24. Find a basis for the solution space W of

$$\begin{bmatrix} 1 & 2 & 1 & 2 & 1 \\ 1 & 2 & 2 & 1 & 2 \\ 2 & 4 & 3 & 3 & 3 \\ 0 & 0 & 1 & -1 & -1 \end{bmatrix} \begin{bmatrix} x_1 \\ x_2 \\ x_3 \\ x_4 \\ x_5 \end{bmatrix} = \begin{bmatrix} 0 \\ 0 \\ 0 \\ 0 \end{bmatrix}.$$

What is the dimension of W?

25. Find the dimension of the solution space of

$$\begin{bmatrix} 1 & 0 & 2 \\ 2 & 1 & 3 \\ 3 & 1 & 2 \end{bmatrix} \begin{bmatrix} x_1 \\ x_2 \\ x_3 \end{bmatrix} = \begin{bmatrix} 0 \\ 0 \\ 0 \end{bmatrix}.$$

26. Find a basis for the solution space W of

$$\begin{bmatrix} 1 & 2 & 2 & -1 & 1 \\ 0 & 2 & 2 & -2 & -1 \\ 2 & 6 & 2 & -4 & 1 \\ 1 & 4 & 0 & -3 & 0 \end{bmatrix} \begin{bmatrix} x_1 \\ x_2 \\ x_3 \\ x_4 \\ x_5 \end{bmatrix} = \begin{bmatrix} 0 \\ 0 \\ 0 \\ 0 \end{bmatrix}.$$

What is the dimension of W?

Theoretical Exercises

T.1. Prove that the vectors $\mathbf{E}_1, \mathbf{E}_2, \ldots, \mathbf{E}_n$ in R^n are linearly independent.

T.2. Let S_1 and S_2 be finite subsets of a vector space and let S_1 be a subset of S_2. Prove:
(a) If S_1 is linearly dependent, so is S_2.
(b) If S_2 is linearly independent, so is S_1.

T.3. Let $S = \{\mathbf{X}_1, \mathbf{X}_2, \ldots, \mathbf{X}_k\}$ be a set of vectors in a vector space. Prove that S is linearly dependent if and only if one of the vectors in S is a linear combination of all the other vectors in S.

T.4. Prove Theorem 3.11.

T.5. Suppose that in the nonzero vector space V, the largest number of vectors in a linearly independent set is m. Show that any set of m linearly independent vectors in V is a basis for V.

T.6. Show that if V is a finite-dimensional vector space, then every nonzero subspace W of V has a finite basis and $\dim W \leqslant \dim V$.

T.7. Prove that if $\dim V = n$, then any $n + 1$ vectors in V are linearly dependent.

T.8. Prove that if $\dim V = n$, then no set of $n - 1$ vectors in V can span V.

T.9. Prove Theorem 3.13.

T.10. Prove Theorem 3.14.

T.11. Prove that if W is a subspace of a finite-dimensional vector space V and $\dim W = \dim V$, then $W = V$.

T.12. Classify all the subspaces of R^3.

3.6 Orthonormal Bases in R^n

A given subspace W of R^n need not contain any of the natural basis vectors, but we want to show that it has a basis with the same properties. That is, we want to show that W contains a basis S such that every vector in S is of unit length and every two vectors in S are perpendicular. The method used to obtain such a basis is the **Gram–Schmidt process**.

DEFINITION A set $S = \{\mathbf{X}_1, \mathbf{X}_2, \ldots, \mathbf{X}_k\}$ in R^n is called **orthogonal** if any two distinct vectors in S are orthogonal, that is, if $\mathbf{X}_i \cdot \mathbf{X}_j = 0$ for $i \neq j$. An **orthonormal** set of vectors is an orthogonal set of unit vectors.

Example 1. If $\mathbf{X}_1 = (1, 0, 2)$, $\mathbf{X}_2 = (-2, 0, 1)$, and $\mathbf{X}_3 = (0, 1, 0)$, then $\{\mathbf{X}_1, \mathbf{X}_2, \mathbf{X}_3\}$ is an orthogonal set in R^3. The vectors

$$\mathbf{Y}_1 = \left(\frac{1}{\sqrt{5}}, 0, \frac{2}{\sqrt{5}}\right) \quad \text{and} \quad \mathbf{Y}_2 = \left(\frac{-2}{\sqrt{5}}, 0, \frac{1}{\sqrt{5}}\right)$$

are unit vectors in the directions of $\mathbf{X}_1$ and $\mathbf{X}_2$, respectively. Since $\mathbf{X}_3$ is also of unit length, it follows that $\{\mathbf{Y}_1, \mathbf{Y}_2, \mathbf{X}_3\}$ is an orthonormal set.

Example 2. The natural basis

$$\{(1, 0, 0), (0, 1, 0), (0, 0, 1)\}$$

is an orthonormal set in R^3. More generally, the natural basis in R^n is an orthonormal set.

An important result about orthonormal sets is the following theorem.

THEOREM 3.15. *Let $S = \{\mathbf{X}_1, \mathbf{X}_2, \ldots, \mathbf{X}_k\}$ be an orthogonal set of nonzero vectors in R^n. Then S is linearly independent.*

proof. Consider the equation

$$c_1\mathbf{X}_1 + c_2\mathbf{X}_2 + \cdots + c_k\mathbf{X}_k = \mathbf{0}. \tag{1}$$

Taking the inner product of both sides of (1) with $\mathbf{X}_i$, $1 \leqslant i \leqslant n$, we have

$$(c_1\mathbf{X}_1 + c_2\mathbf{X}_2 + \cdots + c_k\mathbf{X}_k) \cdot \mathbf{X}_i = \mathbf{0} \cdot \mathbf{X}_i. \tag{2}$$

By properties (c) and (d) of Theorem 3.3, Section 3.2, the left side of (2) is

$$c_1(\mathbf{X}_1 \cdot \mathbf{X}_i) + c_2(\mathbf{X}_2 \cdot \mathbf{X}_i) + \cdots + c_k(\mathbf{X}_k \cdot \mathbf{X}_i),$$

and the right side is 0. Since $\mathbf{X}_j \cdot \mathbf{X}_i = 0$ if $i \neq j$, (2) becomes

$$0 = c_i(\mathbf{X}_i \cdot \mathbf{X}_i) = c_i\|\mathbf{X}_i\|^2. \tag{3}$$

By (a) of Theorem 3.3, Section 3.2, $\|\mathbf{X}_i\| \neq 0$, since $\mathbf{X}_i \neq \mathbf{0}$. Hence (3) implies that $c_i = 0$, $1 \leqslant i \leqslant n$, and S is linearly independent.

COROLLARY 3.2. *An orthonormal set of vectors in R^n is linearly independent.*

proof. Exercise T.2.

From Theorem 3.14 of Section 3.5, it follows that an orthonormal set of n vectors in R^n is a basis for R^n (Exercise T.3). An **orthogonal (orthonormal) basis** for a vector space is a basis that is an orthogonal (orthonormal) set.

THEOREM 3.16 (Gram–Schmidt Process). *Let W be a nonzero subspace of R^n with basis $S = \{\mathbf{X}_1, \mathbf{X}_2, \ldots, \mathbf{X}_m\}$. Then there exists an orthonormal basis $T = \{\mathbf{Z}_1, \mathbf{Z}_2, \ldots, \mathbf{Z}_m\}$ for W.*

proof. The proof is constructive; that is, we develop the desired basis T. However, we first find an orthogonal basis $T^* = \{\mathbf{Y}_1, \mathbf{Y}_2, \ldots, \mathbf{Y}_m\}$ for W.

First, we pick any one of the vectors in S, say $\mathbf{X}_1$, and call it $\mathbf{Y}_1$. Thus $\mathbf{Y}_1 = \mathbf{X}_1$. We now look for a vector $\mathbf{Y}_2$ in the subspace W_1 of W spanned by $\{\mathbf{X}_1, \mathbf{X}_2\}$, which is orthogonal to $\mathbf{Y}_1$. Since $\mathbf{Y}_1 = \mathbf{X}_1$, W_1 is also the subspace spanned by $\{\mathbf{Y}_1, \mathbf{X}_2\}$. Thus

$$\mathbf{Y}_2 = c_1\mathbf{Y}_1 + c_2\mathbf{X}_2.$$

We try to determine c_1 and c_2 so that $\mathbf{Y}_1 \cdot \mathbf{Y}_2 = 0$. Now

$$\begin{aligned} 0 = \mathbf{Y}_2 \cdot \mathbf{Y}_1 &= (c_1\mathbf{Y}_1 + c_2\mathbf{X}_2) \cdot \mathbf{Y}_1 \\ &= c_1(\mathbf{Y}_1 \cdot \mathbf{Y}_1) + c_2\mathbf{X}_2 \cdot \mathbf{Y}_1. \end{aligned} \tag{4}$$

Since $\mathbf{Y}_1 \neq \mathbf{0}$ (why?), $\mathbf{Y}_1 \cdot \mathbf{Y}_1 \neq 0$, and solving for c_1 and c_2 in (4), we have

$$c_1 = -c_2 \frac{\mathbf{X}_2 \cdot \mathbf{Y}_1}{\mathbf{Y}_1 \cdot \mathbf{Y}_1}.$$

We may assign an arbitrary nonzero value to c_2. Thus, letting $c_2 = 1$, we obtain

$$c_1 = \frac{-\mathbf{X}_2 \cdot \mathbf{Y}_1}{\mathbf{Y}_1 \cdot \mathbf{Y}_1}.$$

Hence

$$\mathbf{Y}_2 = c_1\mathbf{Y}_1 + c_2\mathbf{X}_2 = \mathbf{X}_2 - \left(\frac{\mathbf{X}_2 \cdot \mathbf{Y}_1}{\mathbf{Y}_1 \cdot \mathbf{Y}_1}\right)\mathbf{Y}_1.$$

Notice that at this point we have an orthogonal subset $\{\mathbf{Y}_1, \mathbf{Y}_2\}$ of W.

Next, we look for a vector $\mathbf{Y}_3$ in the subspace W_2 of W spanned by $\{\mathbf{X}_1, \mathbf{X}_2, \mathbf{X}_3\}$, which is orthogonal to both $\mathbf{Y}_1$ and $\mathbf{Y}_2$. Of course, W_2 is also the subspace spanned by $\{\mathbf{Y}_1, \mathbf{Y}_2, \mathbf{X}_3\}$ (why?). Thus

$$\mathbf{Y}_3 = d_1\mathbf{Y}_1 + d_2\mathbf{Y}_2 + d_3\mathbf{X}_3.$$

We let $d_3 = 1$ and try to find d_1 and d_2 so that

$$\mathbf{Y}_3 \cdot \mathbf{Y}_1 = 0 \quad \text{and} \quad \mathbf{Y}_3 \cdot \mathbf{Y}_2 = 0.$$

Now

$$0 = \mathbf{Y}_3 \cdot \mathbf{Y}_1 = (d_1\mathbf{Y}_1 + d_2\mathbf{Y}_2 + \mathbf{X}_3) \cdot \mathbf{Y}_1 = d_1(\mathbf{Y}_1 \cdot \mathbf{Y}_1) + \mathbf{X}_3 \cdot \mathbf{Y}_1, \quad (5)$$

$$0 = \mathbf{Y}_3 \cdot \mathbf{Y}_2 = (d_1\mathbf{Y}_1 + d_2\mathbf{Y}_2 + \mathbf{X}_3) \cdot \mathbf{Y}_2 = d_2(\mathbf{Y}_2 \cdot \mathbf{Y}_2) + \mathbf{X}_3 \cdot \mathbf{Y}_2. \quad (6)$$

In obtaining the right sides of (5) and (6), we have used the fact that $\mathbf{Y}_1 \cdot \mathbf{Y}_2 = 0$. Now observe that $\mathbf{Y}_2 \neq \mathbf{0}$ (why?). Solving (5) and (6) for d_1 and d_2, respectively, we obtain

$$d_1 = -\frac{\mathbf{X}_3 \cdot \mathbf{Y}_1}{\mathbf{Y}_1 \cdot \mathbf{Y}_1} \quad \text{and} \quad d_2 = -\frac{\mathbf{X}_3 \cdot \mathbf{Y}_2}{\mathbf{Y}_2 \cdot \mathbf{Y}_2}.$$

Thus

$$\mathbf{Y}_3 = \mathbf{X}_3 - \left(\frac{\mathbf{X}_3 \cdot \mathbf{Y}_1}{\mathbf{Y}_1 \cdot \mathbf{Y}_1}\right)\mathbf{Y}_1 - \left(\frac{\mathbf{X}_3 \cdot \mathbf{Y}_2}{\mathbf{Y}_2 \cdot \mathbf{Y}_2}\right)\mathbf{Y}_2.$$

Notice that at this point we have an orthogonal subset $\{\mathbf{Y}_1, \mathbf{Y}_2, \mathbf{Y}_3\}$ of W.

We next seek a vector $\mathbf{Y}_4$ in the subspace W_3 of W spanned by $\{\mathbf{X}_1, \mathbf{X}_2, \mathbf{X}_3, \mathbf{X}_4\}$, and thus by $\{\mathbf{Y}_1, \mathbf{Y}_2, \mathbf{Y}_3, \mathbf{X}_4\}$, which is orthogonal to $\mathbf{Y}_1$, $\mathbf{Y}_2$, and $\mathbf{Y}_3$. We can then write

$$\mathbf{Y}_4 = \mathbf{X}_4 - \left(\frac{\mathbf{X}_4 \cdot \mathbf{Y}_1}{\mathbf{Y}_1 \cdot \mathbf{Y}_1}\right)\mathbf{Y}_1 - \left(\frac{\mathbf{X}_4 \cdot \mathbf{Y}_2}{\mathbf{Y}_2 \cdot \mathbf{Y}_2}\right)\mathbf{Y}_2 - \left(\frac{\mathbf{X}_4 \cdot \mathbf{Y}_3}{\mathbf{Y}_3 \cdot \mathbf{Y}_3}\right)\mathbf{Y}_3.$$

Continue in this manner until we have an orthogonal set $T^* = \{\mathbf{Y}_1, \mathbf{Y}_2, \ldots, \mathbf{Y}_m\}$ of m vectors. It then follows that T^* is a basis for W. If we now normalize the $\mathbf{Y}_i$, that is, let

$$\mathbf{Z}_i = \frac{\mathbf{Y}_i}{\|\mathbf{Y}_i\|} \qquad (1 \leqslant i \leqslant m),$$

then $T = \{\mathbf{Z}_1, \mathbf{Z}_2, \ldots, \mathbf{Z}_m\}$ is a basis for W.

Example 3. Consider the basis $S = \{\mathbf{X}_1, \mathbf{X}_2, \mathbf{X}_3\}$ for R^3, where

$$\mathbf{X}_1 = (1, 1, 1), \qquad \mathbf{X}_2 = (-1, 0, -1), \qquad \text{and} \qquad \mathbf{X}_3 = (-1, 2, 3).$$

We transform S to an orthonormal basis T as follows.

First, let $\mathbf{Y}_1 = \mathbf{X}_1$. Then

$$\mathbf{Y}_2 = \mathbf{X}_2 - \left(\frac{\mathbf{X}_2 \cdot \mathbf{Y}_1}{\mathbf{Y}_1 \cdot \mathbf{Y}_1}\right)\mathbf{Y}_1 = (-1, 0, -1) + \tfrac{2}{3}(1, 1, 1) = \left(-\tfrac{1}{3}, \tfrac{2}{3}, -\tfrac{1}{3}\right)$$

and

$$\mathbf{Y}_3 = \mathbf{X}_3 - \left(\frac{\mathbf{X}_3 \cdot \mathbf{Y}_1}{\mathbf{Y}_1 \cdot \mathbf{Y}_1}\right)\mathbf{Y}_1 - \left(\frac{\mathbf{X}_3 \cdot \mathbf{Y}_2}{\mathbf{Y}_2 \cdot \mathbf{Y}_2}\right)\mathbf{Y}_2$$

$$= (-1, 2, 3) - \left(\tfrac{4}{3}\right)(1, 1, 1) - \left(\frac{\frac{2}{3}}{\frac{6}{9}}\right)\left(-\tfrac{1}{3}, \tfrac{2}{3}, -\tfrac{1}{3}\right)$$

$$= (-2, 0, 2).$$

Then $\{\mathbf{Y}_1, \mathbf{Y}_2, \mathbf{Y}_3\}$ is an orthogonal basis for R^3. Of course, if this is an orthogonal set, we can clear the fractions in each $\mathbf{Y}_i$ by multiplying by a scalar and the resulting set is also an orthogonal basis for R^3. Hence

$$T^* = \{(1, 1, 1), (-1, 2, -1), (-2, 0, 2)\}$$

is an orthogonal basis for R^3. An orthonormal basis, obtained by normal-iz-ing, is

$$T = \left\{\left(\frac{1}{\sqrt{3}}, \frac{1}{\sqrt{3}}, \frac{1}{\sqrt{3}}\right), \left(-\frac{1}{\sqrt{6}}, \frac{2}{\sqrt{6}}, -\frac{1}{\sqrt{6}}\right), \left(-\frac{1}{\sqrt{2}}, 0, \frac{1}{\sqrt{2}}\right)\right\}.$$

Example 4. Let W be the subspace of R^4 with basis $S = \{\mathbf{X}_1, \mathbf{X}_2\}$, where $\mathbf{X}_1 = (1, -2, 0, 1)$ and $\mathbf{X}_2 = (-1, 0, 0, -1)$. To transform S to an orthonormal basis T, we proceed as follows. First, let $\mathbf{Y}_1 = \mathbf{X}_1$. Then

$$\mathbf{Y}_2 = \mathbf{X}_2 - \left(\frac{\mathbf{X}_2 \cdot \mathbf{Y}_2}{\mathbf{Y}_1 \cdot \mathbf{Y}_1}\right)\mathbf{Y}_1 = (-1, 0, 0, -1) - \left(-\tfrac{2}{6}\right)(1, -2, 0, 1)$$

$$= \left(-\tfrac{2}{3}, -\tfrac{2}{3}, 0, -\tfrac{2}{3}\right).$$

Thus $\{\mathbf{Y}_1, \mathbf{Y}_2\}$ is an orthogonal basis for W. Clearing fractions, we find that

$$\{(1, -2, 0, 1), (-2, -2, 0, -2)\}$$

is an orthogonal basis. Normalizing these vectors, we obtain the following orthonormal basis for

$$\left\{\left(\frac{1}{\sqrt{6}}, -\frac{2}{\sqrt{6}}, 0, \frac{1}{\sqrt{6}}\right), \left(-\frac{1}{\sqrt{3}}, -\frac{1}{\sqrt{3}}, 0, -\frac{1}{\sqrt{3}}\right)\right\}.$$

In the proof of Theorem 3.16 we have also established the following result. At each stage of the Gram—Schmidt process, the ordered set $\{\mathbf{Z}_1, \mathbf{Z}_2, \ldots, \mathbf{Z}_k\}$ is an orthonormal basis for the subspace spanned by $\{\mathbf{X}_1, \mathbf{X}_2, \ldots, \mathbf{X}_k\}$, $1 \leqslant k \leqslant m$. Also, the final orthonormal basis T depends upon the order of the vectors in the given basis S. Thus, if we change the order of the vectors in S, we might obtain a different orthonormal basis T for W.

We make one final observation with regard to the Gram–Schmidt process. In our proof of Theorem 3.16, we first obtained an orthogonal basis T^* and then normalized all the vectors in T^* to obtain the orthonormal basis T. Of course, an alternative course of action is to normalize each vector as soon as we produce it.

Suppose that $S = \{\mathbf{X}_1, \mathbf{X}_2, \ldots, \mathbf{X}_n\}$ is a basis for R^n. In Section 3.5 we had to solve a linear system of n equations in n unknowns to write a given vector $\mathbf{X}$ of R^n as a linear combination of the vectors in S. However, if S is orthonormal, we can obtain the same results with much less work. This is the content of the following theorem.

THEOREM 3.17. *Let* $S = \{\mathbf{X}_1, \mathbf{X}_2, \ldots, \mathbf{X}_n\}$ *be an orthonormal basis for* R^n *and* $\mathbf{X}$ *any vector in* R^n. *Then*

$$\mathbf{X} = c_1\mathbf{X}_1 + c_2\mathbf{X}_2 + \cdots + c_n\mathbf{X}_n,$$

where

$$c_i = \mathbf{X} \cdot \mathbf{X}_i \qquad (1 \leqslant i \leqslant n).$$

proof. Exercise T.4.

3.6 Exercises

1. Which of the following are orthogonal sets of vectors?

(a) $(1, -1, 2), (0, 2, -1), (-1, 1, 1)$.

(b) $(1, 2, -1, 1), (0, -1, -2, 0), (1, 0, 0, -1)$.

(c) $(0, 1, 0, -1), (1, 0, 1, 1), (-1, 1, -1, 2)$.

2. Which of the following are orthonormal sets of vectors?

(a) $(\frac{1}{3}, \frac{2}{3}, \frac{2}{3}), (\frac{2}{3}, \frac{1}{3}, -\frac{2}{3}), (\frac{2}{3}, -\frac{2}{3}, \frac{1}{3})$.

(b) $(1/\sqrt{2}, 0, -1/\sqrt{2}), (1/\sqrt{3}, 1/\sqrt{3}, 1/\sqrt{3}), (0, 1, 0)$.

(c) $(0, 2, 2, 1), (1, 1, -2, 2), (0, -2, 1, 2)$.

3. Find an orthonormal basis for the subspace of R^2 with basis $\{(1, -1), (2, 1)\}$.

4. Find an orthonormal basis for the subspace of R^4 with basis $\{(1, 1, -1, 0), (0, 2, 0, 1)\}$.

5. Use the Gram–Schmidt process to transform the basis $\{(1, 2), (-3, 4)\}$ for R^2 into (a) an orthogonal basis; (b) an orthonormal basis.

6. Use the Gram–Schmidt process to transform the basis $\{(1, 1, 1), (0, 1, 1), (1, 2, 3)\}$ for R^3 into an orthonormal basis for R^3.

7. Find an orthonormal basis for R^3 containing the vectors $(\frac{2}{3}, -\frac{2}{3}, \frac{1}{3})$ and $(\frac{2}{3}, \frac{1}{3}, -\frac{2}{3})$.

8. Construct an orthonormal basis for the subspace W of R^3 *spanned* by $\{(1, 1, 1), (2, 2, 2), (0, 0, 1), (1, 2, 3)\}$.

9. Construct an orthonormal basis for the subspace W of R^4 *spanned* by $\{(1, 1, 0, 0), (2, -1, 0, 1), (3, -3, 0, -2), (1, -2, 0, -3)\}$.

Theoretical Exercises

T.1. Verify that the natural basis for R^n is an orthonormal set.

T.2. Prove Corollary 3.2.

T.3. Show that an orthonormal set of n vectors in R^n is a basis for R^n.

T.4. Prove Theorem 3.17.

4 Linear Transformations and Matrices

4.1 Definition and Examples

Functions occur in almost every endeavor of life. In this chapter we shall study the properties of certain functions called linear transformations, mapping one vector space into another vector space. Linear transformations play an important role in many areas of mathematics, as well as in numerous applied problems in the physical sciences, the social sciences, and economics. Although we talk in terms of general vector spaces, the examples considered will be R^m and R^n.

DEFINITION Let V and W be vector spaces. A **linear transformation L of V into W** is a function assigning a unique vector $L(\mathbf{X})$ in W to each $\mathbf{X}$ in V such that:

(a) $\mathbf{L}(\mathbf{X} + \mathbf{Y}) = L(\mathbf{X}) + \mathbf{L}(\mathbf{Y})$, for every $\mathbf{X}$ and $\mathbf{Y}$ in V.

(b) $\mathbf{L}(c\mathbf{X}) = c\mathbf{L}(\mathbf{X})$, for every $\mathbf{X}$ in V and every scalar c.

We shall write the fact that $\mathbf{L}$ maps V into W, even if it is not a linear transformation, as

$$\mathbf{L} : V \to W.$$

If $V = W$, the linear transformation $\mathbf{L} : V \to W$ is also called a **linear operator on V.**

Example 1. Let $L : R^3 \to R^2$ be defined by

$$\mathbf{L}\left(\begin{bmatrix} x \\ y \\ z \end{bmatrix}\right) = \begin{bmatrix} x \\ y \end{bmatrix}.$$

To verify that $\mathbf{L}$ is a linear transformation, we let

$$\mathbf{X} = \begin{bmatrix} x_1 \\ y_1 \\ z_1 \end{bmatrix} \quad \text{and} \quad \mathbf{Y} = \begin{bmatrix} x_2 \\ y_2 \\ z_2 \end{bmatrix}.$$

Then

$$\mathbf{L}(\mathbf{X} + \mathbf{Y}) = \mathbf{L}\left(\begin{bmatrix} x_1 \\ y_1 \\ z_1 \end{bmatrix} + \begin{bmatrix} x_2 \\ y_2 \\ z_2 \end{bmatrix}\right) = \mathbf{L}\left(\begin{bmatrix} x_1 + x_2 \\ y_1 + y_2 \\ z_1 + z_2 \end{bmatrix}\right)$$

$$= \begin{bmatrix} x_1 + x_2 \\ y_1 + y_2 \end{bmatrix} = \begin{bmatrix} x_1 \\ y_1 \end{bmatrix} + \begin{bmatrix} x_2 \\ y_2 \end{bmatrix} = \mathbf{L}(\mathbf{X}) + \mathbf{L}(\mathbf{Y}).$$

Also, if c is a real number, then

$$\mathbf{L}(c\mathbf{X}) = \mathbf{L}\left(\begin{bmatrix} cx_1 \\ cy_1 \\ cz_1 \end{bmatrix}\right) = \begin{bmatrix} cx_1 \\ cy_1 \end{bmatrix} = c\begin{bmatrix} x_1 \\ y_1 \end{bmatrix} = c\mathbf{L}(\mathbf{X}).$$

Hence $\mathbf{L}$ is a linear transformation, which is called **projection**. It is a simple and helpful matter to describe geometrically the effect of $\mathbf{L}$. The image under $\mathbf{L}$ of a vector $\mathbf{X}$ in R^3 with end point $P(a, b, c)$ is found by drawing a line through P and perpendicular to R^2, the (x, y)-plane. We obtain the point $Q(a, b)$ of intersection of this line with the (x, y)-plane. The vector $\mathbf{Y}$ in R^2 with end point Q is the image of $\mathbf{X}$ under $\mathbf{L}$ (Figure 4.1).

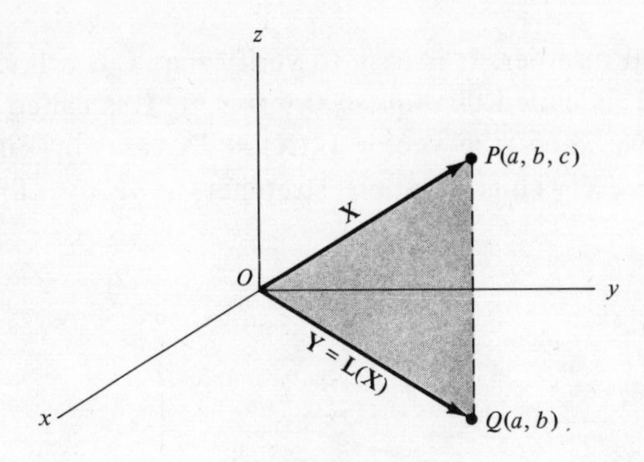

Figure 4.1

Example 2. Let $\mathbf{L} : R^3 \to R^3$ be defined by

$$\mathbf{L}\left(\begin{bmatrix} x \\ y \\ z \end{bmatrix}\right) = \begin{bmatrix} x + 1 \\ 2y \\ z \end{bmatrix}.$$

To determine whether $\mathbf{L}$ is a linear transformation, let

$$\mathbf{X} = \begin{bmatrix} x_1 \\ y_1 \\ z_1 \end{bmatrix} \quad \text{and} \quad \mathbf{Y} = \begin{bmatrix} x_2 \\ y_2 \\ z_2 \end{bmatrix}.$$

Then

$$\mathbf{L}(\mathbf{X} + \mathbf{Y}) = \mathbf{L}\left(\begin{bmatrix} x_1 \\ y_1 \\ z_1 \end{bmatrix} + \begin{bmatrix} x_2 \\ y_2 \\ z_2 \end{bmatrix}\right) = \mathbf{L}\left(\begin{bmatrix} x_1 + x_2 \\ y_1 + y_2 \\ z_1 + z_2 \end{bmatrix}\right) = \begin{bmatrix} (x_1 + x_2) + 1 \\ 2(y_1 + y_2) \\ z_1 + z_2 \end{bmatrix}.$$

On the other hand,

$$\mathbf{L}(\mathbf{X}) + \mathbf{L}(\mathbf{Y}) = \begin{bmatrix} x_1 + 1 \\ 2y_1 \\ z_1 \end{bmatrix} + \begin{bmatrix} x_2 + 1 \\ 2y_2 \\ z_2 \end{bmatrix} = \begin{bmatrix} (x_1 + x_2) + 2 \\ 2(y_1 + y_2) \\ z_1 + z_2 \end{bmatrix}.$$

Since, $\mathbf{L}(\mathbf{X} + \mathbf{Y}) \neq \mathbf{L}(\mathbf{X}) + \mathbf{L}(\mathbf{Y})$, we conclude that the function $\mathbf{L}$ is not a linear transformation.

Example 3. Let $\mathbf{L} : R^3 \to R^3$ be defined by

$$\mathbf{L}\left(\begin{bmatrix} x \\ y \\ z \end{bmatrix}\right) = \begin{bmatrix} rx \\ ry \\ rz \end{bmatrix}.$$

where r is a real number. It is easy to verify that **L** is a linear transformation. If $r > 1$, **L** is called **dilation**; if $0 < r < 1$, **L** is called **contraction**. In Figure 4.2(a), we show the vector $\mathbf{L}_1(\mathbf{X}) = 2\mathbf{X}$, and in Figure 4.2(b) the vector $\mathbf{L}_2(\mathbf{X}) = \frac{1}{2}\mathbf{X}$. Thus dilation stretches a vector, and contraction shrinks it.

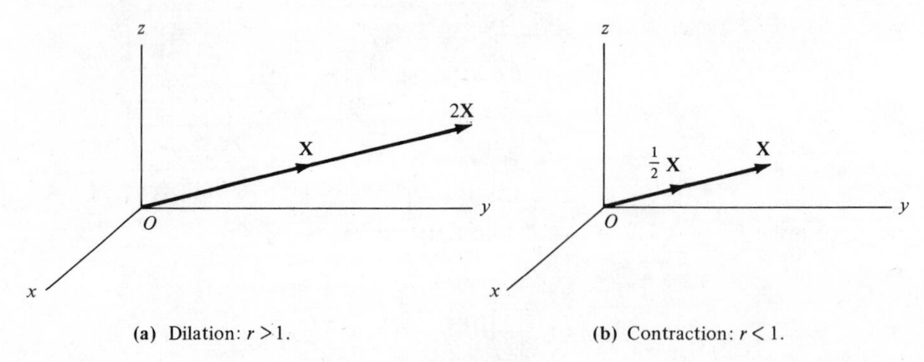

(a) Dilation: $r > 1$.

(b) Contraction: $r < 1$.

Figure 4.2

Example 4. Let $\mathbf{L} : R^2 \to R^3$ be defined by

$$L\left(\begin{bmatrix} x \\ y \end{bmatrix}\right) = \begin{bmatrix} 1 & 0 \\ 0 & 1 \\ 1 & -1 \end{bmatrix} \begin{bmatrix} x \\ y \end{bmatrix}.$$

Then **L** is a linear transformation (verify).

Example 5. Let $\mathbf{L} : R^n \to R^m$ be defined by

$$\mathbf{L}(\mathbf{X}) = \mathbf{AX} \qquad (\mathbf{X} \in R^n),$$

where **A** is an $m \times n$ matrix. It then follows that **L** is a linear transformation (Exercise T.1).

Example 6. Let $\mathbf{L} : R^2 \to R^2$ be defined by

$$\mathbf{L}\left(\begin{bmatrix} x \\ y \end{bmatrix}\right) = \begin{bmatrix} x \\ -y \end{bmatrix}.$$

Then **L** is a linear transformation, which is shown in Figure 4.3. This linear transformation is called **reflection with respect to the x-axis**. The reader may consider **reflection with respect to the y-axis**.

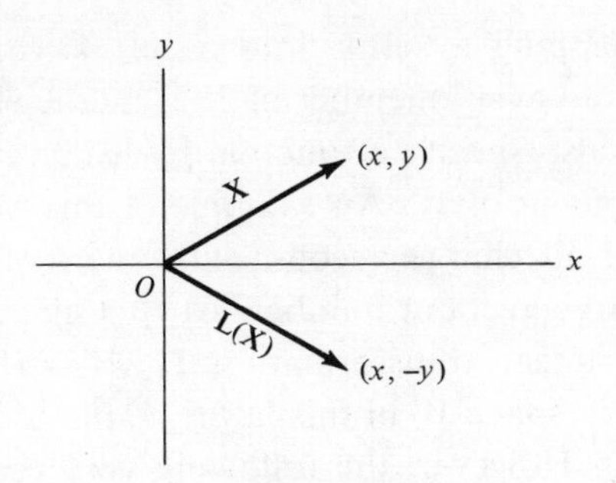

Figure 4.3

THEOREM 4.1. *If* $\mathbf{L} : V \to W$ *is a linear transformation, then*

$$\mathbf{L}(c_1\mathbf{X}_1 + c_2\mathbf{X}_2 + \cdots + c_k\mathbf{X}_k) = c_1\mathbf{L}(\mathbf{X}_1) + c_2\mathbf{L}(\mathbf{X}_2) + \cdots + c_k\mathbf{L}(\mathbf{X}_k).$$

for any vectors $\mathbf{X}_1, \mathbf{X}_2, \ldots, \mathbf{X}_k$ *in* V *and any scalars* $c_1, c_2, \ldots, c_k$.

proof. Exercise T.2.

The following theorem gives some basic properties of linear transformations.

THEOREM 4.2. *Let* $\mathbf{L} : V \to W$ *be a linear transformation. Then*:

(a) $\mathbf{L}(\mathbf{0}_V) = \mathbf{0}_W$, *where* $\mathbf{0}_V$ *and* $\mathbf{0}_W$ *are the zero vectors in* V *and* W, *respectively*.

(b) $\mathbf{L}(\mathbf{X} - \mathbf{Y}) = \mathbf{L}(\mathbf{X}) - \mathbf{L}(\mathbf{Y})$.

proof

(a) We have

$$\mathbf{0}_V = \mathbf{0}_V + \mathbf{0}_V.$$

Then

$$\mathbf{L}(\mathbf{0}_V) = \mathbf{L}(\mathbf{0}_V + \mathbf{0}_V) = \mathbf{L}(\mathbf{0}_V) + \mathbf{L}(\mathbf{0}_V). \tag{1}$$

Subtracting $\mathbf{L}(\mathbf{0}_V)$ from both sides of (1), we obtain

$$\mathbf{L}(\mathbf{0}_V) = \mathbf{0}_W.$$

(b) Exercise T.3.

A function f mapping a set V into a set W can be specified by a formula that assigns to every member of V a unique element of W. On the other hand, we can also specify a function by listing next to each member of V its assigned element of W. An example of this would be provided by listing the names of all charge account customers of a department store along with their charge account number. At first glance, it appears impossible to describe a linear transformation $\mathbf{L} : V \to W$ of a vector space $V \neq \{\mathbf{0}\}$ into a vector space W in this latter manner, since V has infinitely many members in it. However, the following very useful theorem tells us that once we know what $\mathbf{L}$ does to a basis of V, then we have completely determined $\mathbf{L}$. Thus for a finite-dimensional vector space, it is possible to describe $\mathbf{L}$ by giving only the images of a finite number of vectors in the vector space.

THEOREM 4.3. *Let* $\mathbf{L} : V \to W$ *be a linear transformation of an* n-*dimensional vector space* V *into a vector space* W. *Also let* $S = \{\mathbf{X}_1, \mathbf{X}_2, \ldots, \mathbf{X}_n\}$ *be a basis for* V. *If* $\mathbf{X}$ *is any vector in* V, *then* $\mathbf{L}(\mathbf{X})$ *is completely determined by* $\{\mathbf{L}(\mathbf{X}_1), \mathbf{L}(\mathbf{X}_2), \ldots, \mathbf{L}(\mathbf{X}_n)\}$.

proof. Since $\mathbf{X}$ is in V, we can write

$$\mathbf{X} = c_1\mathbf{X}_1 + c_2\mathbf{X}_2 + \cdots + c_n\mathbf{X}_n,$$

where $c_1, c_2, \ldots, c_n$ are uniquely determined real numbers. Then

$$\begin{aligned} \mathbf{L}(\mathbf{X}) &= \mathbf{L}(c_1\mathbf{X}_1 + c_2\mathbf{X}_2 + \cdots + c_n\mathbf{X}_n) \\ &= c_1\mathbf{L}(\mathbf{X}_1) + c_2\mathbf{L}(\mathbf{X}_2) + \cdots + c_n\mathbf{L}(\mathbf{X}_n), \end{aligned}$$

by Theorem 4.1. Thus $\mathbf{L}(\mathbf{X})$ has been completely determined by the elements $\mathbf{L}(\mathbf{X}_1), \mathbf{L}(\mathbf{X}_2), \ldots, \mathbf{L}(\mathbf{X}_n)$.

4.1 Exercises

1. Which of the following are linear transformations?

(a) $\mathbf{L}\left(\begin{bmatrix} x \\ y \end{bmatrix}\right) = \begin{bmatrix} x + 1 \\ y \\ x + y \end{bmatrix}$.

(b) $\mathbf{L}\left(\begin{bmatrix} x \\ y \\ z \end{bmatrix}\right) = \begin{bmatrix} x + y \\ y \\ x - z \end{bmatrix}$.

(c) $\mathbf{L}\left(\begin{bmatrix} x \\ y \end{bmatrix}\right) = \begin{bmatrix} x^2 + x \\ y - y^2 \end{bmatrix}$.

2. Which of the following are linear transformations?

(a) $\mathbf{L}\left(\begin{bmatrix} x \\ y \\ z \end{bmatrix}\right) = \begin{bmatrix} x - y \\ x^2 \\ 2z \end{bmatrix}$.

(b) $\mathbf{L}\left(\begin{bmatrix} x \\ y \\ z \end{bmatrix}\right) = \begin{bmatrix} 2x - 3y \\ 3y - 2z \\ 2z \end{bmatrix}$.

(c) $\mathbf{L}\left(\begin{bmatrix} x \\ y \end{bmatrix}\right) = \begin{bmatrix} x - y \\ 2x + 2 \end{bmatrix}$.

3. Which of the following are linear transformations?

(a) $\mathbf{L}\left(\begin{bmatrix} x \\ y \\ z \end{bmatrix}\right) = \begin{bmatrix} x + y \\ 0 \\ 2x - z \end{bmatrix}$.

(b) $\mathbf{L}\left(\begin{bmatrix} x \\ y \end{bmatrix}\right) = \begin{bmatrix} x^2 - y^2 \\ x^2 + y^2 \end{bmatrix}$.

(c) $\mathbf{L}\left(\begin{bmatrix} x \\ y \end{bmatrix}\right) = \begin{bmatrix} x - y \\ 0 \\ 2x + 3 \end{bmatrix}$.

4. Let $L : R^3 \to R^2$ be a linear transformation for which we know that

$$\mathbf{L}\left(\begin{bmatrix} 1 \\ 0 \\ 0 \end{bmatrix}\right) = \begin{bmatrix} 2 \\ -4 \end{bmatrix}, \quad \mathbf{L}\left(\begin{bmatrix} 0 \\ 1 \\ 0 \end{bmatrix}\right) = \begin{bmatrix} 3 \\ -5 \end{bmatrix}, \quad \mathbf{L}\left(\begin{bmatrix} 0 \\ 0 \\ 1 \end{bmatrix}\right) = \begin{bmatrix} 2 \\ 3 \end{bmatrix}.$$

(a) What is $\mathbf{L}\left(\begin{bmatrix} 1 \\ -2 \\ 3 \end{bmatrix}\right)$?

(b) What is $\mathbf{L}\left(\begin{bmatrix} a \\ b \\ c \end{bmatrix}\right)$?

5. Verify that the functions in Examples 3, 4, and 6 are linear transformations.

6. Describe the following linear transformations geometrically.

(a) $\mathbf{L}\left(\begin{bmatrix} x \\ y \end{bmatrix}\right) = \begin{bmatrix} -x \\ y \end{bmatrix}$.

(b) $\mathbf{L}\left(\begin{bmatrix} x \\ y \end{bmatrix}\right) = \begin{bmatrix} -x \\ -y \end{bmatrix}$.

(c) $\mathbf{L}\left(\begin{bmatrix} x \\ y \end{bmatrix}\right) = \begin{bmatrix} -y \\ x \end{bmatrix}$.

7. Let $\mathbf{L} : R^2 \to R^2$ be a linear transformation for which we know that

$$\mathbf{L}\left(\begin{bmatrix} 1 \\ 0 \end{bmatrix}\right) = \begin{bmatrix} 2 \\ -3 \end{bmatrix} \quad \text{and} \quad \mathbf{L}\left(\begin{bmatrix} 0 \\ 1 \end{bmatrix}\right) = \begin{bmatrix} 1 \\ 2 \end{bmatrix}.$$

(a) What is $\mathbf{L}\left(\begin{bmatrix} 3 \\ -2 \end{bmatrix}\right)$?

(b) What is $\mathbf{L}\left(\begin{bmatrix} a \\ b \end{bmatrix}\right)$?

Theoretical Exercises

T.1. Verify that the function in Example 5 is a linear transformation.

T.2. Prove Theorem 4.1.

T.3. Prove (b) of Theorem 4.2.

T.4. Prove that $\mathbf{L} : V \to W$ is a linear transformation if and only if

$$\mathbf{L}(a\mathbf{X} + b\mathbf{Y}) = a\mathbf{L}(\mathbf{X}) + b\mathbf{L}(\mathbf{Y}),$$

for any scalars a and b and any vectors $\mathbf{X}$ and $\mathbf{Y}$ in V.

4.2 The Kernel and Range of a Linear Transformation

In this section we study special types of linear transformations; we formulate the notions of one-to-one linear transformations and onto linear transformations. We also develop methods for determining when a linear transformation is one-to-one or onto. In Section 4.4 we examine some applications of these notions.

DEFINITION A linear transformation $\mathbf{L} : V \to W$ is said to be **one-to-one** if $\mathbf{X}_1 \neq \mathbf{X}_2$ implies that $\mathbf{L}(\mathbf{X}_1) \neq \mathbf{L}(\mathbf{X}_2)$. An equivalent statement is that $\mathbf{L}$ is one-to-one if $\mathbf{L}(\mathbf{X}_1) = \mathbf{L}(\mathbf{X}_2)$ implies that $\mathbf{X}_1 = \mathbf{X}_2$.

This definition says that $\mathbf{L}$ is one-to-one if $\mathbf{L}(\mathbf{X}_1)$ and $\mathbf{L}(\mathbf{X}_2)$ are distinct whenever $\mathbf{X}_1$ and $\mathbf{X}_2$ are distinct (Figure 4.4).

Example 1. Let $\mathbf{L} : R^2 \to R^2$ be defined by

$$\mathbf{L}\left(\begin{bmatrix} x \\ y \end{bmatrix}\right) = \begin{bmatrix} x + y \\ x - y \end{bmatrix}.$$

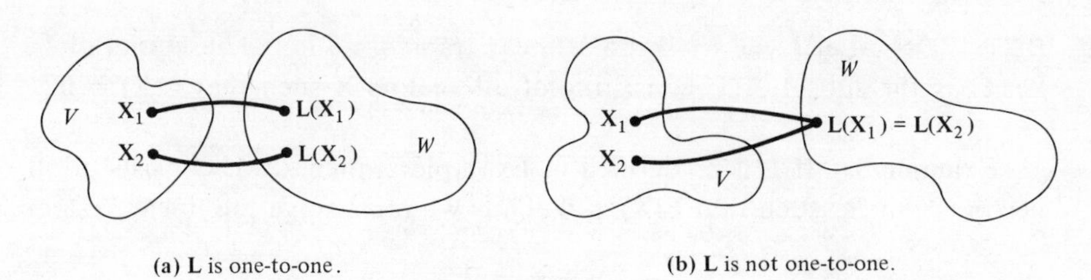

(a) L is one-to-one.

(b) L is not one-to-one.

Figure 4.4

To determine whether **L** is one-to-one, we let

$$\mathbf{X}_1 = \begin{bmatrix} x_1 \\ y_1 \end{bmatrix} \quad \text{and} \quad \mathbf{X}_2 = \begin{bmatrix} x_2 \\ y_2 \end{bmatrix}.$$

Then if

$$\mathbf{L}(\mathbf{X}_1) = \mathbf{L}(\mathbf{X}_2),$$

we have

$$x_1 + y_1 = x_2 + y_2$$
$$x_1 - y_1 = x_2 - y_2.$$

Adding these equations, we obtain $2x_1 = 2x_2$, or $x_1 = x_2$, which implies that $y_1 = y_2$. Hence $\mathbf{X}_1 = \mathbf{X}_2$ and **L** is one-to-one.

Example 2. Let $\mathbf{L} : R^3 \to R^2$ be the linear transformation defined in Example 1 of Section 4.1 (the projection map) by

$$\mathbf{L}\left(\begin{bmatrix} x \\ y \\ z \end{bmatrix}\right) = \begin{bmatrix} x \\ y \end{bmatrix}.$$

Since

$$\mathbf{L}\left(\begin{bmatrix} 1 \\ 3 \\ 3 \end{bmatrix}\right) = \mathbf{L}\left(\begin{bmatrix} 1 \\ 3 \\ -2 \end{bmatrix}\right),$$

we conclude that **L** is not one-to-one.

We shall now develop some more efficient ways of determining whether or not a linear transformation is one-to-one.

DEFINITION Let $\mathbf{L} : V \to W$ be a linear transformation. The **kernel** of **L**, ker **L**, is the subset of V consisting of all vectors **X** such that $\mathbf{L}(\mathbf{X}) = \mathbf{0}_W$.

Example 3. If **L** is as defined in Example 1, then ker **L** consists of all vectors **X** in R^2 such that $\mathbf{L}(\mathbf{X}) = \mathbf{0}$. Thus we must solve the linear system

$$x + y = 0$$
$$x - y = 0$$

for x and y. The only solution is $\mathbf{X} = \mathbf{0}$, so ker $\mathbf{L} = \{\mathbf{0}\}$.

Example 4. If **L** is as defined in Example 2, then ker **L** consists of all vectors **X** in R^3 such that $\mathbf{L}(\mathbf{X}) = \mathbf{0}$. This leads to the linear system

$$x = 0$$
$$y = 0,$$

and z can be any real number. Thus ker **L** consists of all vectors of the form

$$\begin{bmatrix} 0 \\ 0 \\ r \end{bmatrix},$$

where r is any real number.

Example 5. If $\mathbf{L} : R^4 \to R^2$ is defined by

$$\mathbf{L}\left(\begin{bmatrix} x \\ y \\ z \\ w \end{bmatrix}\right) = \begin{bmatrix} x + y \\ z + w \end{bmatrix},$$

then ker (**L**) consists of all vectors **X** in R^3 such that $L(\mathbf{X}) = \mathbf{0}$. This leads to the linear system

$$x + y = 0$$
$$z + w = 0.$$

Thus ker **L** consists of all vectors of the form

$$\begin{bmatrix} r \\ -r \\ s \\ -s \end{bmatrix},$$

where r and s are any real numbers.

We observe that property (a) of Theorem 4.2 of Section 4.1 assures us that ker **L** is never an empty set, since $\mathbf{0}_V$ is in ker **L**.

THEOREM 4.4. *If* $\mathbf{L} : V \to W$ *is a linear transformation, then* ker **L** *is a subspace of* V.

proof. Let **X** and **Y** be in ker **L**. Then since **L** is a linear transformation,

$$\mathbf{L}(\mathbf{X} + \mathbf{Y}) = \mathbf{L}(\mathbf{X}) + \mathbf{L}(\mathbf{Y}) = \mathbf{0}_W + \mathbf{0}_W = \mathbf{0}_W.$$

Also, if c is a scalar, then since **L** is a linear transformation,

$$\mathbf{L}(c\mathbf{X}) = c\mathbf{L}(\mathbf{X}) = c\mathbf{0}_W = \mathbf{0}_W.$$

Hence ker **L** is a subspace of V.

Example 6. If **L** is as in Examples 1 and 3, then ker **L** is the subspace $\{\mathbf{0}\}$; its dimension is zero.

Example 7. If **L** is as in Examples 2 and 4, then a basis for ker **L** is

$$\begin{bmatrix} 0 \\ 0 \\ 1 \end{bmatrix},$$

and dim (ker **L**) = 1. Thus ker L consists of the z-axis in (x, y, z) three-dimensional space R^3.

Example 8. If **L** is as in Example 5, then a basis for ker **L** consists of the vectors

$$\begin{bmatrix} 1 \\ -1 \\ 0 \\ 0 \end{bmatrix} \quad \text{and} \quad \begin{bmatrix} 0 \\ 0 \\ 1 \\ -1 \end{bmatrix};$$

thus dim (ker **L**) = 2.

The following theorem shows that a linear transformation **L** is one-to-one if and only if the kernel of **L** consists only of the zero vector.

THEOREM 4.5. *A linear transformation* $\mathbf{L} : V \to W$ *is one-to-one if and only if* ker $\mathbf{L} = \{\mathbf{0}_V\}$.

proof. Let $\mathbf{L}$ be one-to-one. We show that ker $\mathbf{L} = \{\mathbf{0}_V\}$. Let $\mathbf{X}$ be in ker $\mathbf{L}$. Then $\mathbf{L}(\mathbf{X}) = \mathbf{0}_W$. Also, we already know that $\mathbf{L}(\mathbf{0}_V) = \mathbf{0}_W$. Thus $\mathbf{L}(\mathbf{X}) = \mathbf{L}(\mathbf{0}_V)$. Since $\mathbf{L}$ is one-to-one, we conclude that $\mathbf{X} = \mathbf{0}_V$. Hence ker $\mathbf{L} = \{\mathbf{0}_V\}$.

Conversely, suppose that ker $\mathbf{L} = \{\mathbf{0}_V\}$. We wish to show that $\mathbf{L}$ is one-to-one. Assume that $\mathbf{L}(\mathbf{X}_1) = \mathbf{L}(\mathbf{X}_2)$, for $\mathbf{X}_1$ and $\mathbf{X}_2$ in V. Then

$$\mathbf{L}(\mathbf{X}_1) - \mathbf{L}(\mathbf{X}_2) = \mathbf{0}_W,$$

so $\mathbf{L}(\mathbf{X}_1 - \mathbf{X}_2) = \mathbf{0}_W$, which means that $\mathbf{X}_1 - \mathbf{X}_2$ is in ker $\mathbf{L}$. Hence $\mathbf{X}_1 - \mathbf{X}_2 = \mathbf{0}_V$, so $\mathbf{X}_1 = \mathbf{X}_2$. Thus $\mathbf{L}$ is one-to-one.

Example 9. The linear transformation of Examples 1 and 3 is one-to-one; the one in Examples 2 and 4 is not.

DEFINITION If $\mathbf{L} : V \to W$ is a linear transformation, then the **range** of $\mathbf{L}$, denoted by range $\mathbf{L}$, is the set of all vectors in W that are images, under $\mathbf{L}$, of vectors in V. Thus a vector $\mathbf{Y}$ is in range $\mathbf{L}$ if we can find some vector $\mathbf{X}$ in V such that $\mathbf{L}(\mathbf{X}) = \mathbf{Y}$. If range $\mathbf{L} = W$, we say that $\mathbf{L}$ is **onto**.

THEOREM 4.6. *If $\mathbf{L} : V \to W$ is a linear transformation, then range $\mathbf{L}$ is a subspace of W.*

proof. Let $\mathbf{Y}_1$ and $\mathbf{Y}_2$ be in range $\mathbf{L}$. Then $\mathbf{Y}_1 = \mathbf{L}(\mathbf{X}_1)$ and $\mathbf{Y}_2 = \mathbf{L}(\mathbf{X}_2)$ for some $\mathbf{X}_1$ and $\mathbf{X}_2$ in V. Now

$$\mathbf{Y}_1 + \mathbf{Y}_2 = \mathbf{L}(\mathbf{X}_1) + \mathbf{L}(\mathbf{X}_2) = \mathbf{L}(\mathbf{X}_1 + \mathbf{X}_2),$$

which implies that $\mathbf{Y}_1 + \mathbf{Y}_2$ is in range $\mathbf{L}$. Also, if c is a scalar, then $c\mathbf{Y}_1 = c\mathbf{L}(\mathbf{X}_1) = \mathbf{L}(c\mathbf{X}_1)$, so $c\mathbf{Y}_1$ is in range $\mathbf{L}$. Hence range $\mathbf{L}$ is a subspace of W.

Example 10. Let $\mathbf{L}$ be the linear transformation defined in Example 2. To find out whether $\mathbf{L}$ is onto, we choose any vector $\mathbf{Y} = \begin{bmatrix} y_1 \\ y_2 \end{bmatrix}$ in R^2 and seek a vector $\mathbf{X} = \begin{bmatrix} x_1 \\ x_2 \\ x_3 \end{bmatrix}$ in R^3 such that $\mathbf{L}(\mathbf{X}) = \mathbf{Y}$. Since $\mathbf{L}(\mathbf{X}) = \begin{bmatrix} y_1 \\ y_2 \end{bmatrix}$, we find that if $x_1 = y_1$ and $x_2 = y_2$, then $\mathbf{L}(\mathbf{X}) = \mathbf{Y}$. Therefore, $\mathbf{L}$ is onto and dimension of range $\mathbf{L}$ is 2.

Example 11. Let $\mathbf{L} : R^3 \to R^3$ be defined by

$$\mathbf{L}\left(\begin{bmatrix} x_1 \\ x_2 \\ x_3 \end{bmatrix}\right) = \begin{bmatrix} 1 & 0 & 1 \\ 1 & 1 & 2 \\ 2 & 1 & 3 \end{bmatrix}\begin{bmatrix} x_1 \\ x_2 \\ x_3 \end{bmatrix}.$$

Is $\mathbf{L}$ onto? Given any $\mathbf{Y} = \begin{bmatrix} y_1 \\ y_2 \\ y_3 \end{bmatrix}$ in R^3, where y_1, y_2, and y_3 are any real numbers, can we find $\mathbf{X} = \begin{bmatrix} x_1 \\ x_2 \\ x_3 \end{bmatrix}$ in R^3 so that $\mathbf{L}(\mathbf{X}) = \mathbf{Y}$? We are seeking a solution to the linear system

$$\begin{bmatrix} 1 & 0 & 1 \\ 1 & 1 & 2 \\ 2 & 1 & 3 \end{bmatrix}\begin{bmatrix} x_1 \\ x_2 \\ x_3 \end{bmatrix} = \begin{bmatrix} y_1 \\ y_2 \\ y_3 \end{bmatrix}.$$

The reduced row echelon form of the augmented matrix is

$$\left[\begin{array}{ccc|c} 1 & 0 & 1 & y_1 \\ 0 & 1 & 1 & y_2 - y_1 \\ 0 & 0 & 0 & y_3 - y_2 - y_1 \end{array}\right].$$

Thus a solution exists only for $y_3 - y_2 - y_1 = 0$, and so $\mathbf{L}$ is not onto.

To find a basis for range L, we note that

$$\mathbf{L}\left(\begin{bmatrix} x_1 \\ x_2 \\ x_3 \end{bmatrix}\right) = \begin{bmatrix} 1 & 0 & 1 \\ 1 & 1 & 2 \\ 2 & 1 & 3 \end{bmatrix}\begin{bmatrix} x_1 \\ x_2 \\ x_3 \end{bmatrix} = \begin{bmatrix} x_1 + x_3 \\ x_1 + x_2 + 2x_3 \\ 2x_1 + x_2 + 3x_3 \end{bmatrix}$$

$$= x_1\begin{bmatrix} 1 \\ 1 \\ 2 \end{bmatrix} + x_2\begin{bmatrix} 0 \\ 1 \\ 1 \end{bmatrix} + x_3\begin{bmatrix} 1 \\ 2 \\ 3 \end{bmatrix}.$$

Since

$$\mathbf{Y}_1 = \begin{bmatrix} 1 \\ 1 \\ 2 \end{bmatrix} = \mathbf{L}\left(\begin{bmatrix} 1 \\ 0 \\ 0 \end{bmatrix}\right), \qquad \mathbf{Y}_2 = \begin{bmatrix} 0 \\ 1 \\ 1 \end{bmatrix} = \mathbf{L}\left(\begin{bmatrix} 0 \\ 1 \\ 0 \end{bmatrix}\right),$$

and

$$\mathbf{Y}_3 = \begin{bmatrix} 1 \\ 2 \\ 3 \end{bmatrix} = \mathbf{L}\left(\begin{bmatrix} 0 \\ 0 \\ 1 \end{bmatrix}\right),$$

we observe, from the last equation that $\{\mathbf{Y}_1, \mathbf{Y}_2, \mathbf{Y}_3\}$ spans range **L**. Now $\{\mathbf{Y}_1, \mathbf{Y}_2\}$ is linearly independent, while $\mathbf{Y}_3 = \mathbf{Y}_1 + \mathbf{Y}_2$. Thus $\mathbf{Y}_1$ and $\mathbf{Y}_2$ form a basis for range **L**, and dim (range **L**) = 2.

It is also interesting to find dim (ker **L**) for this example; that is, we wish to find all **X** in R^3 so that $\mathbf{L}(\mathbf{X}) = \mathbf{0}$. From the reduced echelon form of the augmented matrix we find that $x_1 = -x_3$ and $x_2 = -x_3$. Thus ker **L** consists of all vectors of the form $\begin{bmatrix} -r \\ -r \\ r \end{bmatrix}$, where r is any real number. A basis for ker **L** is given by the vector $\begin{bmatrix} -1 \\ -1 \\ 1 \end{bmatrix}$; **L** is not one-to-one and dim (ker **L**) = 1.

To find out if a linear transformation is one-to-one or onto, we must solve a linear system of equations. This is one further demonstration of the frequency with which linear systems must be solved to answer many questions in linear algebra.

We also observe that in Example 11 we have

$$\dim(\ker \mathbf{L}) + \dim(\text{range } \mathbf{L}) = \dim V.$$

This very important result is always true, and we now prove it in the following theorem.

THEOREM 4.7. *If* $\mathbf{L} : V \to W$ *is a linear transformation, then*

$$\dim(\ker \mathbf{L}) + \dim(\text{range } \mathbf{L}) = \dim V. \tag{1}$$

proof. If $\mathbf{L} = \mathbf{0}$, the zero linear transformation defined by $\mathbf{0}(\mathbf{X}) = \mathbf{0}_W$, for **X** in V, then ker $\mathbf{L} = V$ and range $\mathbf{L} = \mathbf{0}_W$. In this case dim (ker **L**) = dim V, and dim (range **L**) = 0, so (1) holds. Now suppose that $\mathbf{L} \neq \mathbf{0}$ and let $n = \dim V$ and $k = \dim(\ker \mathbf{L})$. If $n = k$, then ker $\mathbf{L} = V$ (Exercise T.11, Section 3.5), which implies that $\mathbf{L}(\mathbf{X}) = \mathbf{0}_W$ for every **X** in V. Hence range $\mathbf{L} = \{\mathbf{0}_W\}$, dim (range **L**) = 0, and again (1) holds. Next, suppose that $1 \leqslant k < n$. We shall prove that dim (range **L**) = $n - k$. Let

$\mathbf{X}_1, \mathbf{X}_2, \ldots, \mathbf{X}_k$ be a basis for ker $\mathbf{L}$. By Theorem 3.13 of Section 3.5 we can extend this basis to a basis $S = \{\mathbf{X}_1, \mathbf{X}_2, \ldots, \mathbf{X}_k, \mathbf{X}_{k+1}, \ldots, \mathbf{X}_n\}$ for V. We now prove that the set $T = \{\mathbf{L}(\mathbf{X}_{k+1}), \mathbf{L}(\mathbf{X}_{k+2}), \ldots, \mathbf{L}(\mathbf{X}_n)\}$ is a basis for range $\mathbf{L}$, which will prove that (1) holds.

First, we show that T spans range $\mathbf{L}$. Let $\mathbf{Y}$ be any vector in range $\mathbf{L}$. Then $\mathbf{Y} = \mathbf{L}(\mathbf{X})$ for some $\mathbf{X}$ in V. Since S is a basis for V, we can write

$$\mathbf{X} = c_1\mathbf{X}_1 + c_2\mathbf{X}_2 + \cdots + c_n\mathbf{X}_n,$$

where $c_1, c_2, \ldots, c_n$ are real numbers. Then

$$\begin{aligned}
\mathbf{Y} &= \mathbf{L}(\mathbf{X}) \\
&= \mathbf{L}(c_1\mathbf{X}_1 + c_2\mathbf{X}_2 + \cdots + c_k\mathbf{X}_k + c_{k+1}\mathbf{X}_{k+1} + \cdots + c_n\mathbf{X}_n) \\
&= c_1\mathbf{L}(\mathbf{X}_1) + c_2\mathbf{L}(\mathbf{X}_2) + \cdots + c_k\mathbf{L}(\mathbf{X}_k) \\
&\quad + c_{k+1}\mathbf{L}(\mathbf{X}_{k+1}) + \cdots + c_n\mathbf{L}(\mathbf{X}_n) \\
&= c_{k+1}\mathbf{L}(\mathbf{X}_{k+1}) + \cdots + c_n\mathbf{L}(\mathbf{X}_n),
\end{aligned}$$

because $\mathbf{X}_1, \mathbf{X}_2, \ldots, \mathbf{X}_k$ are in ker $\mathbf{L}$. Hence T spans range $\mathbf{L}$.

Now we show that T is linearly independent. Suppose that

$$c_{k+1}\mathbf{L}(\mathbf{X}_{k+1}) + c_{k+2}\mathbf{L}(\mathbf{X}_{k+2}) + \cdots + c_n\mathbf{L}(\mathbf{X}_n) = \mathbf{0}_W; \tag{2}$$

then

$$\mathbf{L}(c_{k+1}\mathbf{X}_{k+1} + c_{k+2}\mathbf{X}_{k+2} + \cdots + c_n\mathbf{X}_n) = \mathbf{0}_W.$$

Hence the vector $c_{k+1}\mathbf{X}_{k+1} + c_{k+2}\mathbf{X}_{k+2} + \cdots + c_n\mathbf{X}_n$ is in ker $\mathbf{L}$, and we can write it as a linear combination of the vectors in the basis for ker $\mathbf{L}$:

$$c_{k+1}\mathbf{X}_{k+1} + c_{k+2}\mathbf{X}_{k+2} + \cdots + c_n\mathbf{X}_n = d_1\mathbf{X}_1 + d_2\mathbf{X}_2 + \cdots + d_k\mathbf{X}_k,$$

where $d_1, d_2, \ldots, d_k$ are real numbers. Then

$$d_1\mathbf{X}_1 + d_2\mathbf{X}_2 + \cdots + d_k\mathbf{X}_k - c_{k+1}\mathbf{X}_{k+1} - c_{k+2}\mathbf{X}_{k+2} - \cdots - c_n\mathbf{X}_n = \mathbf{0}_V.$$

Since S is linearly independent, we conclude that

$$d_1 = d_2 = \cdots = d_k = c_{k+1} = \cdots = c_n = 0.$$

Referring back to Equation (2), we find that this means that T is linearly independent and is a basis for range $\mathbf{L}$.

We have seen examples of linear transformations that are onto and not one-to-one, or neither one-to-one nor onto. In general, one of these properties has nothing to do with the other one. However, in a special case we can establish the following corollary, whose proof we leave to the reader.

COROLLARY 4.1. *Let* $\mathbf{L} : V \to W$ *be a linear transformation and let* $\dim V = \dim W$.

(a) *If* $\mathbf{L}$ *is one-to-one, then it is onto.*
(b) *If* $\mathbf{L}$ *is onto, then it is one-to-one.*

proof. Exercise T.1.

4.2 Exercises

1. Let $\mathbf{L} : R^2 \to R^3$ be defined by

$$\mathbf{L}\left(\begin{bmatrix} x \\ y \end{bmatrix}\right) = \begin{bmatrix} x \\ x + y \\ y \end{bmatrix}.$$

(a) Find ker $\mathbf{L}$.
(b) Is $\mathbf{L}$ one-to-one?
(c) Is $\mathbf{L}$ onto?

2. Let $\mathbf{L} : R^4 \to R^3$ be defined by

$$\mathbf{L}\left(\begin{bmatrix} x \\ y \\ z \\ w \end{bmatrix}\right) = \begin{bmatrix} x + y \\ z + w \\ x + z \end{bmatrix}.$$

(a) Find a basis for ker $\mathbf{L}$.
(b) Find a basis for range $\mathbf{L}$.
(c) Verify Theorem 4.7.

3. Let $\mathbf{L} : R^5 \to R^4$ be defined by

$$\mathbf{L}\left(\begin{bmatrix} x_1 \\ x_2 \\ x_3 \\ x_4 \\ x_5 \end{bmatrix}\right) = \begin{bmatrix} 1 & 0 & -1 & 3 & -1 \\ 1 & 0 & 0 & 2 & -1 \\ 2 & 0 & -1 & 5 & -1 \\ 0 & 0 & -1 & 1 & 0 \end{bmatrix}\begin{bmatrix} x_1 \\ x_2 \\ x_3 \\ x_4 \\ x_5 \end{bmatrix}.$$

(a) Find a basis for ker $\mathbf{L}$.
(b) Find a basis for range $\mathbf{L}$.
(c) Verify Theorem 4.7.

4. Let $\mathbf{L} : R^3 \to R^3$ be defined by

$$\mathbf{L}\left(\begin{bmatrix} x \\ y \\ z \end{bmatrix}\right) = \begin{bmatrix} 4 & 2 & 2 \\ 2 & 3 & -1 \\ -1 & 1 & -2 \end{bmatrix}\begin{bmatrix} x \\ y \\ z \end{bmatrix}.$$

(a) Is $\mathbf{L}$ one-to-one?
(b) Find the dimension of range $\mathbf{L}$.

5. Let $\mathbf{L} : R^4 \to R^3$ be defined by

$$\mathbf{L}\left(\begin{bmatrix} x \\ y \\ z \\ w \end{bmatrix}\right) = \begin{bmatrix} x + y \\ y - z \\ z - w \end{bmatrix}.$$

(a) Is $\mathbf{L}$ onto?
(b) Find the dimension of ker $\mathbf{L}$.
(c) Verify Theorem 4.7.

6. Let $\mathbf{L} : R^3 \to R^3$ be defined by

$$\mathbf{L}\left(\begin{bmatrix} x \\ y \\ z \end{bmatrix}\right) = \begin{bmatrix} x - y \\ x + 2y \\ z \end{bmatrix}.$$

(a) Find a basis for ker $\mathbf{L}$.
(b) Find a basis for range $\mathbf{L}$.
(c) Verify Theorem 4.7.

7. Verify Theorem 4.7 for the following linear transformations.

(a) $\mathbf{L}\left(\begin{bmatrix} x \\ y \end{bmatrix}\right) = \begin{bmatrix} x + y \\ y \end{bmatrix}.$

(b) $\mathbf{L}\left(\begin{bmatrix} x \\ y \\ z \end{bmatrix}\right) = \begin{bmatrix} 4 & -1 & -1 \\ 2 & 2 & 3 \\ 2 & -3 & -4 \end{bmatrix}\begin{bmatrix} x \\ y \\ z \end{bmatrix}.$

(c) $\mathbf{L}\left(\begin{bmatrix} x \\ y \\ z \end{bmatrix}\right) = \begin{bmatrix} x + y - z \\ x + y \\ y + z \end{bmatrix}.$

8. Let $\mathbf{L} : R^4 \to R^4$ be defined by

$$\mathbf{L}\left(\begin{bmatrix} x \\ y \\ z \\ w \end{bmatrix}\right) = \begin{bmatrix} 1 & 2 & 1 & 3 \\ 2 & 1 & -1 & 2 \\ 1 & 0 & 0 & -1 \\ 4 & 1 & -1 & 0 \end{bmatrix}\begin{bmatrix} x \\ y \\ z \\ w \end{bmatrix}.$$

(a) Find a basis for ker $\mathbf{L}$.
(b) Find a basis for range $\mathbf{L}$.
(c) Verify Theorem 4.7.

9. Let $\mathbf{L} : R^4 \to R^6$ be a linear transformation.
(a) If dim (ker $\mathbf{L}$) = 2, what is dim (range $\mathbf{L}$)?
(b) If dim (range $\mathbf{L}$) = 3, what is dim (ker $\mathbf{L}$)?

10. Let $\mathbf{L} : V \to R^5$ be a linear transformation.
(a) If $\mathbf{L}$ is onto and dim (ker $\mathbf{L}$) = 2, what is dim V?
(b) If $\mathbf{L}$ is one-to-one and onto, what is dim V?

Theoretical Exercises

T.1. Prove Corollary 4.1.

T.2. Let $\mathbf{L} : V \to W$ be a linear transformation. If $\{\mathbf{X}_1, \mathbf{X}_2, \ldots, \mathbf{X}_k\}$ spans V, show that $\{\mathbf{L}(\mathbf{X}_1), \mathbf{L}(\mathbf{X}_2), \ldots, \mathbf{L}(\mathbf{X}_k)\}$ spans range $\mathbf{L}$.

T.3. Let $\mathbf{L} : V \to W$ be a linear transformation.
(a) Show that dim (range $\mathbf{L}$) $\leqslant$ dim V.
(b) Prove that if $\mathbf{L}$ is onto, then dim $W \leqslant$ dim V.

T.4. Let $\mathbf{L} : V \to W$ be a linear transformation, and let $S = \{\mathbf{X}_1, \mathbf{X}_2, \ldots, \mathbf{X}_n\}$ be a set of vectors in S. Prove that if $T = \{\mathbf{L}(\mathbf{X}_1), \mathbf{L}(\mathbf{X}_2), \ldots, \mathbf{L}(\mathbf{X}_n)\}$ is linearly independent, then so is S. (*Hint*: Assume that S is linearly dependent.) What can we say about the converse?

T.5. Let $\mathbf{L} : V \to W$ be a linear transformation. Prove that $\mathbf{L}$ is one-to-one if and only if dim (range $\mathbf{L}$) = dim V.

T.6. Let $\mathbf{L} : V \to W$ be a linear transformation. Show that $\mathbf{L}$ is one-to-one if and only if the image of every linearly independent set of vectors in V is a linearly independent set of vectors in W.

T.7. Let $\mathbf{L} : V \to W$ be a linear transformation, and let dim V = dim W. Show that $\mathbf{L}$ is one-to-one if and only if the image under $\mathbf{L}$ of a basis for V is a basis for W.

4.3 The Matrix of a Linear Transformation

In Example 5 of Section 4.1 we saw that if $\mathbf{A}$ is an $m \times n$ matrix, then we can define a linear transformation $\mathbf{L} : R^n \to R^m$ by $\mathbf{L}(\mathbf{X}) = \mathbf{AX}$, for $\mathbf{X}$ in R^n. Now let $\mathbf{L} : V \to W$ be a linear transformation of a finite-dimensional vector space V into a finite-dimensional vector space W. In this section we shall show how to attach a unique matrix to $\mathbf{L}$ which will enable us to find $\mathbf{L}(\mathbf{X})$, for $\mathbf{X}$ in V, by merely performing matrix multiplication. We first discuss the notion of a coordinate vector.

Coordinate Vectors

DEFINITION Let V be an n-dimensional vector space with basis $S = \{\mathbf{X}_1, \mathbf{X}_2, \ldots, \mathbf{X}_n\}$. If

$$\mathbf{X} = a_1\mathbf{X}_1 + a_2\mathbf{X}_2 + \cdots + a_n\mathbf{X}_n \tag{1}$$

is any vector in V, then the vector

$$[\mathbf{X}]_S = \begin{bmatrix} a_1 \\ a_2 \\ \vdots \\ a_n \end{bmatrix}$$

in R^n is called the **coordinate vector of X with respect to the basis** S. The components of $[\mathbf{X}]_S$ are called the **coordinates of X with respect to** S.

Example 1. Let $S = \{\mathbf{X}_1, \mathbf{X}_2, \mathbf{X}_3\}$ be a basis for R^3, where

$$\mathbf{X}_1 = \begin{bmatrix} 1 \\ -1 \\ 0 \end{bmatrix}, \quad \mathbf{X}_2 = \begin{bmatrix} 0 \\ 1 \\ 1 \end{bmatrix}, \quad \text{and} \quad \mathbf{X}_3 = \begin{bmatrix} 1 \\ -1 \\ 1 \end{bmatrix}.$$

If $\mathbf{X} = \begin{bmatrix} 1 \\ 2 \\ 2 \end{bmatrix}$, then to find $[\mathbf{X}]_S$ we must find c_1, c_2, and c_3 such that

$$\begin{bmatrix} 1 \\ 2 \\ 2 \end{bmatrix} = c_1\mathbf{X}_1 + c_2\mathbf{X}_2 + c_3\mathbf{X}_3.$$

This leads to the linear system

$$\begin{bmatrix} 1 & 0 & 1 \\ -1 & 1 & -1 \\ 0 & 1 & 1 \end{bmatrix}\begin{bmatrix} c_1 \\ c_2 \\ c_3 \end{bmatrix} = \begin{bmatrix} 1 \\ 2 \\ 2 \end{bmatrix}.$$

The solution is

$$c_1 = 2, \qquad c_2 = 3, \qquad \text{and} \qquad c_3 = -1.$$

Thus

$$[\mathbf{X}]_S = \begin{bmatrix} 2 \\ 3 \\ -1 \end{bmatrix}.$$

Example 2. Let S be the natural basis for R^n and let $\mathbf{X}$ be a vector in R^n. Then

$$[\mathbf{X}]_S = \mathbf{X}.$$

Example 3. Let $S = \{\mathbf{X}_1, \mathbf{X}_2, \ldots, \mathbf{X}_n\}$ be a basis for an n-dimensional vector space V. Then since

$$\mathbf{X}_1 = 1\mathbf{X}_1 + 0\mathbf{X}_2 + \cdots + 0\mathbf{X}_n,$$

we have

$$[\mathbf{X}_1]_S = \begin{bmatrix} 1 \\ 0 \\ \vdots \\ 0 \end{bmatrix}.$$

Similarly,

$$[\mathbf{X}_j]_S = \mathbf{E}_j \qquad (1 \leqslant j \leqslant n),$$

where $\{\mathbf{E}_1, \mathbf{E}_2, \ldots, \mathbf{E}_n\}$ is the natural basis for R^n.

THEOREM 4.8. *Let* $\mathbf{L} : V \to W$ *be a linear transformation of an* n-*dimensional vector space* V *into an* m-*dimensional vector space* W ($n \neq 0$ *and* $m \neq 0$) *and let* $S = \{\mathbf{X}_1, \mathbf{X}_2, \ldots, \mathbf{X}_n\}$ *and* $T = \{\mathbf{Y}_1, \mathbf{Y}_2, \ldots, \mathbf{Y}_m\}$ *be bases for* V *and* W, *respectively. Then the* $m \times n$ *matrix* $\mathbf{A}$ *whose* jth *column is the coordinate vector* $[\mathbf{L}(\mathbf{X}_j)]_T$ *of* $\mathbf{L}(\mathbf{X}_j)$ *with respect to* T *is associated with* $\mathbf{L}$ *and has the following property: If* $\mathbf{Y} = \mathbf{L}(\mathbf{X})$ *for some* $\mathbf{X}$ *in* V, *then*

$$[\mathbf{Y}]_T = \mathbf{A}[\mathbf{X}]_S, \tag{2}$$

where $[\mathbf{X}]_S$ *and* $[\mathbf{Y}]_T$ *are the coordinate vectors of* $\mathbf{X}$ *and* $\mathbf{Y}$ *with respect to the respective bases* S *and* T. *Moreover,* $\mathbf{A}$ *is the only matrix with this property.*

proof. The proof is a constructive one; that is, we show how to construct the matrix **A**. Consider the vector $\mathbf{X}_j$ in V for $j = 1, 2, \ldots, n$. Then $\mathbf{L}(\mathbf{X}_j)$ is a vector in W, and since T is a basis for W we can express this vector as a linear combination of the vectors in T in a unique manner. Thus

$$\mathbf{L}(\mathbf{X}_j) = c_{1j}\mathbf{Y}_1 + c_{2j}\mathbf{Y}_2 + \cdots + c_{mj}\mathbf{Y}_m \qquad (1 \leqslant j \leqslant n). \tag{3}$$

This means that the coordinate vector of $\mathbf{L}(\mathbf{X}_j)$ with respect to T is

$$[\mathbf{L}(\mathbf{X}_j)]_T = \begin{bmatrix} c_{1j} \\ c_{2j} \\ \vdots \\ c_{mj} \end{bmatrix}.$$

We now define the $m \times n$ matrix **A** by choosing $[\mathbf{L}(\mathbf{X}_j)]_T$ as the jth column of **A** and show that this matrix satisfies the properties stated in the theorem. We leave the rest of the proof as Exercise T.1 and amply illustrate it in the following examples.

Example 4. Let $\mathbf{L} : R^3 \to R^2$ be defined by

$$\mathbf{L}\left(\begin{bmatrix} x \\ y \\ z \end{bmatrix}\right) = \begin{bmatrix} x + y \\ y - z \end{bmatrix}. \tag{5}$$

Let

$$S = \{\mathbf{X}_1, \mathbf{X}_2, \mathbf{X}_3\} \qquad \text{and} \qquad T = \{\mathbf{Y}_1, \mathbf{Y}_2\}$$

be bases for R^3 and R^2, respectively, where

$$\mathbf{X}_1 = \begin{bmatrix} 1 \\ 0 \\ 0 \end{bmatrix}, \quad \mathbf{X}_2 = \begin{bmatrix} 0 \\ 1 \\ 0 \end{bmatrix}, \quad \mathbf{X}_3 = \begin{bmatrix} 0 \\ 0 \\ 1 \end{bmatrix}, \quad \mathbf{Y}_1 = \begin{bmatrix} 1 \\ 0 \end{bmatrix}, \quad \text{and} \quad \mathbf{Y}_2 = \begin{bmatrix} 0 \\ 1 \end{bmatrix}.$$

We now find the matrix **A** associated with **L**. We have

$$\mathbf{L}(\mathbf{X}_1) = \begin{bmatrix} 1 + 0 \\ 0 - 0 \end{bmatrix} = \begin{bmatrix} 1 \\ 0 \end{bmatrix} = 1\mathbf{Y}_1 + 0\mathbf{Y}_2, \text{ so } [\mathbf{L}(\mathbf{X}_1)]_T = \begin{bmatrix} 1 \\ 0 \end{bmatrix};$$

$$\mathbf{L}(\mathbf{X}_2) = \begin{bmatrix} 0 + 1 \\ 1 - 0 \end{bmatrix} = \begin{bmatrix} 1 \\ 1 \end{bmatrix} = 1\mathbf{Y}_1 + 1\mathbf{Y}_2, \text{ so } [\mathbf{L}(\mathbf{X}_2)]_T = \begin{bmatrix} 1 \\ 1 \end{bmatrix};$$

$$\mathbf{L}(\mathbf{X}_3) = \begin{bmatrix} 0 + 0 \\ 0 - 1 \end{bmatrix} = \begin{bmatrix} 0 \\ -1 \end{bmatrix} = 0\mathbf{Y}_1 - 1\mathbf{Y}_2, \text{ so } [\mathbf{L}(\mathbf{X}_3)]_T = \begin{bmatrix} 0 \\ -1 \end{bmatrix}.$$

Hence

$$\mathbf{A} = \begin{bmatrix} 1 & 1 & 0 \\ 0 & 1 & -1 \end{bmatrix}.$$

DEFINITION The matrix **A** of Theorem 4.8 is called the **matrix of L with respect to the bases** S **and** T. Equation (2) is called the **representation of L with respect to** S **and** T. We also say that Equation (2) **represents L with respect to** S **and** T.

We observe that having **A** enables us to replace **L** by **A**, **X** by $[\mathbf{X}]_S$, and **Y** by $[\mathbf{Y}]_T$ in the expression $\mathbf{L}(\mathbf{X}) = \mathbf{Y}$ to obtain $\mathbf{A}[\mathbf{X}]_S = [\mathbf{Y}]_T$. We can thus work with matrices rather than with linear transformations. Physicists and others who deal at great length with linear transformations perform most of their computations with the matrices of the linear transformations.

Example 5. Let $\mathbf{L} : R^3 \to R^2$ be defined as in Example 4. Now let

$$S = \{\mathbf{X}_1, \mathbf{X}_2, \mathbf{X}_3\} \qquad \text{and} \qquad T = \{\mathbf{Y}_1, \mathbf{Y}_2\}$$

be bases for R^3 and R^2, respectively, where

$$\mathbf{X}_1 = \begin{bmatrix} 1 \\ 0 \\ 1 \end{bmatrix}, \quad \mathbf{X}_2 = \begin{bmatrix} 0 \\ 1 \\ 1 \end{bmatrix}, \quad \mathbf{X}_3 = \begin{bmatrix} 1 \\ 1 \\ 1 \end{bmatrix}, \quad \mathbf{Y}_1 = \begin{bmatrix} 1 \\ 2 \end{bmatrix}, \quad \text{and} \quad \mathbf{Y}_2 = \begin{bmatrix} -1 \\ 1 \end{bmatrix}.$$

Then

$$\mathbf{L}(\mathbf{X}_1) = \begin{bmatrix} 1 \\ -1 \end{bmatrix} = 0\mathbf{Y}_1 - 1\mathbf{Y}_2, \text{ so } \left[\mathbf{L}(\mathbf{X}_1)\right]_T = \begin{bmatrix} 0 \\ -1 \end{bmatrix};$$

$$\mathbf{L}(\mathbf{X}_2) = \begin{bmatrix} 1 \\ 0 \end{bmatrix} = \tfrac{1}{3}\mathbf{Y}_1 - \tfrac{2}{3}\mathbf{Y}_2, \text{ so } \left[\mathbf{L}(\mathbf{X}_2)\right]_T = \begin{bmatrix} \frac{1}{3} \\ -\frac{2}{3} \end{bmatrix};$$

$$\mathbf{L}(\mathbf{X}_3) = \begin{bmatrix} 2 \\ 0 \end{bmatrix} = \tfrac{2}{3}\mathbf{Y}_1 - \tfrac{4}{3}\mathbf{Y}_2, \text{ so } \left[\mathbf{L}(\mathbf{X}_3)\right]_T = \begin{bmatrix} \frac{2}{3} \\ -\frac{4}{3} \end{bmatrix}.$$

Hence the matrix of **L** is

$$\mathbf{A} = \begin{bmatrix} 0 & \frac{1}{3} & \frac{2}{3} \\ -1 & -\frac{2}{3} & -\frac{4}{3} \end{bmatrix}.$$

The representation of $\mathbf{L}$ with respect to S and T is

$$[L(\mathbf{X})]_T = \begin{bmatrix} 0 & \frac{1}{3} & \frac{2}{3} \\ -1 & -\frac{2}{3} & -\frac{4}{3} \end{bmatrix} [\mathbf{X}]_S. \tag{6}$$

To verify this equation, let

$$\mathbf{X} = \begin{bmatrix} 1 \\ 6 \\ 3 \end{bmatrix}.$$

Then from the definition of $\mathbf{L}$, Equation (4), we have

$$\mathbf{L}(\mathbf{X}) = \begin{bmatrix} 1+6 \\ 6-3 \end{bmatrix} = \begin{bmatrix} 7 \\ 3 \end{bmatrix}.$$

Now

$$[\mathbf{X}]_S = \begin{bmatrix} -3 \\ 2 \\ 4 \end{bmatrix} = \begin{bmatrix} x_1 \\ x_2 \\ x_3 \end{bmatrix}.$$

Then from (6),

$$[L(\mathbf{X})]_T = \mathbf{A}[\mathbf{X}]_S = \begin{bmatrix} \frac{10}{3} \\ -\frac{11}{3} \end{bmatrix} = \begin{bmatrix} y_1 \\ y_2 \end{bmatrix}.$$

Hence

$$\mathbf{L}(\mathbf{X}) = y_1\mathbf{Y}_1 + y_2\mathbf{Y}_2 = \tfrac{10}{3}\begin{bmatrix} 1 \\ 2 \end{bmatrix} - \tfrac{11}{3}\begin{bmatrix} -1 \\ 1 \end{bmatrix} = \begin{bmatrix} 7 \\ 3 \end{bmatrix},$$

which agrees with the previous value for $\mathbf{L}(\mathbf{X})$.

Example 6. Let $\mathbf{L} : R^3 \to R^2$ be as defined in Example 5. Now let

$$S = \{\mathbf{X}_3, \mathbf{X}_2, \mathbf{X}_1\} \qquad \text{and} \qquad T = \{\mathbf{Y}_1, \mathbf{Y}_2\},$$

where $\mathbf{X}_1$, $\mathbf{X}_2$, $\mathbf{X}_3$, $\mathbf{Y}_1$, and $\mathbf{Y}_2$ are as in Example 5. Then the matrix of $\mathbf{L}$ is

$$\mathbf{A} = \begin{bmatrix} \frac{2}{3} & \frac{1}{3} & 0 \\ -\frac{4}{3} & -\frac{2}{3} & -1 \end{bmatrix}.$$

Notice that if we change the order of the vectors in the bases S and T, then the matrix $\mathbf{A}$ of $\mathbf{L}$ may change.

Example 7. Let $\mathbf{L} : R^3 \to R^2$ be defined by

$$\mathbf{L}\left(\begin{bmatrix} x \\ y \\ z \end{bmatrix}\right) = \begin{bmatrix} 1 & 1 & 1 \\ 1 & 2 & 3 \end{bmatrix}\begin{bmatrix} x \\ y \\ z \end{bmatrix}.$$

Let

$$S = \{\mathbf{X}_1, \mathbf{X}_2, \mathbf{X}_3\} \quad \text{and} \quad T = \{\mathbf{Y}_1, \mathbf{Y}_2\}$$

be the natural bases for R^3 and R^2, respectively. Now

$$\mathbf{L}(\mathbf{X}_1) = \begin{bmatrix} 1 & 1 & 1 \\ 1 & 2 & 3 \end{bmatrix}\begin{bmatrix} 1 \\ 0 \\ 0 \end{bmatrix} = \begin{bmatrix} 1 \\ 1 \end{bmatrix} = 1\mathbf{Y}_1 + 1\mathbf{Y}_2, \text{ so } [\mathbf{L}(\mathbf{X}_1)]_T = \begin{bmatrix} 1 \\ 1 \end{bmatrix};$$

$$\mathbf{L}(\mathbf{X}_2) = \begin{bmatrix} 1 & 1 & 1 \\ 1 & 2 & 3 \end{bmatrix}\begin{bmatrix} 0 \\ 1 \\ 0 \end{bmatrix} = \begin{bmatrix} 1 \\ 2 \end{bmatrix} = 1\mathbf{Y}_1 + 2\mathbf{Y}_2, \text{ so } [\mathbf{L}(\mathbf{X}_2)]_T = \begin{bmatrix} 1 \\ 2 \end{bmatrix}.$$

Also,

$$[\mathbf{L}(\mathbf{X}_3)]_T = \begin{bmatrix} 1 \\ 3 \end{bmatrix} \text{ (verify).}$$

Then the matrix of $\mathbf{L}$ is

$$\mathbf{A} = \begin{bmatrix} 1 & 1 & 1 \\ 1 & 2 & 3 \end{bmatrix}.$$

Of course, the reason that $\mathbf{A}$ is the same matrix as the one involved in the definition of $\mathbf{L}$ is that the natural bases are being used for R^3 and R^2.

Example 8. Let $\mathbf{L} : R^3 \to R^2$ be defined as in Example 7. Now let

$$S = \{\mathbf{X}_1, \mathbf{X}_2, \mathbf{X}_3\} \quad \text{and} \quad T = \{\mathbf{Y}_1, \mathbf{Y}_2\},$$

where

$$\mathbf{X}_1 = \begin{bmatrix} 1 \\ 1 \\ 0 \end{bmatrix}, \quad \mathbf{X}_2 = \begin{bmatrix} 0 \\ 1 \\ 1 \end{bmatrix}, \quad \mathbf{X}_3 = \begin{bmatrix} 0 \\ 0 \\ 1 \end{bmatrix}, \quad \mathbf{Y}_1 = \begin{bmatrix} 1 \\ 2 \end{bmatrix}, \quad \text{and} \quad \mathbf{Y}_2 = \begin{bmatrix} 1 \\ 3 \end{bmatrix}.$$

Then the matrix **A** of **L** is (verify)

$$\mathbf{A} = \begin{bmatrix} 3 & 1 & 0 \\ -1 & 1 & 1 \end{bmatrix}.$$

This matrix is, of course, far different from the one that defined **L**. Thus, although a matrix **A** may be involved in the definition of a linear transformation **L**, we cannot conclude that it is necessarily the matrix representing **L** with respect to two given bases S and T.

If $\mathbf{L} : V \to V$ is a linear transformation and V is an n-dimensional vector space, then to obtain a representation of **L**, we fix bases S and T for V and obtain the matrix of **L** with respect to S and T. However, it is often convenient in this case to choose $S = T$. To avoid redundancy in this case, we refer to **A** as the **matrix** of **L** **with respect to** S. We refer to the representation as the **representation of L with respect to** S.

It is, of course, clear that if $\mathbf{I} : S \to S$ is the identity linear transformation defined by $\mathbf{I}(\mathbf{X}) = \mathbf{X}$, for every **X** in V, then the matrix of **I** with respect to any basis S for V is the identity matrix $\mathbf{I}_n$ (Exercise T.2), where $n = \dim V$.

4.3 Exercises

1. Let

$$S = \left\{ \begin{bmatrix} 1 \\ -1 \end{bmatrix}, \begin{bmatrix} 2 \\ 3 \end{bmatrix} \right\}$$

be a basis for R^2. Find the coordinate vectors of the following vectors with respect to S.

(a) $\begin{bmatrix} -3 \\ -7 \end{bmatrix}$. (b) $\begin{bmatrix} 3 \\ 7 \end{bmatrix}$. (c) $\begin{bmatrix} 12 \\ 13 \end{bmatrix}$. (d) $\begin{bmatrix} 2 \\ 3 \end{bmatrix}$.

2. Let

$$S = \left\{ \begin{bmatrix} 1 \\ -1 \\ 2 \end{bmatrix}, \begin{bmatrix} 0 \\ 2 \\ 1 \end{bmatrix}, \begin{bmatrix} 1 \\ 0 \\ 0 \end{bmatrix} \right\}$$

be a basis for R^3. Find the coordinate vectors of the following vectors with respect to S.

(a) $\begin{bmatrix} 1 \\ 4 \\ 2 \end{bmatrix}$. (b) $\begin{bmatrix} 3 \\ -4 \\ 3 \end{bmatrix}$. (c) $\begin{bmatrix} 2 \\ -5 \\ 5 \end{bmatrix}$. (d) $\begin{bmatrix} 1 \\ -2 \\ 4 \end{bmatrix}$.

3. Repeat Exercise 2 for

$$S = \left\{ \begin{bmatrix} 1 \\ 1 \\ 3 \end{bmatrix}, \begin{bmatrix} -1 \\ 0 \\ 1 \end{bmatrix}, \begin{bmatrix} 1 \\ 1 \\ 0 \end{bmatrix} \right\}.$$

(a) $\begin{bmatrix} 2 \\ 2 \\ 3 \end{bmatrix}$. (b) $\begin{bmatrix} -2 \\ 0 \\ -1 \end{bmatrix}$. (c) $\begin{bmatrix} -1 \\ 0 \\ 1 \end{bmatrix}$. (d) $\begin{bmatrix} 0 \\ 1 \\ -5 \end{bmatrix}$.

4. Let $\mathbf{L} : R^2 \to R^2$ be defined by

$$\mathbf{L}\left(\begin{bmatrix} x \\ y \end{bmatrix}\right) = \begin{bmatrix} x + 2y \\ 2x - y \end{bmatrix}.$$

Let S be the natural basis for R^2 and let $T = \left\{ \begin{bmatrix} -1 \\ 2 \end{bmatrix}, \begin{bmatrix} 2 \\ 0 \end{bmatrix} \right\}$ be another basis for R^2. Find the matrix of $\mathbf{L}$ with respect to

(a) S.
(b) S and T.
(c) T and S.
(d) T.
(e) Compute $\mathbf{L}\left(\begin{bmatrix} 1 \\ 2 \end{bmatrix}\right)$ using the definition of $\mathbf{L}$ and also using the matrices obtained in (a), (b), (c), and (d).

5. Let $\mathbf{L} : R^2 \to R^3$ be defined by

$$\mathbf{L}\left(\begin{bmatrix} x \\ y \end{bmatrix}\right) = \begin{bmatrix} x - 2y \\ 2x + y \\ x + y \end{bmatrix}.$$

Let S and T be the natural bases for R^2 and R^3, respectively. Also, let

$$S' = \left\{ \begin{bmatrix} 1 \\ -1 \end{bmatrix}, \begin{bmatrix} 0 \\ 1 \end{bmatrix} \right\} \quad \text{and} \quad T' = \left\{ \begin{bmatrix} 1 \\ 1 \\ 0 \end{bmatrix}, \begin{bmatrix} 0 \\ 1 \\ 1 \end{bmatrix}, \begin{bmatrix} 1 \\ -1 \\ 1 \end{bmatrix} \right\}$$

be bases for R^2 and R^3, respectively. Find the matrix of $\mathbf{L}$ with respect to

(a) S and T.
(b) S' and T'.
(c) Compute $\mathbf{L}\left(\begin{bmatrix} 1 \\ 2 \end{bmatrix}\right)$ using the matrices obtained in (a) and (b).

6. Let $\mathbf{L} : R^3 \to R^3$ be defined by

$$\mathbf{L}\left(\begin{bmatrix} x \\ y \\ z \end{bmatrix}\right) = \begin{bmatrix} x + 2y + z \\ 2x - y \\ 2y + z \end{bmatrix}.$$

Let S be the natural basis for R^3 and let $T = \left\{\begin{bmatrix} 1 \\ 0 \\ 1 \end{bmatrix}, \begin{bmatrix} 0 \\ 1 \\ 1 \end{bmatrix}, \begin{bmatrix} 0 \\ 0 \\ 1 \end{bmatrix}\right\}$ be another basis for R^3. Find the matrix of $\mathbf{L}$ with respect to

(a) S.
(b) S and T.
(c) T and S.
(d) T.
(e) Compute $\mathbf{L}\left(\begin{bmatrix} 1 \\ 1 \\ -2 \end{bmatrix}\right)$ using the representations obtained in (a), (b), (c), and (d).

7. Let $\mathbf{L} : R^3 \to R^2$ be defined by

$$\mathbf{L}\left(\begin{bmatrix} x \\ y \\ z \end{bmatrix}\right) = \begin{bmatrix} x + y \\ y - z \end{bmatrix}.$$

Let S and T be the natural bases for R^3 and R^2, respectively. Also, let

$$S' = \left\{\begin{bmatrix} 1 \\ 1 \\ 0 \end{bmatrix}, \begin{bmatrix} 0 \\ 1 \\ 0 \end{bmatrix}, \begin{bmatrix} -1 \\ 1 \\ 1 \end{bmatrix}\right\} \quad \text{and} \quad T' = \left\{\begin{bmatrix} -1 \\ 1 \end{bmatrix}, \begin{bmatrix} 1 \\ 2 \end{bmatrix}\right\}$$

be bases for R^3 and R^2, respectively.
Find the representation of $\mathbf{L}$ with respect to

(a) S and T.
(b) S' and T'.
(c) Compute $\mathbf{L}\left(\begin{bmatrix} 1 \\ 2 \\ 3 \end{bmatrix}\right)$ using both representations.

8. Let $\mathbf{L} : R^2 \to R^3$ be defined by

$$\mathbf{L}\left(\begin{bmatrix} x \\ y \end{bmatrix}\right) = \begin{bmatrix} 1 & 1 \\ 1 & -1 \\ 1 & 2 \end{bmatrix}\begin{bmatrix} x \\ y \end{bmatrix}.$$

(a) Represent $\mathbf{L}$ with respect to the natural bases for R^2 and R^3.
(b) Represent $\mathbf{L}$ with respect to the bases S' and T' of Exercise 5.
(c) Compute $\mathbf{L}\left(\begin{bmatrix} 2 \\ -3 \end{bmatrix}\right)$ using both representations.

9. Let $\mathbf{L} : R^2 \to R^2$ be a linear transformation. Suppose that the matrix of $\mathbf{L}$ with respect to the basis

$$\mathbf{S} = \{\mathbf{X}_1, \mathbf{X}_2\}$$

is

$$\mathbf{A} = \begin{bmatrix} 2 & -3 \\ -1 & 4 \end{bmatrix},$$

where

$$\mathbf{X}_1 = \begin{bmatrix} 1 \\ 2 \end{bmatrix} \quad \text{and} \quad \mathbf{X}_2 = \begin{bmatrix} 1 \\ -1 \end{bmatrix}.$$

(a) Compute $[\mathbf{L}(\mathbf{X}_1)]_S$ and $[\mathbf{L}(\mathbf{X}_2)]_S$.
(b) Compute $\mathbf{L}(\mathbf{X}_1)$ and $\mathbf{L}(\mathbf{X}_2)$.
(c) Compute $\mathbf{L}\left(\begin{bmatrix} -2 \\ 3 \end{bmatrix}\right)$.

Find the matrix of $\mathbf{L}$ with respect to the natural bases for R^3 and R^2.

10. Let the matrix of $\mathbf{L} : R^2 \to R^2$ with respect to the basis

$$S = \left\{\begin{bmatrix} 1 \\ -1 \end{bmatrix}, \begin{bmatrix} 0 \\ 1 \end{bmatrix}\right\}$$

be

$$\begin{bmatrix} 1 & 2 \\ -2 & 3 \end{bmatrix}.$$

Find the matrix of $\mathbf{L}$ with respect to the natural basis for R^2.
[*Hint:* Let $\mathbf{X}_1 = \begin{bmatrix} 1 \\ -1 \end{bmatrix}$, $\mathbf{X}_2 = \begin{bmatrix} 0 \\ 1 \end{bmatrix}$. Find $\mathbf{L}(\mathbf{X}_1)$ and $\mathbf{L}(\mathbf{X}_2)$.]

11. Let $\mathbf{L} : R^3 \to R^3$ be defined by

$$\mathbf{L}\left(\begin{bmatrix} 1 \\ 0 \\ 0 \end{bmatrix}\right) = \begin{bmatrix} 1 \\ 1 \\ 0 \end{bmatrix}, \quad \mathbf{L}\left(\begin{bmatrix} 0 \\ 1 \\ 0 \end{bmatrix}\right) = \begin{bmatrix} 2 \\ 0 \\ 1 \end{bmatrix}, \quad \mathbf{L}\left(\begin{bmatrix} 0 \\ 0 \\ 1 \end{bmatrix}\right) = \begin{bmatrix} 1 \\ 0 \\ 1 \end{bmatrix}.$$

(a) Find the matrix of $\mathbf{L}$ with respect to the natural basis S for R^3.
(b) Find $\mathbf{L}\left(\begin{bmatrix} 1 \\ 2 \\ 3 \end{bmatrix}\right)$ using the definition of $\mathbf{L}$ and also using the matrix obtained in (a).

12. Let the matrix of $\mathbf{L} : R^3 \to R^2$ with respect to the bases

$$S = \{\mathbf{X}_1, \mathbf{X}_2, \mathbf{X}_3\} \quad \text{and} \quad T = \{\mathbf{Y}_1, \mathbf{Y}_2\}$$

be

$$\mathbf{A} = \begin{bmatrix} 1 & 2 & 1 \\ -1 & 1 & 0 \end{bmatrix},$$

where

$$\mathbf{X}_1 = \begin{bmatrix} -1 \\ 1 \\ 0 \end{bmatrix}, \quad \mathbf{X}_2 = \begin{bmatrix} 0 \\ 1 \\ 1 \end{bmatrix}, \quad \text{and} \quad \mathbf{X}_3 = \begin{bmatrix} 1 \\ 0 \\ 0 \end{bmatrix}$$

and

$$\mathbf{Y}_1 = \begin{bmatrix} 1 \\ 2 \end{bmatrix}, \quad \mathbf{Y}_2 = \begin{bmatrix} 1 \\ -1 \end{bmatrix}.$$

(a) Compute $[\mathbf{L}(\mathbf{X}_1)]_T$, $[\mathbf{L}(\mathbf{X}_2)]_T$, and $[\mathbf{L}(\mathbf{X}_3)]_T$.
(b) Compute $\mathbf{L}(\mathbf{X}_1)$ and $\mathbf{L}(\mathbf{X}_2)$ and $\mathbf{L}(\mathbf{X}_3)$.
(c) Compute $\mathbf{L}\left(\begin{bmatrix} 2 \\ 1 \\ -1 \end{bmatrix}\right)$.

Theoretical Exercises

T.1. Complete the proof of Theorem 4.8.

T.2. If $\mathbf{I} : V \to V$ is the identity transformation, show that the matrix of $\mathbf{I}$ with respect to any basis S for V is the identity matrix $\mathbf{I}_n$, where $n = \dim V$.

T.3. Let $\mathbf{0} : V \to W$ be the zero linear transformation, defined by $\mathbf{0}(\mathbf{X}) = \mathbf{0}_W$, for any $\mathbf{X}$ in S. Prove that the matrix of $\mathbf{0}$ with respect to any bases for V and W is the $m \times n$ zero matrix (where $n = \dim V$, $m = \dim W$).

4.4 The Rank of a Matrix

In this section we formulate the notion of the rank of a matrix and use it to compute the dimension of the solution space of a homogeneous system. Several other applications are also discussed.

DEFINITION Let

$$\mathbf{A} = \begin{bmatrix} a_{11} & a_{12} & \cdots & a_{1n} \\ a_{21} & a_{22} & \cdots & a_{2n} \\ \vdots & \vdots & & \vdots \\ a_{m1} & a_{m2} & \cdots & a_{mn} \end{bmatrix}$$

be an $m \times n$ matrix. The columns of **A**,

$$\begin{bmatrix} a_{11} \\ a_{21} \\ \vdots \\ a_{m1} \end{bmatrix}, \begin{bmatrix} a_{12} \\ a_{22} \\ \vdots \\ a_{m2} \end{bmatrix}, \ldots, \begin{bmatrix} a_{1n} \\ a_{2n} \\ \vdots \\ a_{mn} \end{bmatrix},$$

considered as vectors in R^m, span a subspace of R^m called the **column space of A**. The dimension of the column space of **A** is called the **column rank of A.**

Example 1. Let

$$\mathbf{A} = \begin{bmatrix} 1 & -2 & 0 & 3 & -4 \\ 3 & 2 & 8 & 1 & 4 \\ 2 & 3 & 7 & 2 & 3 \\ -1 & 2 & 0 & 4 & -3 \end{bmatrix},$$

and let the columns of **A** be

$$\mathbf{X}_1 = \begin{bmatrix} 1 \\ 3 \\ 2 \\ -1 \end{bmatrix}, \quad \mathbf{X}_2 = \begin{bmatrix} -2 \\ 2 \\ 3 \\ 2 \end{bmatrix}, \quad \mathbf{X}_3 = \begin{bmatrix} 0 \\ 8 \\ 7 \\ 0 \end{bmatrix},$$

$$\mathbf{X}_4 = \begin{bmatrix} 3 \\ 1 \\ 2 \\ 4 \end{bmatrix}, \quad \text{and} \quad \mathbf{X}_5 = \begin{bmatrix} -4 \\ 4 \\ 3 \\ -3 \end{bmatrix}.$$

Proceeding as in Example 12 of Section 3.5, we find that a basis for the column space of **A** is $\{\mathbf{X}_1, \mathbf{X}_2, \mathbf{X}_4\}$. Hence the column rank of **A** is 3.

DEFINITION The $m \times n$ matrix **A** is row equivalent to a matrix **B** in reduced row echelon form. The number of nonzero rows of B is called the **row rank of A.**

Example 2. Let **A** be as in Example 1. Then **A** is row equivalent to

$$\mathbf{B} = \begin{bmatrix} 1 & 0 & 2 & 0 & 1 \\ 0 & 1 & 1 & 0 & 1 \\ 0 & 0 & 0 & 1 & -1 \\ 0 & 0 & 0 & 0 & 0 \end{bmatrix},$$

which is in reduced row echelon form. Hence the row rank of **A** is 3.

We omit the proof of the following theorem.

THEOREM 4.9. *The row and column ranks of an $m \times n$ matrix are equal.*

We now merely refer to the **rank** of an $m \times n$ matrix, and write rank **A**. Thus to find the rank of an $m \times n$ matrix we use elementary row operations to obtain the matrix **B**, in reduced row echelon form, which is row equivalent to **A**. The number of nonzero rows of **B** is rank **A**.

Consider now the homogeneous system

$$\mathbf{AX} = \mathbf{0}, \tag{1}$$

where **A** is $m \times n$. The solution space W is a subspace of R^n; we now study the dimension of W. We examine the linear transformation $\mathbf{L} : R^n \to R^m$ defined by

$$\mathbf{L}(\mathbf{X}) = \mathbf{AX},$$

for **X** in R^n. It is easy to show (Exercise T.1) that the column space of **A** is range **L**. Also, the solution space W of (1) is the kernel of **L**. Now recall Equation (1) in Theorem 4.7 of Section 4.2,

$$n = \dim(\ker \mathbf{L}) + \dim(\text{range } \mathbf{L}).$$

Since dim (range **L**) = dim (column space of **A**) = rank **A**, we conclude that

$$\dim W = \dim(\ker \mathbf{L}) = n - \text{rank } \mathbf{A}. \tag{2}$$

Thus the dimension of the solution space of (1) is the number of columns of **A** minus its rank.

Example 3. Consider the homogeneous system of Example 11 of Section 1.3. The coefficient matrix is

$$\mathbf{A} = \begin{bmatrix} 1 & 1 & 1 & 1 \\ 1 & 0 & 0 & 1 \\ 1 & 2 & 1 & 0 \end{bmatrix},$$

which is row equivalent to

$$\mathbf{B} = \begin{bmatrix} 1 & 0 & 0 & 1 \\ 0 & 1 & 0 & -1 \\ 0 & 0 & 1 & 1 \end{bmatrix}.$$

The matrix $\mathbf{B}$ is in reduced row echelon form and rank $\mathbf{A} = 3$. Hence the dimension of the solution space is $4 - 3 = 1$. In Example 11 of Section 1.3 we found that the solution to the system is of the form

$$\mathbf{X} = \begin{bmatrix} -r \\ r \\ -r \\ r \end{bmatrix},$$

where r is any real number. Thus a basis for the solution space is the vector

$$\begin{bmatrix} -1 \\ 1 \\ -1 \\ 1 \end{bmatrix},$$

which confirms the fact that the dimension of the solution space is 1.

For a square matrix, its rank can be used to determine whether the matrix is singular or nonsingular, as the following theorem shows.

THEOREM 4.10. *An $n \times n$ matrix is nonsingular if and only if rank $\mathbf{A} = n$.*

proof. Suppose that $\mathbf{A}$ is nonsingular. Then $\mathbf{A}$ is row equivalent to $\mathbf{I}_n$ (Theorem 1.10, Section 1.4), and so rank $\mathbf{A} = n$.

Conversely, if rank $\mathbf{A} = n$, then by definition, $\mathbf{A}$ is row equivalent to $\mathbf{I}_n$ and so $\mathbf{A}$ is nonsingular (Theorem 1.10, Section 1.4).

From a practical point of view, this result is not too useful, since most of the time we want to know not only whether A is nonsingular, but also its inverse. The method developed in Chapter 1 enables us to find A^{-1}, if it exists, and tells us if it does not exist. Thus we do not have to learn first if A^{-1} exists, and then go through another procedure to obtain it.

An immediate consequence of Theorem 4.10 is the following corollary, which is a criterion for the rank of an $n \times n$ matrix to be n.

COROLLARY 4.2. *If $\mathbf{A}$ is an $n \times n$ matrix, then rank $\mathbf{A} = n$ if and only if $|\mathbf{A}| \neq 0$.*

proof. Exercise T.2.

The following corollary gives another method of testing whether n given vectors in R^n are linearly dependent or linearly independent.

COROLLARY 4.3. *Let $S = \{\mathbf{X}_1, \mathbf{X}_2, \ldots, \mathbf{X}_n\}$ be a set of n vectors in R^n and let $\mathbf{A}$ be the matrix whose columns are the vectors in S. Then S is linearly independent if and only if $|\mathbf{A}| \neq 0$.*

proof. Exercise T.3.

For $n > 4$, the method of Corollary 4.3 for testing linear dependence is not as efficient as the direct method of Section 3.5, calling for the solution of a linear system.

COROLLARY 4.4. *The homogeneous system $\mathbf{AX} = \mathbf{0}$ of n linear equations in n unknowns has a nontrivial solution if and only if rank $\mathbf{A} < n$.*

proof. This follows from Corollary 4.2 and from the fact that $\mathbf{AX} = \mathbf{0}$ has a nontrivial solution if and only if $\mathbf{A}$ is singular (Theorem 1.12, Section 1.4).

An alternative proof follows from Equation (2). There is a nontrivial solution to the system $\mathbf{AX} = \mathbf{0}$ if and only if the dimension of the solution space is positive. Now from (2), the dimension of the solution space is

$$n - \text{rank } \mathbf{A},$$

which is positive if and only if rank $\mathbf{A} < n$.

Example 4. Let

$$\mathbf{A} = \begin{bmatrix} 1 & 2 & 0 \\ 0 & 1 & 3 \\ 2 & 1 & 3 \end{bmatrix}.$$

If we transform $\mathbf{A}$ to reduced row echelon form $\mathbf{B}$, we find that $\mathbf{B} = \mathbf{I}_3$ (verify). Thus rank $\mathbf{A} = 3$ and $\mathbf{A}$ is nonsingular. Moreover, the homogeneous linear system $\mathbf{AX} = \mathbf{0}$ has only the trivial solution.

Example 5. Let

$$\mathbf{A} = \begin{bmatrix} 1 & 2 & 0 \\ 1 & 1 & -3 \\ 1 & 3 & 3 \end{bmatrix}.$$

Then **A** is row equivalent to

$$\begin{bmatrix} 1 & 0 & -6 \\ 0 & 1 & 3 \\ 0 & 0 & 0 \end{bmatrix},$$

a matrix in reduced row echelon form. Hence rank $\mathbf{A} < 3$, and **A** is singular. Moreover, $\mathbf{AX} = \mathbf{0}$ has a nontrivial solution.

Our final application of rank is to linear systems. If we consider the linear system $\mathbf{AX} = \mathbf{B}$ of m linear equations in n unknowns, where $\mathbf{A} = [a_{ij}]$, then we observe that the system can also be written as the equation

$$x_1 \begin{bmatrix} a_{11} \\ a_{21} \\ \vdots \\ a_{m1} \end{bmatrix} + x_2 \begin{bmatrix} a_{12} \\ a_{22} \\ \vdots \\ a_{m2} \end{bmatrix} + \cdots + x_n \begin{bmatrix} a_{1n} \\ a_{2n} \\ \vdots \\ a_{mn} \end{bmatrix} = \begin{bmatrix} b_1 \\ b_2 \\ \vdots \\ b_m \end{bmatrix}.$$

Thus $\mathbf{AX} = \mathbf{B}$ has a solution if and only if **B** is a linear combination of the columns of **A**; that is, if and only if **B** belongs to the column space of **A**. This means that if $\mathbf{AX} = \mathbf{B}$ has a solution, then rank $\mathbf{A}$ = rank $[\mathbf{A} \mid \mathbf{B}]$, where $[\mathbf{A} \mid \mathbf{B}]$ is the augmented matrix of the system. Conversely, if rank $[\mathbf{A} \mid \mathbf{B}]$ = rank **A**, then **B** is in the column space of **A**, which means that the system has a solution. Thus $\mathbf{AX} = \mathbf{B}$ has a solution if and only if rank $\mathbf{A}$ = rank $[\mathbf{A} \mid \mathbf{B}]$. This result, while of interest, is not of great computational value, since we usually are interested in finding a solution rather than in knowing whether or not a solution exists.

Example 6. Consider the linear system

$$\begin{bmatrix} 2 & 1 & 3 \\ 1 & -2 & 2 \\ 0 & 1 & 3 \end{bmatrix} \begin{bmatrix} x_1 \\ x_2 \\ x_3 \end{bmatrix} = \begin{bmatrix} 1 \\ 2 \\ 3 \end{bmatrix}.$$

Since rank $\mathbf{A}$ = rank $[\mathbf{A} \mid \mathbf{B}] = 3$, the linear system has a solution.

Example 7. The linear system

$$\begin{bmatrix} 1 & 2 & 3 \\ 1 & -3 & 4 \\ 2 & -1 & 7 \end{bmatrix} \begin{bmatrix} x_1 \\ x_2 \\ x_3 \end{bmatrix} = \begin{bmatrix} 4 \\ 5 \\ 6 \end{bmatrix}$$

has no solution, because rank $\mathbf{A} = 2$ and rank $[\mathbf{A} \mid \mathbf{B}] = 3$.

4.4 Exercises

1. Find the row and column ranks of

$$\mathbf{A} = \begin{bmatrix} 1 & 2 & 3 & 2 & 1 \\ 3 & 1 & -5 & -2 & 1 \\ 7 & 8 & -1 & 2 & 5 \end{bmatrix}.$$

2. Find the row and column ranks of

$$\mathbf{A} = \begin{bmatrix} 1 & 3 & 2 & 0 & 0 & 1 \\ 2 & 1 & -5 & 1 & 2 & 0 \\ 3 & 2 & 5 & 1 & -2 & 1 \\ 5 & 8 & 9 & 1 & -2 & 2 \\ 9 & 9 & 4 & 2 & 0 & 2 \end{bmatrix}.$$

In Exercises 3 through 7 compute the rank of each matrix.

3. $\begin{bmatrix} 1 & 2 & 1 & 3 \\ 2 & 1 & -4 & -5 \\ 7 & 8 & -5 & -1 \\ 10 & 14 & -2 & 8 \end{bmatrix}.$

4. $\begin{bmatrix} 1 & 2 & 1 & 3 \\ 2 & 1 & -4 & -5 \\ 1 & 1 & 0 & 0 \\ 0 & 0 & 1 & 1 \end{bmatrix}.$

5. $\begin{bmatrix} 1 & 2 & 3 \\ -1 & 2 & 1 \\ 3 & 1 & 2 \end{bmatrix}.$

6. $\begin{bmatrix} 1 & -2 & -1 \\ 2 & -1 & 3 \\ 7 & -8 & 3 \end{bmatrix}.$

7. $\begin{bmatrix} 1 & -2 & -1 \\ 2 & -1 & 3 \\ 7 & -8 & 3 \\ 5 & -7 & 0 \end{bmatrix}.$

In Exercises 8 through 10 determine the dimension of the solution space to the homogeneous system $\mathbf{AX} = \mathbf{0}$ for the given matrix $\mathbf{A}$.

8. $\mathbf{A} = \begin{bmatrix} 1 & 1 & -2 \\ 1 & 2 & 3 \\ 0 & 1 & 3 \end{bmatrix}.$

9. $\mathbf{A} = \begin{bmatrix} 1 & 1 & -2 & 0 & 0 \\ 1 & 2 & 3 & 6 & 7 \\ 2 & 1 & 3 & 6 & 5 \end{bmatrix}.$

10. $\mathbf{A} = \begin{bmatrix} 1 & 1 & -2 \\ 1 & 2 & 3 \\ 3 & 4 & -1 \end{bmatrix}.$

In Exercises 11 through 13 use Theorem 4.10 to determine whether each matrix is singular or nonsingular.

11. $\begin{bmatrix} 1 & 2 & -3 \\ -1 & 2 & 3 \\ 0 & 8 & 0 \end{bmatrix}$.

12. $\begin{bmatrix} 1 & 2 & -3 \\ -1 & 2 & 3 \\ 0 & 1 & 1 \end{bmatrix}$.

13. $\begin{bmatrix} 1 & 1 & 4 & -1 \\ 1 & 2 & 3 & 2 \\ -1 & 3 & 2 & 1 \\ -2 & 6 & 12 & -4 \end{bmatrix}$.

Use Corollary 4.3 to do Exercises 14 and 15.

14. Is

$$S = \left\{ \begin{bmatrix} 2 \\ 2 \\ 3 \end{bmatrix}, \begin{bmatrix} 1 \\ 0 \\ 2 \end{bmatrix}, \begin{bmatrix} 0 \\ 1 \\ 3 \end{bmatrix} \right\}$$

a linearly independent set of vectors in R^3?

15. Is

$$S = \left\{ \begin{bmatrix} 4 \\ 1 \\ 2 \end{bmatrix}, \begin{bmatrix} 2 \\ 5 \\ -5 \end{bmatrix}, \begin{bmatrix} 2 \\ -1 \\ 3 \end{bmatrix} \right\}$$

a linearly independent set of vectors in R^3?

In Exercises 16 through 18 find which homogeneous systems have a nontrivial solution for the given matrix **A** by using Corollary 4.4.

16. $\mathbf{A} = \begin{bmatrix} 1 & 1 & 2 & -1 \\ 1 & 3 & -1 & 2 \\ 1 & 1 & 1 & 3 \\ 1 & 2 & 1 & 1 \end{bmatrix}$.

17. $\mathbf{A} = \begin{bmatrix} 1 & 2 & 3 \\ 0 & 1 & 0 \\ 1 & 0 & 3 \end{bmatrix}$.

18. $\mathbf{A} = \begin{bmatrix} 1 & 2 & -1 \\ 2 & -1 & 3 \\ 5 & -4 & 3 \end{bmatrix}$.

In Exercises 19 through 22 determine which of the linear systems have a solution by comparing the ranks of the coefficient and augmented matrices.

19. $\begin{bmatrix} 1 & 2 & 5 & -2 \\ 2 & 3 & -2 & 4 \\ 5 & 1 & 0 & 2 \end{bmatrix} \begin{bmatrix} x_1 \\ x_2 \\ x_3 \\ x_4 \end{bmatrix} = \begin{bmatrix} 0 \\ 0 \\ 0 \end{bmatrix}.$

20. $\begin{bmatrix} 1 & 2 & 5 & -2 \\ 2 & 3 & -2 & 4 \\ 5 & 1 & 0 & 2 \end{bmatrix} \begin{bmatrix} x_1 \\ x_2 \\ x_3 \\ x_4 \end{bmatrix} = \begin{bmatrix} 1 \\ -13 \\ 3 \end{bmatrix}.$

21. $\begin{bmatrix} 1 & -2 & -3 & 4 \\ 4 & -1 & -5 & 6 \\ 2 & 3 & 1 & -2 \end{bmatrix} \begin{bmatrix} x_1 \\ x_2 \\ x_3 \\ x_4 \end{bmatrix} = \begin{bmatrix} 1 \\ 2 \\ 2 \end{bmatrix}.$

22. $\begin{bmatrix} 1 & 1 & 1 \\ 1 & -1 & 1 \\ 5 & 1 & 5 \end{bmatrix} \begin{bmatrix} x_1 \\ x_2 \\ x_3 \end{bmatrix} = \begin{bmatrix} 6 \\ 2 \\ 5 \end{bmatrix}.$

Theoretical Exercises

T.1. Let **A** be an $m \times n$ matrix and let $\mathbf{L} : R^m \to R^n$ be defined by $\mathbf{L}(\mathbf{X}) = \mathbf{AX}$ for **X** in R^m. Show that the column space of **A** is the range of **L**.

T.2. Prove Corollary 4.2.

T.3. Prove Corollary 4.3.

5 Eigenvalues and Eigenvectors

5.1 Diagonalization

In this chapter every matrix considered is a square matrix. If **A** is an $n \times n$ matrix, then we can define the linear transformation $\mathbf{L} : R^n \to R^n$ by $\mathbf{L}(\mathbf{X}) = \mathbf{AX}$, for **X** in R^n. A question of considerable importance in a great many applied problems is the determination of vectors **X**, if there are any, such that **X** and **L(X)** are parallel. Such questions arise in all applications involving vibrations; they arise in aerodynamics, elasticity, nuclear physics, mechanics, chemical engineering, biology, differential equations, and so on. In this section we shall formulate this problem precisely; we also define some pertinent terminology. In the next section we settle the question for symmetric matrices and briefly discuss the situation in the general case.

DEFINITION Let **A** be an $n \times n$ matrix. The real number λ is called an **eigenvalue** of **A** if there exists a nonzero vector **X** in R^n such that

$$\mathbf{AX} = \lambda\mathbf{X}. \tag{1}$$

Every nonzero vector $\mathbf{X}$ satisfying (1) is called an **eigenvector of A associated with the eigenvalue** λ. We might mention that the word "eigenvalue" is a hybrid one ("eigen" in German means "proper"). Eigenvalues are also called **proper values, characteristic values,** and **latent values**; and eigenvectors are also called **proper vectors**, and so on, accordingly.

Note that $\mathbf{X} = \mathbf{0}$ always satisfies Equation (1), but we insist that an eigenvector $\mathbf{X}$ be a nonzero vector. In some practical applications one encounters matrices with complex entries and vector spaces with scalars that are complex numbers. In such a setting the above definition of eigenvalue is modified so that an eigenvalue can be a real *or* complex number. Such treatments are presented in more advanced books. In this book we require an eigenvalue to be a real number.

Example 1. If $\mathbf{A}$ is the identity matrix $\mathbf{I}_n$, then the only eigenvalue is $\lambda = 1$; every nonzero vector in R^n is an eigenvector of $\mathbf{A}$ associated with the eigenvalue $\lambda = 1$:

$$\mathbf{I}_n\mathbf{X} = 1\mathbf{X}.$$

Example 2. Let

$$\mathbf{A} = \begin{bmatrix} 0 & \frac{1}{2} \\ \frac{1}{2} & 0 \end{bmatrix}.$$

Then

$$\mathbf{A}\begin{bmatrix} 1 \\ 1 \end{bmatrix} = \begin{bmatrix} 0 & \frac{1}{2} \\ \frac{1}{2} & 0 \end{bmatrix}\begin{bmatrix} 1 \\ 1 \end{bmatrix} = \begin{bmatrix} \frac{1}{2} \\ \frac{1}{2} \end{bmatrix} = \frac{1}{2}\begin{bmatrix} 1 \\ 1 \end{bmatrix},$$

so that $\mathbf{X}_1 = \begin{bmatrix} 1 \\ 1 \end{bmatrix}$ is an eigenvector of $\mathbf{A}$ associated with the eigenvalue $\lambda_1 = \frac{1}{2}$. Also,

$$\mathbf{A}\begin{bmatrix} 1 \\ -1 \end{bmatrix} = \begin{bmatrix} 0 & \frac{1}{2} \\ \frac{1}{2} & 0 \end{bmatrix}\begin{bmatrix} 1 \\ -1 \end{bmatrix} = \begin{bmatrix} -\frac{1}{2} \\ \frac{1}{2} \end{bmatrix} = -\frac{1}{2}\begin{bmatrix} 1 \\ -1 \end{bmatrix},$$

so that $\mathbf{X}_2 = \begin{bmatrix} 1 \\ -1 \end{bmatrix}$ is an eigenvector of $\mathbf{A}$ associated with the eigenvalue $\lambda_2 = -\frac{1}{2}$.

If we let $\mathbf{L} : R^2 \to R^2$ be defined by

$$\mathbf{L}(\mathbf{X}) = \mathbf{AX} = \begin{bmatrix} 0 & \frac{1}{2} \\ \frac{1}{2} & 0 \end{bmatrix} \begin{bmatrix} x_1 \\ x_2 \end{bmatrix},$$

then Figure 5.1 shows that $\mathbf{X}_1$ and $\mathbf{L}(\mathbf{X}_1)$ are parallel, and $\mathbf{X}_2$ and $\mathbf{L}(\mathbf{X}_2)$ are parallel also. This illustrates the fact that if $\mathbf{X}$ is an eigenvector of $\mathbf{A}$, then $\mathbf{X}$ and $\mathbf{AX}$ are parallel.

In Figure 5.2 we show $\mathbf{X}$ and $\mathbf{AX}$ for the cases $\lambda > 1$, $0 < \lambda < 1$, and $\lambda < 0$. Thus, if $\lambda > 1$, then $\mathbf{L}$ is a dilation and if $0 < \lambda < 1$, then $\mathbf{L}$ is a contraction; if $\lambda < 0$, then $\mathbf{L}$ reverses the direction of $\mathbf{X}$.

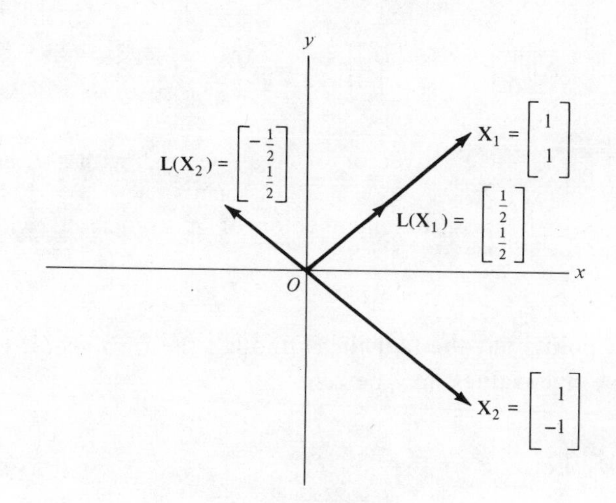

Figure 5.1

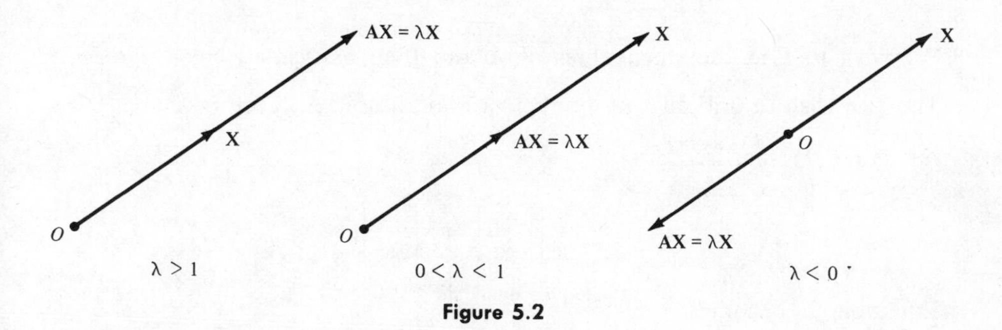

Figure 5.2

An eigenvalue λ of $\mathbf{A}$ can have associated with it many different eigenvectors. In fact, if $\mathbf{X}$ is an eigenvector of $\mathbf{A}$ associated with λ (that is,

$\mathbf{AX} = \lambda\mathbf{X}$) and r is any nonzero real number, then

$$\mathbf{A}(r\mathbf{X}) = r(\mathbf{AX}) = r(\lambda\mathbf{X}) = \lambda(r\mathbf{X}).$$

Thus $r\mathbf{X}$ is also an eigenvector of $\mathbf{A}$ associated with λ.

Example 3. Let

$$\mathbf{A} = \begin{bmatrix} 0 & 0 \\ 0 & 1 \end{bmatrix}.$$

Then

$$\mathbf{A}\begin{bmatrix} 1 \\ 0 \end{bmatrix} = \begin{bmatrix} 0 & 0 \\ 0 & 1 \end{bmatrix}\begin{bmatrix} 1 \\ 0 \end{bmatrix} = \begin{bmatrix} 0 \\ 0 \end{bmatrix} = 0\begin{bmatrix} 1 \\ 0 \end{bmatrix},$$

so that $\mathbf{X}_1 = \begin{bmatrix} 1 \\ 0 \end{bmatrix}$ is an eigenvector of $\mathbf{A}$ associated with the eigenvalue $\lambda_1 = 0$. Also, $\mathbf{X}_2 = \begin{bmatrix} 0 \\ 1 \end{bmatrix}$ is an eigenvector of $\mathbf{A}$ associated with $\lambda_2 = 1$ (verify).

Example 3 points out the fact that although the zero vector cannot be an eigenvector, eigenvalues may be zero.

Example 4. Let

$$\mathbf{A} = \begin{bmatrix} 1 & 1 \\ -2 & 4 \end{bmatrix}.$$

We wish to find the eigenvalues of $\mathbf{A}$ and their associated eigenvectors. Thus we wish to find all real numbers λ and all nonzero vectors $\mathbf{X} = \begin{bmatrix} x_1 \\ x_2 \end{bmatrix}$ satisfying (1), that is,

$$\begin{bmatrix} 1 & 1 \\ -2 & 4 \end{bmatrix}\begin{bmatrix} x_1 \\ x_2 \end{bmatrix} = \lambda\begin{bmatrix} x_1 \\ x_2 \end{bmatrix}. \tag{2}$$

Equation (2) becomes

$$\begin{aligned} x_1 + x_2 &= \lambda x_1 \\ -2x_1 + 4x_2 &= \lambda x_2, \end{aligned}$$

or

$$\begin{aligned} (\lambda - 1)x_1 - \qquad\quad x_2 &= 0 \\ 2x_1 + (\lambda - 4)x_2 &= 0. \end{aligned} \tag{3}$$

Equation (3) is a homogeneous system of two equations in two unknowns. From Corollary 2.3 of Section 2.2 it follows that the linear system in (3) has a nontrivial solution if and only if the determinant of its coefficient matrix is zero: thus if and only if

$$\begin{vmatrix} \lambda - 1 & 1 \\ 2 & \lambda - 4 \end{vmatrix} = 0.$$

This means that

$$(\lambda - 1)(\lambda - 4) + 2 = 0,$$

or

$$\lambda^2 - 5\lambda + 6 = 0 = (\lambda - 3)(\lambda - 2).$$

Hence

$$\lambda_1 = 2 \qquad \text{and} \qquad \lambda_2 = 3$$

are the eigenvalues of **A**. To find an eigenvector of **A** associated with $\lambda_1 = 2$, we form the linear system

$$\mathbf{AX} = 2\mathbf{X},$$

or

$$\begin{bmatrix} 1 & 1 \\ -2 & 4 \end{bmatrix} \begin{bmatrix} x_1 \\ x_2 \end{bmatrix} = 2 \begin{bmatrix} x_1 \\ x_2 \end{bmatrix}.$$

This gives

$$\begin{aligned} x_1 + \ x_2 &= 2x_1 \\ -2x_1 + 4x_2 &= 2x_2 \end{aligned} \quad \text{or} \quad \begin{aligned} (2 - 1)x_1 - \qquad\quad x_2 &= 0 \\ 2x_1 + (2 - 4)x_2 &= 0 \end{aligned} \quad \text{or} \quad \begin{aligned} x_1 - \ x_2 &= 0 \\ 2x_1 - 2x_2 &= 0. \end{aligned}$$

Note that we could have obtained this last linear system by merely substituting $\lambda = 2$ in (3). A nontrivial solution to this last system is

$$\mathbf{X}_1 = \begin{bmatrix} 1 \\ 1 \end{bmatrix},$$

which is an eigenvector of **A** associated with $\lambda_1 = 2$. Similarly, for $\lambda_2 = 3$ we obtain, from (3),

$$\begin{aligned} (3-1)x_1 - \qquad x_2 &= 0 \\ 2x_1 + (3-4)x_2 &= 0 \end{aligned} \quad \text{or} \quad \begin{aligned} 2x_1 - x_2 &= 0 \\ 2x_1 - x_2 &= 0. \end{aligned}$$

Thus $\mathbf{X}_2 = \begin{bmatrix} 1 \\ 2 \end{bmatrix}$ is an eigenvector of **A** associated with $\lambda_2 = 3$.

In Examples 1, 2, and 3 we found eigenvalues and eigenvectors by inspection, whereas in Example 4 we proceeded in a more systematic fashion. We use the procedure of Example 4 as our standard method, as follows.

DEFINITION Let $\mathbf{A} = [a_{ij}]$ be an $n \times n$ matrix. The determinant

$$f(\lambda) = |\lambda \mathbf{I}_n - \mathbf{A}| = \begin{vmatrix} \lambda - a_{11} & -a_{12} & & -a_{1n} \\ -a_{21} & \lambda - a_{22} & & -a_{2n} \\ \vdots & \vdots & & \vdots \\ -a_{n1} & -a_{n2} & \cdots & \lambda - a_{nn} \end{vmatrix} \tag{4}$$

is called the **characteristic polynomial of A**. The equation

$$f(\lambda) = |\lambda \mathbf{I}_n - \mathbf{A}| = 0$$

is called the **characteristic equation of A**.

Recall from Chapter 2 that each term in the expansion of the determinant of an $n \times n$ matrix is a product of n elements of the matrix, containing exactly one element from each row and exactly one element from each column. Thus if we expand $f(\lambda) = |\lambda \mathbf{I}_n - \mathbf{A}|$, we obtain a polynomial of degree n. The expression involving λ^n in the characteristic polynomial of **A** comes from the product $(\lambda - a_{11})(\lambda - a_{22}) \cdots (\lambda - a_{nn})$, and so the coefficient of λ^n is 1. We can then write

$$f(\lambda) = |\lambda \mathbf{I}_n - \mathbf{A}| = \lambda^n + a_1\lambda^{n-1} + a_2\lambda^{n-2} + \cdots + a_{n-1}\lambda + a_n.$$

If we let $\lambda = 0$, then we get $|-\mathbf{A}| = a_n$, which shows that the constant term a_n is $(-1)^n|\mathbf{A}|$. A polynomial of degree n with real coefficients has n roots, some of which may be complex numbers.

Example 5. Let

$$\mathbf{A} = \begin{bmatrix} 1 & 2 & -1 \\ 1 & 0 & 1 \\ 4 & -4 & 5 \end{bmatrix}.$$

The characteristic polynomial of **A** is

$$f(\lambda) = |\lambda\mathbf{I}_3 - \mathbf{A}| = \begin{vmatrix} \lambda - 1 & -2 & 1 \\ -1 & \lambda & -1 \\ -4 & 4 & \lambda - 5 \end{vmatrix} = \lambda^3 - 6\lambda^2 + 11\lambda - 6.$$

We now connect the characteristic polynomial of a matrix with its eigenvalues in the following theorem.

THEOREM 5.1. *The eigenvalues of* **A** *are the real roots of the characteristic polynomial of* **A**.

proof. Let λ be an eigenvalue of **A** with associated eigenvector **X**. Then

$$\mathbf{AX} = \lambda\mathbf{X},$$

which can be rewritten as

$$\mathbf{AX} = (\lambda\mathbf{I}_n)\mathbf{X}$$

or

$$(\lambda\mathbf{I}_n - \mathbf{A})\mathbf{X} = \mathbf{0}, \tag{5}$$

a homogeneous system of n equations in n unknowns. This system has a nontrivial solution if and only if the determinant of its coefficient matrix vanishes (Corollary 2.3, Section 2.2), that is, if and only if $|\lambda\mathbf{I}_n - \mathbf{A}| = 0$.

Conversely, if λ is a real root of the characteristic polynomial of A, then $|\lambda\mathbf{I}_n - \mathbf{A}| = 0$, so the homogeneous system (5) has a nontrivial solution **X**. Hence λ is an eigenvalue of **A**.

Thus to find the eigenvalues of a given matrix **A**, we must find the real roots of its characteristic polynomial $f(\lambda)$. There are many methods for finding approximations to the roots of a polynomial, some of them more effective than others; indeed, many computer codes are available to find the roots of a polynomial. Two results that are sometimes useful in this connection are: (1) the product of all the roots of the polynomial $f(\lambda) = \lambda^n + a_1\lambda^{n-1} + \cdots + a_{n-1}\lambda + a_n$ is $(-1)^n a_n$, and (2) if $a_1, a_2, \ldots, a_n$ are integers, then $f(\lambda)$ cannot have a rational root that is not already an integer. Thus as possible real roots of $f(\lambda)$ one can try the factors of a_n. Of course, $f(\lambda)$ might well have irrational roots. However, all the characteristic polynomials to be solved in the rest of this chapter have only integer roots, and each of these roots is a factor of the constant term of the characteristic polynomial of **A**. The corresponding eigenvectors are obtained by substituting for λ in Equation (5) and solving the resulting homogeneous system.

Example 6. Consider the matrix of Example 5. The characteristic polynomial is

$$f(\lambda) = \lambda^3 - 6\lambda^2 + 11\lambda - 6 = (\lambda - 1)(\lambda - 2)(\lambda - 3).$$

The eigenvalues of **A** are then

$$\lambda_1 = 1, \qquad \lambda_2 = 2, \qquad \lambda_3 = 3.$$

To find an eigenvector $\mathbf{X}_1$ associated with $\lambda_1 = 1$, we form the system

$$(\lambda_1\mathbf{I}_3 - \mathbf{A})\mathbf{X} = \mathbf{0},$$

$$\begin{bmatrix} 1-1 & -2 & 1 \\ -1 & 1 & -1 \\ -4 & 4 & 1-5 \end{bmatrix}\begin{bmatrix} x_1 \\ x_2 \\ x_3 \end{bmatrix} = \begin{bmatrix} 0 \\ 0 \\ 0 \end{bmatrix}$$

or

$$\begin{bmatrix} 0 & -2 & 1 \\ -1 & 1 & -1 \\ -4 & 4 & -4 \end{bmatrix}\begin{bmatrix} x_1 \\ x_2 \\ x_3 \end{bmatrix} = \begin{bmatrix} 0 \\ 0 \\ 0 \end{bmatrix}.$$

A solution is $\begin{bmatrix} -\frac{r}{2} \\ \frac{r}{2} \\ r \end{bmatrix}$ for any real number r. Thus $\mathbf{X}_1 = \begin{bmatrix} -1 \\ 1 \\ 2 \end{bmatrix}$ is an eigenvector of **A** associated with $\lambda_1 = 1$.

To find an eigenvector $\mathbf{X}_2$ associated with $\lambda_2 = 2$, we form the system

$$(\lambda_2\mathbf{I}_3 - \mathbf{A})\mathbf{X} = \mathbf{0},$$

that is,

$$\begin{bmatrix} 2-1 & -2 & 1 \\ -1 & 2 & -1 \\ -4 & 4 & 2-5 \end{bmatrix} \begin{bmatrix} x_1 \\ x_2 \\ x_3 \end{bmatrix} = \begin{bmatrix} 0 \\ 0 \\ 0 \end{bmatrix}$$

or

$$\begin{bmatrix} 1 & -2 & 1 \\ -1 & 2 & -1 \\ -4 & 4 & -3 \end{bmatrix} \begin{bmatrix} x_1 \\ x_2 \\ x_3 \end{bmatrix} = \begin{bmatrix} 0 \\ 0 \\ 0 \end{bmatrix}.$$

A solution is $\begin{bmatrix} -\frac{r}{2} \\ \frac{r}{4} \\ r \end{bmatrix}$, for any real number r. Thus $\mathbf{X}_2 = \begin{bmatrix} -2 \\ 1 \\ 4 \end{bmatrix}$ is an eigenvector of $\mathbf{A}$ associated with $\lambda_2 = 2$.

To find an eigenvector $\mathbf{X}_3$ associated with $\lambda_3 = 3$, we form the system

$$(\lambda_3 \mathbf{I}_3 - \mathbf{A})\mathbf{X} = \mathbf{0},$$

and find that a solution is $\begin{bmatrix} -\frac{r}{4} \\ \frac{r}{4} \\ r \end{bmatrix}$, for any real number r. Thus $\mathbf{X}_3 = \begin{bmatrix} -1 \\ 1 \\ 4 \end{bmatrix}$ is an eigenvector of $\mathbf{A}$ associated with $\lambda_3 = 3$.

Of course, the characteristic polynomial of a given matrix may have complex roots and it may even have no real roots. However, for the matrices that we are most interested in, symmetric matrices, all the roots of the characteristic polynomial are real. We shall state this in Section 5.2.

Example 7. Let $\mathbf{A} = \begin{bmatrix} 0 & 1 \\ -1 & 0 \end{bmatrix}$. Then the characteristic polynomial of $\mathbf{A}$ is

$$f(\lambda) = \lambda^2 + 1,$$

which has no real roots. Thus $\mathbf{A}$ has no eigenvalues.

Similar Matrices

We shall now develop an equivalent formulation for the eigenvalue–eigenvector problem, which will shed further light on the problem.

DEFINITION A matrix **B** is said to be **similar** to a matrix **A** if there is a nonsingular matrix **P** such that

$$\mathbf{B} = \mathbf{P}^{-1}\mathbf{A}\mathbf{P}.$$

Example 8. Let

$$\mathbf{A} = \begin{bmatrix} 1 & 1 \\ -2 & 4 \end{bmatrix}$$

be the matrix of Example 4. Let

$$\mathbf{P} = \begin{bmatrix} 1 & 1 \\ 1 & 2 \end{bmatrix}.$$

Then

$$\mathbf{P}^{-1} = \begin{bmatrix} 2 & -1 \\ -1 & 1 \end{bmatrix}$$

and

$$\mathbf{B} = \mathbf{P}^{-1}\mathbf{A}\mathbf{P} = \begin{bmatrix} 2 & -1 \\ -1 & 1 \end{bmatrix}\begin{bmatrix} 1 & 1 \\ -2 & 4 \end{bmatrix}\begin{bmatrix} 1 & 1 \\ 1 & 2 \end{bmatrix} = \begin{bmatrix} 2 & 0 \\ 0 & 3 \end{bmatrix}.$$

Thus **B** is similar to **A**.

We shall let the reader (Exercise T.1) prove that the following elementary properties hold for similarity:

1. **A** is similar to **A**.
2. If **B** is similar to **A**, then **A** is similar to **B**.
3. If **A** is similar to **B** and **B** is similar to **C**, then **A** is similar to **C**.

By property (2) we replace the statements "**A** is similar to **B**" and "**B** is similar to **A**" by "**A** and **B** are similar."

DEFINITION We shall say that the matrix **A** is **diagonalizable** if it is similar to a diagonal matrix. In this case we also say that **A can be diagonalized**.

Example 9. If **A** and **B** are as in Example 8, then **A** is diagonalizable, since it is similar to **B**.

THEOREM 5.2. *An $n \times n$ matrix* $\mathbf{A}$ *is diagonalizable if and only if it has n linearly independent eigenvectors. In this case* $\mathbf{A}$ *is similar to a matrix* $\mathbf{D}$ *whose diagonal elements are the eigenvalues of* $\mathbf{A}$.

proof. Suppose that $\mathbf{A}$ is similar to $\mathbf{D}$. Then

$$\mathbf{P}^{-1}\mathbf{A}\mathbf{P} = \mathbf{D},$$

so that

$$\mathbf{A}\mathbf{P} = \mathbf{P}\mathbf{D}. \tag{6}$$

Let

$$\mathbf{D} = \begin{bmatrix} \lambda_1 & 0 & \cdots & 0 \\ 0 & \lambda_2 & \cdots & 0 \\ \vdots & & & \vdots \\ 0 & \cdots & 0 & \lambda_n \end{bmatrix},$$

and let $\mathbf{X}_j$, $j = 1, 2, \ldots, n$ be the jth column of $\mathbf{P}$. Note that the jth column of the matrix $\mathbf{AP}$ is $\mathbf{AX}_j$ and the jth column of $\mathbf{PD}$ is $\lambda_j\mathbf{X}_j$ (see Exercise T.21 in Section 1.2 and verify). Thus from (6) we have

$$\mathbf{A}\mathbf{X}_j = \lambda_j\mathbf{X}_j. \tag{7}$$

Since $\mathbf{P}$ is a nonsingular matrix, its columns are linearly independent and so are all nonzero. Hence λ_j is an eigenvalue of $\mathbf{A}$ and $\mathbf{X}_j$ is a corresponding eigenvector. Moreover, since $\mathbf{P}$ is nonsingular, its column vectors are linearly independent (Corollary 4.3, Section 4.4).

Conversely, suppose that $\lambda_1, \lambda_2, \ldots, \lambda_n$ are n eigenvalues of $\mathbf{A}$ and that the corresponding eigenvectors $\mathbf{X}_1, \mathbf{X}_2, \ldots, \mathbf{X}_n$ are linearly independent. Let $\mathbf{P}$ be the matrix whose jth column is $\mathbf{X}_j$. From Corollary 4.3 of Section 4.4 it follows that $\mathbf{P}$ is nonsingular. From (7) we obtain (6), which implies that $\mathbf{A}$ is diagonalizable. This completes the proof.

Example 10. Let $\mathbf{A}$ be as in Example 4. The eigenvalues are $\lambda_1 = 2$ and $\lambda_2 = 3$. The corresponding eigenvectors $\mathbf{X}_1 = \begin{bmatrix} 1 \\ 1 \end{bmatrix}$ and $\mathbf{X}_2 = \begin{bmatrix} 1 \\ 2 \end{bmatrix}$ are linearly independent. Hence **A** is diagonalizable. Here

$$\mathbf{P} = \begin{bmatrix} 1 & 1 \\ 1 & 2 \end{bmatrix} \quad \text{and} \quad \mathbf{P}^{-1} = \begin{bmatrix} 2 & -1 \\ -1 & 1 \end{bmatrix}.$$

Thus

$$\mathbf{P}^{-1}\mathbf{A}\mathbf{P} = \begin{bmatrix} 2 & -1 \\ -1 & 1 \end{bmatrix} \begin{bmatrix} 1 & 1 \\ -2 & 4 \end{bmatrix} \begin{bmatrix} 1 & 1 \\ 1 & 2 \end{bmatrix} = \begin{bmatrix} 2 & 0 \\ 0 & 3 \end{bmatrix}.$$

On the other hand, if we let $\lambda_1 = 3$ and $\lambda_2 = 2$, then $\mathbf{X}_1 = \begin{bmatrix} 1 \\ 2 \end{bmatrix}$ and $\mathbf{X}_2 = \begin{bmatrix} 1 \\ 1 \end{bmatrix}$. Then

$$\mathbf{P} = \begin{bmatrix} 1 & 1 \\ 2 & 1 \end{bmatrix} \quad \text{and} \quad \mathbf{P}^{-1} = \begin{bmatrix} -1 & 1 \\ 2 & -1 \end{bmatrix}.$$

Hence

$$\mathbf{P}^{-1}\mathbf{A}\mathbf{P} = \begin{bmatrix} -1 & 1 \\ 2 & -1 \end{bmatrix} \begin{bmatrix} 1 & 1 \\ -2 & 4 \end{bmatrix} \begin{bmatrix} 1 & 1 \\ 2 & 1 \end{bmatrix} = \begin{bmatrix} 3 & 0 \\ 0 & 2 \end{bmatrix}.$$

Example 11. Let

$$\mathbf{A} = \begin{bmatrix} 1 & 1 \\ 0 & 1 \end{bmatrix}.$$

The eigenvalues here are $\lambda_1 = 1$ and $\lambda_2 = 1$. Eigenvectors associated with λ_1 and λ_2 are vectors of the form

$$\begin{bmatrix} r \\ 0 \end{bmatrix},$$

where r is any real number. Since **A** does not have two linearly independent eigenvectors, we conclude that **A** is not diagonalizable.

COROLLARY 5.1. *Consider the linear transformation* $\mathbf{L} : R^n \to R^n$ *defined by* $\mathbf{L}(\mathbf{X}) = \mathbf{AX}$ *for* $\mathbf{X}$ *in* R^n. *Then* **A** *is diagonalizable if and only if there is a basis for* R^n *with respect to which the matrix of* **L** *is diagonal.*

proof. Suppose that **A** is diagonalizable. Then by Theorem 5.2 it has n linearly independent eigenvectors $\mathbf{X}_1, \mathbf{X}_2, \ldots, \mathbf{X}_n$, with corresponding eigenvalues $\lambda_1, \lambda_2, \ldots, \lambda_n$. Since n linearly independent vectors in R^n form a basis (Theorem 3.14 in Section 3.5), we can conclude that $S = \{\mathbf{X}_1, \mathbf{X}_2, \ldots, \mathbf{X}_n\}$ is a basis for R^n. Now

$$\begin{aligned}\mathbf{L}(\mathbf{X}_j) &= \mathbf{A}\mathbf{X}_j \\ &= \lambda_j\mathbf{X}_j = 0\mathbf{X}_1 + \cdots + 0\mathbf{X}_{j-1} + \lambda_j\mathbf{X}_j + 0\mathbf{X}_{j+1} + \cdots + 0\mathbf{X}_n,\end{aligned}$$

so the coordinate vector $[\mathbf{L}(\mathbf{X}_j)]_S$ of $\mathbf{L}(\mathbf{X}_j)$ with respect to S is

$$\begin{bmatrix} 0 \\ \vdots \\ 0 \\ \lambda_j \\ 0 \\ \vdots \\ 0 \end{bmatrix} \text{--- } j\text{th row.} \tag{8}$$

Hence the matrix of $\mathbf{L}$ with respect to S is (Theorem 4.8 of Section 4.3)

$$\begin{bmatrix} \lambda_1 & 0 & \cdots & 0 \\ 0 & \lambda_2 & \cdots & 0 \\ \vdots & & & \vdots \\ 0 & \cdots & 0 & \lambda_n \end{bmatrix}. \tag{9}$$

Conversely, suppose that there is a basis $S = \{\mathbf{X}_1, \mathbf{X}_2, \ldots, \mathbf{X}_n\}$ for R^n with respect to which the matrix of $\mathbf{L}$ is diagonal, say, of the form in (9). Then the coordinate vector of $\mathbf{L}(\mathbf{X}_j)$ with respect to S is (8), so

$$\mathbf{L}(\mathbf{X}_j) = 0\mathbf{X}_1 + \cdots + 0\mathbf{X}_{j-1} + \lambda_j\mathbf{X}_j + 0\mathbf{X}_{j+1} + \cdots + 0\mathbf{X}_n = \lambda_j\mathbf{X}_j.$$

Since $\mathbf{L}(\mathbf{X}_j) = \mathbf{A}\mathbf{X}_j$, we have

$$\mathbf{A}\mathbf{X}_j = \lambda_j\mathbf{X}_j,$$

which means that $\mathbf{X}_1, \mathbf{X}_2, \ldots, \mathbf{X}_n$ are eigenvectors of $\mathbf{A}$. Since they form a basis for R^n, they are linearly independent, and by Theorem 5.2 we conclude that $\mathbf{A}$ is diagonalizable.

The following is a useful theorem because it identifies a large class of matrices that can be diagonalized.

THEOREM 5.3. *A matrix* $\mathbf{A}$ *is diagonalizable if all the roots of its characteristic polynomial are real and distinct.*

proof. Let $\lambda_1, \lambda_2, \ldots, \lambda_n$ be the distinct eigenvalues of $\mathbf{A}$ and let $S = \{\mathbf{X}_1, \mathbf{X}_2, \ldots, \mathbf{X}_n\}$ be the set of associated eigenvectors. We wish to show that S is linearly independent.

Suppose that S is a linearly dependent set of vectors. Then Theorem 3.9 of Section 3.5 implies that some vector $\mathbf{X}_j$ is a linear combination of the preceding vectors in S. We can assume that $S_1 = \{\mathbf{X}_1, \mathbf{X}_2, \ldots, \mathbf{X}_{j-1}\}$ is linearly independent, for otherwise one of the vectors in S_1 is a linear combination of the preceding ones, and we can choose a new set S_2, and so on. We thus have that S_1 is linearly independent and that

$$\mathbf{X}_j = c_1\mathbf{X}_1 + c_2\mathbf{X}_2 + \cdots + c_{j-1}\mathbf{X}_{j-1}, \tag{10}$$

where $c_1, c_2, \ldots, c_{j-1}$ are real numbers. Premultiplying (multiplying on the left) both sides of Equation (10) by $\mathbf{A}$, we obtain

$$\begin{aligned} \mathbf{A}\mathbf{X}_j &= \mathbf{A}(c_1\mathbf{X}_1 + c_2\mathbf{X}_2 + \cdots + c_{j-1}\mathbf{X}_{j-1}) \\ &= c_1\mathbf{A}\mathbf{X}_1 + c_2\mathbf{A}\mathbf{X}_2 + \cdots + c_{j-1}\mathbf{A}\mathbf{X}_{j-1}. \end{aligned} \tag{11}$$

Since $\lambda_1, \lambda_2, \ldots, \lambda_j$ are eigenvalues of $\mathbf{A}$ and $\mathbf{X}_1, \mathbf{X}_2, \ldots, \mathbf{X}_j$, its associated eigenvectors, we know that $\mathbf{A}\mathbf{X}_i = \lambda_i\mathbf{X}_i$ for $i = 1, 2, \ldots, j$. Substituting in (11), we have

$$\lambda_j\mathbf{X}_j = c_1\lambda_1\mathbf{X}_1 + c_2\lambda_2\mathbf{X}_2 + \cdots + c_{j-1}\lambda_{j-1}\mathbf{X}_{j-1}. \tag{12}$$

Multiplying (10) by λ_j, we obtain

$$\lambda_j\mathbf{X}_j = \lambda_j c_1\mathbf{X}_1 + \lambda_j c_2\mathbf{X}_2 + \cdots + \lambda_j c_{j-1}\mathbf{X}_{j-1}. \tag{13}$$

Subtracting (13) from (12), we have

$$\begin{aligned} 0 &= \lambda_j\mathbf{X}_j - \lambda_j\mathbf{X}_j \\ &= c_1(\lambda_1 - \lambda_j)\mathbf{X}_1 + c_2(\lambda_2 - \lambda_j)\mathbf{X}_2 + \cdots + c_{j-1}(\lambda_{j-1} - \lambda_j)\mathbf{X}_{j-1}. \end{aligned}$$

Since S_1 is linearly independent, we must have

$$c_1(\lambda_1 - \lambda_j) = 0, \quad c_2(\lambda_2 - \lambda_j) = 0, \ldots, c_{j-1}(\lambda_{j-1} - \lambda_j) = 0.$$

Now

$$\lambda_1 - \lambda_j \neq 0, \quad \lambda_2 - \lambda_j \neq 0, \ldots, (\lambda_{j-1} - \lambda_j) \neq 0,$$

which implies that

$$c_1 = c_2 = \cdots = c_{j-1} = 0.$$

From (10) we conclude that $\mathbf{X}_j = \mathbf{0}$, which is impossible if $\mathbf{X}_j$ is an eigenvector. Hence S is linearly independent, and from Theorem 5.2 it follows that $\mathbf{A}$ is diagonalizable.

If all the roots of the characteristic polynomial of $\mathbf{A}$ are real and not all distinct, then $\mathbf{A}$ may or may not be diagonalizable. The characteristic polynomial of $\mathbf{A}$ can be written as the product of n factors, each of the form $\lambda - \lambda_0$, where λ_0 is a root of the characteristic polynomial. Now the eigenvalues of $\mathbf{A}$ are the real roots of the characteristic polynomial of $\mathbf{A}$. Thus the characteristic polynomial can be written as

$$(\lambda - \lambda_1)^{k_1}(\lambda - \lambda_2)^{k_2} \cdots (\lambda - \lambda_r)^{k_r},$$

where $\lambda_1, \lambda_2, \ldots, \lambda_r$ are the distinct eigenvalues of $\mathbf{A}$, and $k_1, k_2, \ldots, k_r$ are integers whose sum is n. The integer k_i is called the **multiplicity** of λ_i. Thus in Example 11, $\lambda = 1$ is an eigenvalue of $\mathbf{A} = \begin{bmatrix} 1 & 1 \\ 0 & 1 \end{bmatrix}$ of multiplicity 2. It can be shown that if the roots of the characteristic polynomial of $\mathbf{A}$ are all real, then $\mathbf{A}$ can be diagonalized if, for each eigenvalue of multiplicity k, we can find k linearly independent eigenvectors. This means that the solution space of the linear system $(\lambda\mathbf{I}_n - \mathbf{A})\mathbf{X} = \mathbf{0}$ has dimension k. We consider the following examples.

Example 12. Let

$$\mathbf{A} = \begin{bmatrix} 0 & 0 & 1 \\ 0 & 1 & 2 \\ 0 & 0 & 1 \end{bmatrix}.$$

The characteristic polynomial of $\mathbf{A}$ is $f(\lambda) = \lambda(\lambda - 1)^2$, so the eigenvalues of $\mathbf{A}$ are $\lambda_1 = 0$, $\lambda_2 = 1$, and $\lambda_3 = 1$. Thus $\lambda_2 = 1$ is an eigenvalue of multiplicity 2. We now consider the eigenvectors associated with the eigenvalues $\lambda_2 = \lambda_3 = 1$. They are obtained by solving the linear system

$(1\mathbf{I}_3 - \mathbf{A})\mathbf{X} = \mathbf{0}$:

$$\begin{bmatrix} 1 & 0 & -1 \\ 0 & 0 & -2 \\ 0 & 0 & 0 \end{bmatrix} \begin{bmatrix} x_1 \\ x_2 \\ x_3 \end{bmatrix} = \begin{bmatrix} 0 \\ 0 \\ 0 \end{bmatrix}.$$

A solution is any vector of the form $\begin{bmatrix} 0 \\ r \\ 0 \end{bmatrix}$, where r is any real number, so the dimension of the solution space of $(1\mathbf{I}_3 - \mathbf{A})\mathbf{X} = \mathbf{0}$ is 1, and the set of eigenvectors is not linearly independent; $\mathbf{A}$ cannot be diagonalized.

Example 13. Let

$$\mathbf{A} = \begin{bmatrix} 0 & 0 & 0 \\ 0 & 1 & 0 \\ 1 & 0 & 1 \end{bmatrix}.$$

The characteristic polynomial of $\mathbf{A}$ is $f(\lambda) = \lambda(\lambda - 1)^2$, so the eigenvalues of $\mathbf{A}$ are $\lambda_1 = 0$, $\lambda_2 = 1$, $\lambda_3 = 1$; $\lambda_2 = 1$ is again an eigenvalue of multiplicity 2. Now we again consider the solution space of $(1\mathbf{I}_3 - \mathbf{A})\mathbf{X} = \mathbf{0}$, that is, of

$$\begin{bmatrix} 1 & 0 & 0 \\ 0 & 0 & 0 \\ -1 & 0 & 0 \end{bmatrix} \begin{bmatrix} x_1 \\ x_2 \\ x_3 \end{bmatrix} = \begin{bmatrix} 0 \\ 0 \\ 0 \end{bmatrix}.$$

A solution is any vector of the form $\begin{bmatrix} 0 \\ r \\ s \end{bmatrix}$ for any real numbers r and s. Thus we can take as eigenvectors $\mathbf{X}_2$ and $\mathbf{X}_3$ the vectors $\mathbf{X}_2 = \begin{bmatrix} 0 \\ 1 \\ 0 \end{bmatrix}$ and $\mathbf{X}_3 = \begin{bmatrix} 0 \\ 0 \\ 1 \end{bmatrix}$. Now we look for an eigenvector associated with $\lambda_1 = 0$. We have to solve $(0\mathbf{I}_3 - \mathbf{A})\mathbf{X} = \mathbf{0}$, or

$$\begin{bmatrix} 0 & 0 & 0 \\ 0 & -1 & 0 \\ -1 & 0 & -1 \end{bmatrix} \begin{bmatrix} x_1 \\ x_2 \\ x_3 \end{bmatrix} = \begin{bmatrix} 0 \\ 0 \\ 0 \end{bmatrix}.$$

A solution is any vector of the form $\begin{bmatrix} t \\ 0 \\ -t \end{bmatrix}$ for any real number t. Thus

$\mathbf{X}_1 = \begin{bmatrix} 1 \\ 0 \\ -1 \end{bmatrix}$ is an eigenvector associated with $\lambda_1 = 0$. Since $\mathbf{X}_1$, $\mathbf{X}_2$, and $\mathbf{X}_3$ are linearly independent, **A** can be diagonalized.

Thus an $n \times n$ matrix may fail to be diagonalizable either because all the roots of its characteristic polynomial are not real numbers, or because it does not have n linearly independent eigenvectors.

Eigenvalues and eigenvectors satisfy many important and interesting properties. For example, if **A** is an upper (lower) triangular matrix, then the eigenvalues of **A** are the elements on the main diagonal of **A**, since the determinant of such a matrix is the product of the elements on the main diagonal of **A**. Also, let λ be a fixed eigenvalue of **A**. The set S consisting of all eigenvectors of **A** associated with λ as well as the zero vector is a subspace of R^n (Exercise T.2) called the **eigenspace associated with** λ. Other properties are developed in the exercises of this section.

It must be pointed out that the method for finding the eigenvalues of a matrix by obtaining the roots of the characteristic polynomial is not a practical one for $n > 4$, owing to the need for evaluating a determinant. An efficient numerical method for finding eigenvalues will be presented in Section 8.3.

5.1 Exercises

In Exercises 1 through 3 find the characteristic polynomial of each matrix.

1. $\begin{bmatrix} 1 & 2 & 1 \\ 0 & 1 & 2 \\ -1 & 3 & 2 \end{bmatrix}$. **2.** $\begin{bmatrix} 2 & 1 \\ -1 & 3 \end{bmatrix}$. **3.** $\begin{bmatrix} 4 & -1 & 3 \\ 0 & 2 & 1 \\ 0 & 0 & 3 \end{bmatrix}$.

In Exercises 4 through 11 find the characteristic polynomial and the eigenvalues and the eigenvectors of each matrix.

4. $\begin{bmatrix} 0 & 1 & 2 \\ 0 & 0 & 3 \\ 0 & 0 & 0 \end{bmatrix}$. **5.** $\begin{bmatrix} 1 & 0 & 0 \\ -1 & 3 & 0 \\ 3 & 2 & -2 \end{bmatrix}$. **6.** $\begin{bmatrix} 1 & 1 \\ 1 & 1 \end{bmatrix}$.

7. $\begin{bmatrix} 1 & -1 \\ 2 & 4 \end{bmatrix}$. **8.** $\begin{bmatrix} 2 & -2 & 3 \\ 0 & 3 & -2 \\ 0 & -1 & 2 \end{bmatrix}$. **9.** $\begin{bmatrix} 2 & 2 & 3 \\ 1 & 2 & 1 \\ 2 & -2 & 1 \end{bmatrix}$.

10. $\begin{bmatrix} 2 & 0 & 0 \\ 3 & -1 & 0 \\ 0 & 4 & 3 \end{bmatrix}$. **11.** $\begin{bmatrix} 1 & 2 & 3 & 4 \\ 0 & -1 & 3 & 2 \\ 0 & 0 & 3 & 3 \\ 0 & 0 & 0 & 2 \end{bmatrix}$.

In Exercises 12 through 16 find which of the matrices are diagonalizable.

12. $\begin{bmatrix} 1 & 4 \\ 1 & -2 \end{bmatrix}$. **13.** $\begin{bmatrix} 1 & 0 \\ -2 & 1 \end{bmatrix}$. **14.** $\begin{bmatrix} 1 & 1 & -2 \\ 4 & 0 & 4 \\ 1 & -1 & 4 \end{bmatrix}$.

15. $\begin{bmatrix} 1 & 2 & 3 \\ 0 & -1 & 2 \\ 0 & 0 & 2 \end{bmatrix}$. **16.** $\begin{bmatrix} 3 & 1 & 0 \\ 0 & 3 & 1 \\ 0 & 0 & 3 \end{bmatrix}$.

In Exercises 17 through 21 find for each matrix **A**, if possible, a nonsingular matrix **P** such that $\mathbf{P}^{-1}\mathbf{AP}$ is diagonal.

17. $\begin{bmatrix} 4 & 2 & 3 \\ 2 & 1 & 2 \\ -1 & -2 & 0 \end{bmatrix}$. **18.** $\begin{bmatrix} 1 & 1 & 2 \\ 0 & 1 & 0 \\ 0 & 1 & 3 \end{bmatrix}$. **19.** $\begin{bmatrix} 1 & 2 & 3 \\ 0 & 1 & 0 \\ 2 & 1 & 2 \end{bmatrix}$.

20. $\begin{bmatrix} 0 & -1 \\ 2 & 3 \end{bmatrix}$. **21.** $\begin{bmatrix} 3 & -2 & 1 \\ 0 & 2 & 0 \\ 0 & 0 & 0 \end{bmatrix}$.

In Exercises 22 and 23 find bases for the eigenspaces associated with each eigenvalue.

22. $\begin{bmatrix} 2 & 3 & 0 \\ 0 & 1 & 0 \\ 0 & 0 & 2 \end{bmatrix}$. **23.** $\begin{bmatrix} 2 & 2 & 3 & 4 \\ 0 & 2 & 3 & 2 \\ 0 & 0 & 1 & 1 \\ 0 & 0 & 0 & 1 \end{bmatrix}$.

Theoretical Exercises

T.1. Prove:

(a) **A** is similar to **A**.

(b) If **B** is similar to **A**, then **A** is similar to **B**.

(c) If **A** is similar to **B** and **B** is similar to **C**, then **A** is similar to **C**.

T.2. Let λ be a fixed eigenvalue of **A**. Prove that the set S consisting of all eigenvectors of **A** associated with λ, as well as the zero vector, is a subspace of R^n. This subspace is called the **eigenspace associated with** $\boldsymbol{\lambda}$.

T.3. Prove that if **A** and **B** are similar matrices, they have the same characteristic polynomials and hence the same eigenvalues.

T.4. Prove that if **A** is an upper (lower) triangular matrix, then the eigenvalues of **A** are the elements on the main diagonal of **A**.

T.5. Prove that **A** and $\mathbf{A}^T$ have the same eigenvalues. What, if anything, can we say about the associated eigenvectors of **A** and $\mathbf{A}^T$?

T.6. If λ is an eigenvalue of **A** with associated eigenvector **X**, prove that λ^k is an eigenvalue of $\mathbf{A}^k = \mathbf{A} \cdot \mathbf{A} \cdots \mathbf{A}$ (k factors) with associated eigenvector **X**, where k is a positive integer.

T.7. An $n \times n$ matrix $\mathbf{A}$ is called **nilpotent** if $\mathbf{A}^k = \mathbf{0}$ for some positive integer k. Prove that if $\mathbf{A}$ is nilpotent, then the only eigenvalue of $\mathbf{A}$ is 0. (*Hint:* Use Exercise T.6.)

T.8. Let $\mathbf{A}$ be an $n \times n$ matrix.

(a) Show that $|\mathbf{A}|$ is the product of all the roots of the characteristic polynomial of $\mathbf{A}$.

(b) Show that $\mathbf{A}$ is singular if and only if 0 is an eigenvalue of $\mathbf{A}$.

T.9. Let λ be an eigenvalue of the nonsingular matrix $\mathbf{A}$ with associated eigenvector $\mathbf{X}$. Show that $1/\lambda$ is an eigenvalue of $\mathbf{A}^{-1}$ with associated eigenvector $\mathbf{X}$.

T.10. Let

$$\mathbf{A} = \begin{bmatrix} a & b \\ c & d \end{bmatrix}.$$

Find necessary and sufficient conditions for $\mathbf{A}$ to be diagonalizable.

T.11. If $\mathbf{A}$ and $\mathbf{B}$ are nonsingular, show that $\mathbf{AB}$ and $\mathbf{BA}$ are similar.

T.12. Prove that if $\mathbf{A}$ is diagonalizable, then (a) $\mathbf{A}^T$ is diagonalizable and (b) $\mathbf{A}^k$ is diagonalizable, where k is a positive integer.

5.2 Diagonalization of Symmetric Matrices

In this section we consider the diagonalization of symmetric matrices ($\mathbf{A} = \mathbf{A}^T$). We restrict our attention to this case because it is easier to handle than that of general matrices and also because symmetric matrices arise in many applied problems.

As an example of such a problem, consider the task of identifying the conic represented by the equation

$$2x^2 + 2xy + 2y^2 = 9,$$

which can be written in matrix form as

$$\begin{bmatrix} x & y \end{bmatrix} \begin{bmatrix} 2 & 1 \\ 1 & 2 \end{bmatrix} \begin{bmatrix} x \\ y \end{bmatrix} = 9.$$

Observe that the matrix used here is a symmetric matrix. This problem is handled in Section 7.2. We shall merely remark here that the solution calls for the determination of the eigenvalues and eigenvectors of the matrix

$$\begin{bmatrix} 2 & 1 \\ 1 & 2 \end{bmatrix}.$$

The x- and y-axes are then rotated to a new set of axes, which lie along the eigenvectors of the matrix. In the new set of axes, the given conic can be readily identified.

In Example 7 of Section 5.1 we saw a matrix with the property that the roots of its characteristic polynomial were complex numbers. This cannot happen for a symmetric matrix, as the following theorem asserts. We omit the proof.

THEOREM 5.4. *All the roots of the characteristic polynomial of a symmetric matrix are real numbers.*

COROLLARY 5.2. *If* **A** *is a symmetric matrix all of whose eigenvalues are distinct, then* **A** *is diagonalizable.*

proof. Since **A** is symmetric, all its eigenvalues are real. From Theorem 5.3 it now follows that **A** can be diagonalized.

THEOREM 5.5. *If* **A** *is a symmetric matrix, then eigenvectors that belong to distinct eigenvalues of* **A** *are orthogonal.*

proof. First, we shall let the reader verify (Exercise T.1) the property that if **X** and **Y** are vectors in R^n, then

$$(\mathbf{AX}) \cdot \mathbf{Y} = \mathbf{X} \cdot (\mathbf{A}^T\mathbf{Y}). \tag{1}$$

Now let $\mathbf{X}_1$ and $\mathbf{X}_2$ be eigenvectors of **A** associated with the distinct eigenvalues λ_1 and λ_2 of **A**. We then have

$$\mathbf{AX}_1 = \lambda_1\mathbf{X}_1 \quad \text{and} \quad \mathbf{AX}_2 = \lambda_2\mathbf{X}_2.$$

Now

$$\begin{aligned} \lambda_1(\mathbf{X}_1 \cdot \mathbf{X}_2) &= (\lambda_1\mathbf{X}_1) \cdot \mathbf{X}_2 = (\mathbf{AX}_1) \cdot \mathbf{X}_2 \\ &= \mathbf{X}_1 \cdot (\mathbf{A}^T\mathbf{X}_2) = \mathbf{X}_1 \cdot (\mathbf{AX}_2) \\ &= \mathbf{X}_1 \cdot (\lambda_2\mathbf{X}_2) = \lambda_2(\mathbf{X}_1 \cdot \mathbf{X}_2), \end{aligned} \tag{2}$$

where we have used the fact that $\mathbf{A} = \mathbf{A}^T$. Thus

$$\lambda_1(\mathbf{X}_1 \cdot \mathbf{X}_2) = \lambda_2(\mathbf{X}_1 \cdot \mathbf{X}_2)$$

and subtracting we obtain

$$\begin{aligned} 0 &= \lambda_1(\mathbf{X}_1 \cdot \mathbf{X}_2) - \lambda_2(\mathbf{X}_1 \cdot \mathbf{X}_2) \\ &= (\lambda_1 - \lambda_2)(\mathbf{X}_1 \cdot \mathbf{X}_2). \end{aligned} \tag{3}$$

Since $\lambda_1 \neq \lambda_2$, we conclude that $\mathbf{X}_1 \cdot \mathbf{X}_2 = 0$.

Example 1. Let

$$\mathbf{A} = \begin{bmatrix} 0 & 0 & -2 \\ 0 & -2 & 0 \\ -2 & 0 & 3 \end{bmatrix}.$$

We find that the characteristic polynomial of $\mathbf{A}$ is

$$f(\lambda) = (\lambda - 2)(\lambda - 4)(\lambda + 1),$$

so the eigenvalues of $\mathbf{A}$ are

$$\lambda_1 = -2, \qquad \lambda_2 = 4, \qquad \text{and} \qquad \lambda_3 = -1.$$

We find the associated eigenvectors by solving the linear system $(\lambda \mathbf{I}_3 - \mathbf{A})\mathbf{X} = \mathbf{0}$; and obtain the respective eigenvectors

$$\mathbf{X}_1 = \begin{bmatrix} 0 \\ 1 \\ 0 \end{bmatrix}, \qquad \mathbf{X}_2 = \begin{bmatrix} -1 \\ 0 \\ 2 \end{bmatrix}, \qquad \text{and} \qquad \mathbf{X}_3 = \begin{bmatrix} 2 \\ 0 \\ 1 \end{bmatrix}.$$

It is clear that $\{\mathbf{X}_1, \mathbf{X}_2, \mathbf{X}_3\}$ is an orthogonal set of vectors in R^3 (and is thus linearly independent by Theorem 3.15, Section 3.6). Thus $\mathbf{A}$ is diagonalizable and is similar to

$$\mathbf{D} = \begin{bmatrix} -2 & 0 & 0 \\ 0 & 4 & 0 \\ 0 & 0 & -1 \end{bmatrix}.$$

We recall that if $\mathbf{A}$ can be diagonalized, then there exists a nonsingular matrix $\mathbf{P}$ such that $\mathbf{P}^{-1}\mathbf{A}\mathbf{P}$ is diagonal. Moreover, the columns of $\mathbf{P}$ are eigenvectors of $\mathbf{A}$. Now, if the eigenvectors of $\mathbf{A}$ form an orthogonal set S, as happens when $\mathbf{A}$ is symmetric and the eigenvalues of $\mathbf{A}$ are distinct, then since any scalar multiple of an eigenvector of **A** is also an eigenvector of **A**, we can normalize S to obtain an orthonormal set $T = \{\mathbf{X}_1, \mathbf{X}_2, \ldots, \mathbf{X}_n\}$ of eigenvectors of **A**. The jth column of **P** is the eigenvector $\mathbf{X}_j$ associated with λ_j, and we now examine what type of matrix **P** must be. We can write **P** as

$$\mathbf{P} = [\mathbf{X}_1, \mathbf{X}_2, \ldots, \mathbf{X}_n].$$

Then

$$\mathbf{P}^T = \begin{bmatrix} \mathbf{X}_1^T \\ \mathbf{X}_2^T \\ \cdot \\ \cdot \\ \cdot \\ \mathbf{X}_n^T \end{bmatrix},$$

where $\mathbf{X}_i^T$, $1 \leqslant i \leqslant n$, is the transpose of the $n \times 1$ matrix (or vector) $\mathbf{X}_i$. We find that the i, jth entry in $\mathbf{P}^T\mathbf{P}$ is $\mathbf{X}_i \cdot \mathbf{X}_j$ (verify). Since

$$\begin{aligned} \mathbf{X}_i \cdot \mathbf{X}_j &= 1 \qquad \text{if } i = j \\ &= 0 \qquad \text{if } i \neq j, \end{aligned}$$

then $\mathbf{P}^T\mathbf{P} = \mathbf{I}_n$. Thus $\mathbf{P}^T = \mathbf{P}^{-1}$. Such matrices are important enough to have a special name.

DEFINITION A nonsingular matrix **A** is called **orthogonal** if

$$\mathbf{A}^{-1} = \mathbf{A}^T.$$

Of course, we can also say that a nonsingular matrix is orthogonal if $\mathbf{A}^T\mathbf{A} = \mathbf{I}_n$.

Example 2. Let

$$\mathbf{A} = \begin{bmatrix} \frac{2}{3} & -\frac{2}{3} & \frac{1}{3} \\ \frac{2}{3} & \frac{1}{3} & -\frac{2}{3} \\ \frac{1}{3} & \frac{2}{3} & \frac{2}{3} \end{bmatrix}.$$

It is easy to check that $\mathbf{A}^T\mathbf{A} = \mathbf{I}_n$.

Example 3. Let **A** be the matrix of Example 1. We already know that the set of eigenvectors

$$\left\{\begin{bmatrix}0\\1\\0\end{bmatrix}, \begin{bmatrix}-1\\0\\2\end{bmatrix}, \begin{bmatrix}2\\0\\1\end{bmatrix}\right\}$$

is orthogonal. If we normalize these vectors, we find that

$$T = \left\{\begin{bmatrix}0\\1\\0\end{bmatrix}, \begin{bmatrix}-\frac{1}{\sqrt{5}}\\0\\\frac{2}{\sqrt{5}}\end{bmatrix}, \begin{bmatrix}\frac{2}{\sqrt{5}}\\0\\\frac{1}{\sqrt{5}}\end{bmatrix}\right\}$$

is an orthonormal set of vectors. The matrix **P** such that $\mathbf{P}^{-1}\mathbf{AP}$ is diagonal is the matrix whose columns are the vectors in T. Thus

$$\mathbf{P} = \begin{bmatrix}0 & -\frac{1}{\sqrt{5}} & \frac{2}{\sqrt{5}}\\1 & 0 & 0\\0 & \frac{2}{\sqrt{5}} & \frac{1}{\sqrt{5}}\end{bmatrix}.$$

We leave it to the reader to verify (Exercise 4) that **P** is an orthogonal matrix and that

$$\mathbf{P}^{-1}\mathbf{AP} = \mathbf{P}^T\mathbf{AP} = \begin{bmatrix}-2 & 0 & 0\\0 & 4 & 0\\0 & 0 & -1\end{bmatrix}.$$

The following theorem is not difficult to show, and we leave its proof to the reader (Exercise T.2).

THEOREM 5.6. *The $n \times n$ matrix **A** is orthogonal if and only if the columns of **A** form an orthonormal set of vectors in R^n.*

If **A** is an orthogonal matrix, then it is easy to show that $|\mathbf{A}| = \pm 1$ (Exercise T.3). We now look at the geometrical implications of orthogonal matrices. Thus, if **A** is an orthogonal $n \times n$ matrix, then consider the linear transformation $\mathbf{L} : R^n \to R^n$ defined by $\mathbf{L}(\mathbf{X}) = \mathbf{AX}$, for $\mathbf{X}$ in R^n. We now compute $\mathbf{L}(\mathbf{X}) \cdot \mathbf{L}(\mathbf{Y})$ for any vectors $\mathbf{X}$ and $\mathbf{Y}$ in R^n. We have, using Equation (1) and the fact that $\mathbf{A}^T\mathbf{A} = \mathbf{I}_n$,

$$\mathbf{L}(\mathbf{X}) \cdot \mathbf{L}(\mathbf{Y}) = (\mathbf{AX}) \cdot (\mathbf{AY}) = \mathbf{X} \cdot (\mathbf{A}^T\mathbf{AY}) = \mathbf{X} \cdot (\mathbf{I}_n\mathbf{Y}) = \mathbf{X} \cdot \mathbf{Y}.$$

This means that $\mathbf{L}$ preserves length. It is, of course, clear that if θ is the angle between vectors $\mathbf{X}$ and $\mathbf{Y}$ in R^n, then the angle between $\mathbf{L}(\mathbf{X})$ and $\mathbf{L}(\mathbf{Y})$ is also θ (Exercise T.4). Conversely, let $\mathbf{L} : R^n \to R^n$ be a linear transformation that preserves length; that is,

$$\mathbf{L}(\mathbf{X}) \cdot \mathbf{L}(\mathbf{Y}) = \mathbf{X} \cdot \mathbf{Y},$$

for any $\mathbf{X}$ and $\mathbf{Y}$ in R^n. Let $\mathbf{A}$ be the matrix of $\mathbf{L}$ with respect to the natural basis for R^n. Then

$$\mathbf{L}(\mathbf{X}) = \mathbf{AX}.$$

If $\mathbf{X}$ and $\mathbf{Y}$ are any vectors in R^n, we have, using (1),

$$\mathbf{X} \cdot \mathbf{Y} = \mathbf{L}(\mathbf{X}) \cdot \mathbf{L}(\mathbf{Y}) = (\mathbf{AX}) \cdot (\mathbf{AY}) = \mathbf{X} \cdot (\mathbf{A}^T\mathbf{AY}).$$

Since this holds for all $\mathbf{Y}$ in R^n, then by Exercise T.8 of Section 3.2, we conclude that

$$\mathbf{A}^T\mathbf{AY} = \mathbf{Y}$$

for any $\mathbf{Y}$ in R^n. It then follows that $\mathbf{A}^T\mathbf{A} = \mathbf{I}_n$ (Exercise T.9), so that $\mathbf{A}$ is an orthogonal matrix. Other properties of orthogonal matrices are examined by the theoretical exercises.

We now turn to the general situation for a symmetric matrix; even if $\mathbf{A}$ has eigenvalues whose multiplicities are greater than one, it turns out that we can still diagonalize $\mathbf{A}$. We omit the proof of the following theorem.

THEOREM 5.7. *If* $\mathbf{A}$ *is a symmetric* $n \times n$ *matrix, then there exists an orthogonal matrix* $\mathbf{P}$ *such that* $\mathbf{P}^{-1}\mathbf{AP} = \mathbf{D}$, *a diagonal matrix. The eigenvalues of* $\mathbf{A}$ *lie on the main diagonal of* $\mathbf{D}$.

It can be shown that if $\mathbf{A}$ has an eigenvalue λ of multiplicity k, then the solution space of the linear system $(\lambda\mathbf{I}_n - \mathbf{A})\mathbf{X} = \mathbf{0}$ (the eigenspace of λ) has dimension k. This means that there exist k linearly independent eigenvectors of $\mathbf{A}$ associated with the eigenvalue λ. We can, of course, choose an orthonormal basis for this solution space. Thus we obtain a set of k orthonormal eigenvectors associated with the eigenvalue λ. Since eigenvectors associated with distinct eigenvalues are orthogonal, if we form the set of all eigenvectors we get an orthonormal set. Hence the matrix $\mathbf{P}$ whose columns are the eigenvectors is orthogonal.

Example 4. Let

$$\mathbf{A} = \begin{bmatrix} 0 & 2 & 2 \\ 2 & 0 & 2 \\ 2 & 2 & 0 \end{bmatrix}.$$

The characteristic polynomial of A is

$$f(\lambda) = (\lambda + 2)^2(\lambda - 4),$$

so the eigenvalues are

$$\lambda_1 = -2, \qquad \lambda_2 = -2, \qquad \text{and} \qquad \lambda_3 = 4.$$

That is, -2 is an eigenvalue whose multiplicity is 2. To find the eigenvectors associated with λ_1 and λ_2, we solve the homogeneous linear system $(-2\mathbf{I}_3 - \mathbf{A})\mathbf{X} = \mathbf{0}$:

$$\begin{bmatrix} -2 & -2 & -2 \\ -2 & -2 & -2 \\ -2 & -2 & -2 \end{bmatrix} \begin{bmatrix} x_1 \\ x_2 \\ x_3 \end{bmatrix} = \begin{bmatrix} 0 \\ 0 \\ 0 \end{bmatrix}. \tag{4}$$

A basis for the solution space of (4) consists of the eigenvectors

$$\mathbf{X}_1 = \begin{bmatrix} -1 \\ 1 \\ 0 \end{bmatrix} \qquad \text{and} \qquad \mathbf{X}_2 = \begin{bmatrix} -1 \\ 0 \\ 1 \end{bmatrix}.$$

Now $\mathbf{X}_1$ and $\mathbf{X}_2$ are not orthogonal, since $\mathbf{X}_1 \cdot \mathbf{X}_2 \neq 0$. We can use the Gram–Schmidt process to obtain an orthonormal basis for the solution space of (4) (the eigenspace of $\lambda_1 = -2$) as follows. Let

$$\mathbf{Y}_1 = \mathbf{X}_1 = \begin{bmatrix} -1 \\ 1 \\ 0 \end{bmatrix}$$

and

$$\mathbf{Y}_2 = \mathbf{X}_2 - \left(\frac{\mathbf{X}_2 \cdot \mathbf{Y}_1}{\mathbf{Y}_1 \cdot \mathbf{Y}_1}\right)\mathbf{X}_1 = \begin{bmatrix} -\frac{1}{2} \\ -\frac{1}{2} \\ 1 \end{bmatrix}.$$

Let

$$\mathbf{Y}_2^* = 2\mathbf{Y}_2 = \begin{bmatrix} -1 \\ -1 \\ 2 \end{bmatrix}.$$

The set $\{\mathbf{Y}_1, \mathbf{Y}_2^*\}$ is an orthogonal set of vectors. Normalizing these eigenvectors, we obtain

$$\mathbf{Z}_1 = \frac{\mathbf{Y}_1}{|\mathbf{Y}_1|} = \frac{1}{\sqrt{2}}\begin{bmatrix} -1 \\ 1 \\ 0 \end{bmatrix} \quad \text{and} \quad \mathbf{Z}_2 = \frac{\mathbf{Y}_2^*}{|\mathbf{Y}_2^*|} = \frac{1}{\sqrt{6}}\begin{bmatrix} -1 \\ -1 \\ 2 \end{bmatrix}.$$

The set $\{\mathbf{Z}_1, \mathbf{Z}_2\}$ is an orthonormal basis of eigenvectors of $\mathbf{A}$ for the solution space of (4). Now we find a basis for the solution space of $(4\mathbf{I}_3 - \mathbf{A})\mathbf{X} = \mathbf{0}$,

$$\begin{bmatrix} 4 & -2 & -2 \\ -2 & 4 & -2 \\ -2 & -2 & 4 \end{bmatrix}\begin{bmatrix} x_1 \\ x_2 \\ x_3 \end{bmatrix} = \begin{bmatrix} 0 \\ 0 \\ 0 \end{bmatrix}, \tag{5}$$

to consist of

$$\mathbf{X}_3 = \begin{bmatrix} 1 \\ 1 \\ 1 \end{bmatrix}.$$

Normalizing this vector, we have the eigenvector

$$\mathbf{Z}_3 = \frac{1}{\sqrt{3}}\begin{bmatrix} 1 \\ 1 \\ 1 \end{bmatrix}$$

as an orthonormal basis for the solution space of (5). Since eigenvectors associated with distinct eigenvalues are orthogonal, we observe that $\mathbf{Z}_3$ is orthogonal to both $\mathbf{Z}_1$ and $\mathbf{Z}_2$. Thus the set $\{\mathbf{Z}_1, \mathbf{Z}_2, \mathbf{Z}_3\}$ is an orthonormal basis of R^3 consisting of eigenvectors of $\mathbf{A}$. The matrix $\mathbf{P}$ is the matrix whose jth column is $\mathbf{Z}_j$:

$$\mathbf{P} = \begin{bmatrix} -\frac{1}{\sqrt{2}} & -\frac{1}{\sqrt{6}} & \frac{1}{\sqrt{3}} \\ \frac{1}{\sqrt{2}} & -\frac{1}{\sqrt{6}} & \frac{1}{\sqrt{3}} \\ 0 & \frac{2}{\sqrt{6}} & \frac{1}{\sqrt{3}} \end{bmatrix}.$$

We leave it to the reader to verify that

$$\mathbf{P}^{-1}\mathbf{AP} = \mathbf{P}^T\mathbf{AP} = \begin{bmatrix} -2 & 0 & 0 \\ 0 & -2 & 0 \\ 0 & 0 & 4 \end{bmatrix}.$$

Example 5. Let

$$\mathbf{A} = \begin{bmatrix} 1 & 2 & 0 & 0 \\ 2 & 1 & 0 & 0 \\ 0 & 0 & 1 & 2 \\ 0 & 0 & 2 & 1 \end{bmatrix}.$$

The characteristic polynomial of $\mathbf{A}$ is

$$f(\lambda) = (\lambda + 1)^2(\lambda - 3)^2,$$

so the eigenvalues of $\mathbf{A}$ are

$$\lambda_1 = -1, \quad \lambda_2 = -1, \quad \lambda_3 = 3, \quad \text{and} \quad \lambda_4 = 3.$$

We find (verify) that a basis for the solution space of

$$(-1\mathbf{I}_3 - \mathbf{A})\mathbf{X} = \mathbf{0} \tag{6}$$

consists of the eigenvectors

$$\mathbf{X}_1 = \begin{bmatrix} 1 \\ -1 \\ 0 \\ 0 \end{bmatrix} \quad \text{and} \quad \mathbf{X}_2 = \begin{bmatrix} 0 \\ 0 \\ 1 \\ -1 \end{bmatrix},$$

which are orthogonal. Normalizing these eigenvectors, we obtain

$$\mathbf{Z}_1 = \begin{bmatrix} \frac{1}{\sqrt{2}} \\ -\frac{1}{\sqrt{2}} \\ 0 \\ 0 \end{bmatrix} \quad \text{and} \quad \mathbf{Z}_2 = \begin{bmatrix} 0 \\ 0 \\ \frac{1}{\sqrt{2}} \\ -\frac{1}{\sqrt{2}} \end{bmatrix}$$

as an orthonormal basis of eigenvectors for the solution space of (6). We also find (verify) that a basis for the solution space of

$$(3\mathbf{I}_3 - \mathbf{A})\mathbf{X} = \mathbf{0} \tag{7}$$

consists of the eigenvectors

$$\mathbf{X}_3 = \begin{bmatrix} 1 \\ 1 \\ 0 \\ 0 \end{bmatrix} \quad \text{and} \quad \mathbf{X}_4 = \begin{bmatrix} 0 \\ 0 \\ 1 \\ 1 \end{bmatrix},$$

which are orthogonal. Normalizing these eigenvectors, we obtain

$$\mathbf{Z}_3 = \begin{bmatrix} \frac{1}{\sqrt{2}} \\ \frac{1}{\sqrt{2}} \\ 0 \\ 0 \end{bmatrix} \quad \text{and} \quad \mathbf{Z}_4 = \begin{bmatrix} 0 \\ 0 \\ \frac{1}{\sqrt{2}} \\ \frac{1}{\sqrt{2}} \end{bmatrix}$$

as an orthonormal basis of eigenvectors for the solution space of (7). Since eigenvectors associated with distinct eigenvalues are orthogonal, we conclude that

$$\{\mathbf{Z}_1, \mathbf{Z}_2, \mathbf{Z}_3, \mathbf{Z}_4\}$$

is an orthonormal basis of R^4 consisting of eigenvectors of **A**. The matrix

$\mathbf{P}$ is the matrix whose jth column is $\mathbf{Z}_j$:

$$\mathbf{P} = \begin{bmatrix} \frac{1}{\sqrt{2}} & 0 & \frac{1}{\sqrt{2}} & 0 \\ -\frac{1}{\sqrt{2}} & 0 & \frac{1}{\sqrt{2}} & 0 \\ 0 & \frac{1}{\sqrt{2}} & 0 & \frac{1}{\sqrt{2}} \\ 0 & -\frac{1}{\sqrt{2}} & 0 & \frac{1}{\sqrt{2}} \end{bmatrix}.$$

Suppose now that $\mathbf{A}$ is an $n \times n$ matrix for which we can find an orthogonal matrix $\mathbf{P}$ such that $\mathbf{P}^{-1}\mathbf{AP}$ is a diagonal matrix $\mathbf{D}$. Thus $\mathbf{P}^{-1}\mathbf{AP} = \mathbf{D}$, or $\mathbf{A} = \mathbf{PDP}^{-1}$. Since $\mathbf{P}^{-1} = \mathbf{P}^T$, we can write $\mathbf{A} = \mathbf{PDP}^T$. Then $\mathbf{A}^T = (\mathbf{PDP}^T)^T = (\mathbf{P}^T)^T\mathbf{D}^T\mathbf{P}^T = \mathbf{PDP}^T = \mathbf{A}$ ($\mathbf{D} = \mathbf{D}^T$, since $\mathbf{D}$ is a diagonal matrix). Thus $\mathbf{A}$ is symmetric.

Some remarks about nonsymmetric matrices are in order at this point. Theorem 5.3 assures us that $\mathbf{A}$ is diagonalizable if all the roots of its characteristic polynomial are real and distinct. We also studied examples, in Section 5.1, of nonsymmetric matrices with repeated eigenvalues that were diagonalizable and others that were not diagonalizable. There are some striking differences between the symmetric and nonsymmetric cases, which we now summarize. If $\mathbf{A}$ is nonsymmetric, then the roots of its characteristic polynomial need not all be real numbers; if an eigenvalue λ has multiplicity k, then the solution space of $(\lambda\mathbf{I}_n - \mathbf{A})\mathbf{X} = \mathbf{0}$ may have dimension $< k$; if the roots of the characteristic polynomial of $\mathbf{A}$ are all real, it is still possible for $\mathbf{A}$ not to have n linearly independent eigenvectors (which means that $\mathbf{A}$ cannot be diagonalized); eigenvectors associated with distinct eigenvalues need not be orthogonal. Thus, in Example 13 of Section 5.1, the eigenvectors $\mathbf{X}_1$ and $\mathbf{X}_3$ associated with the eigenvalues $\lambda_1 = 0$ and $\lambda_3 = 1$ are not orthogonal. If a matrix $\mathbf{A}$ cannot be diagonalized, then we can often find a matrix $\mathbf{B}$ similar to $\mathbf{A}$ which is "nearly diagonal." The matrix $\mathbf{B}$ is said to be in **Jordan canonical form**; its treatment lies beyond the scope of this book but is studied in advanced books on linear algebra [for example, K. Hoffman and R. Kunze, *Linear Algebra*, 2nd ed. Prentice-Hall, Inc., (Englewood Cliffs, N. J.: 1971)].

It should also be noted that, in many applications, we need only find a diagonal matrix $\mathbf{D}$ that is similar to the given matrix $\mathbf{A}$; that is, we do not need the orthogonal matrix $\mathbf{P}$ such that $\mathbf{P}^{-1}\mathbf{AP} = \mathbf{D}$. Many of the matrices to be diagonalized in applied problems are either symmetric, or all the roots of their characteristic polynomial are real. Of course, the methods for finding eigenvalues that have been presented in this chapter are not recommended for matrices with more than four rows because of the need to evaluate determinants. In Section 8.3 we shall consider an efficient numerical method for diagonalizing a symmetric matrix.

5.2 Exercises

1. Verify that

$$\mathbf{P} = \begin{bmatrix} \frac{2}{3} & -\frac{2}{3} & \frac{1}{3} \\ \frac{2}{3} & \frac{1}{3} & -\frac{2}{3} \\ \frac{1}{3} & \frac{2}{3} & \frac{2}{3} \end{bmatrix}$$

is an orthogonal matrix.

2. Find the inverse of each of the following orthogonal matrices

$$\text{(a) } \mathbf{A} = \begin{bmatrix} 1 & 0 & 0 \\ 0 & \cos\theta & \sin\theta \\ 0 & -\sin\theta & \cos\theta \end{bmatrix}. \qquad \text{(b) } \mathbf{B} = \begin{bmatrix} 1 & 0 & 0 \\ 0 & \dfrac{1}{\sqrt{2}} & -\dfrac{1}{\sqrt{2}} \\ 0 & -\dfrac{1}{\sqrt{2}} & -\dfrac{1}{\sqrt{2}} \end{bmatrix}.$$

3. Verify Theorem 5.6 for the matrices in Exercise 2.

4. Verify that the matrix $\mathbf{P}$ in Example 3 is an orthogonal matrix.

5. For the orthogonal matrix $\mathbf{A} = \begin{bmatrix} \dfrac{1}{\sqrt{2}} & -\dfrac{1}{\sqrt{2}} \\ -\dfrac{1}{\sqrt{2}} & -\dfrac{1}{\sqrt{2}} \end{bmatrix}$, verify that $(\mathbf{AX}) \cdot (\mathbf{AY}) = \mathbf{X} \cdot \mathbf{Y}$ for any $\mathbf{X}$ and $\mathbf{Y}$ in R^2.

6. Let $\mathbf{L} : R^2 \to R^2$ be the linear transformation performing a counterclockwise rotation through 45°; and let $\mathbf{A}$ be the matrix of $\mathbf{L}$ with respect to the natural basis for R^2. Show that $\mathbf{A}$ is orthogonal.

In Exercises 7 through 12 diagonalize each given matrix $\mathbf{A}$ and find an orthogonal matrix $\mathbf{P}$ such that $\mathbf{P}^{-1}\mathbf{AP}$ is diagonal.

7. $\begin{bmatrix} 2 & 2 \\ 2 & 2 \end{bmatrix}$. **8.** $\begin{bmatrix} 0 & 0 & 1 \\ 0 & 0 & 0 \\ 1 & 0 & 0 \end{bmatrix}$. **9.** $\begin{bmatrix} 0 & 0 & 0 \\ 0 & 2 & 2 \\ 0 & 2 & 2 \end{bmatrix}$.

10. $\begin{bmatrix} 0 & 0 & 0 & 0 \\ 0 & 0 & 0 & 0 \\ 0 & 0 & 0 & 1 \\ 0 & 0 & 1 & 0 \end{bmatrix}$. 11. $\begin{bmatrix} 0 & -1 & -1 \\ -1 & 0 & -1 \\ -1 & -1 & 0 \end{bmatrix}$. 12. $\begin{bmatrix} -1 & 2 & 2 \\ 2 & -1 & 2 \\ 2 & 2 & -1 \end{bmatrix}$.

In Exercises 13 through 20 diagonalize each given matrix.

13. $\begin{bmatrix} 2 & 1 \\ 1 & 2 \end{bmatrix}$. 14. $\begin{bmatrix} 2 & 2 & 0 & 0 \\ 2 & 2 & 0 & 0 \\ 0 & 0 & 2 & 2 \\ 0 & 0 & 2 & 2 \end{bmatrix}$. 15. $\begin{bmatrix} 1 & 1 & 0 \\ 1 & 1 & 0 \\ 0 & 0 & 1 \end{bmatrix}$.

16. $\begin{bmatrix} 1 & 0 & 0 \\ 0 & 3 & -2 \\ 0 & -2 & 3 \end{bmatrix}$. 17. $\begin{bmatrix} 1 & 0 & 0 \\ 0 & 1 & 1 \\ 0 & 1 & 1 \end{bmatrix}$. 18. $\begin{bmatrix} 0 & 0 & 0 & 1 \\ 0 & 0 & 0 & 0 \\ 0 & 0 & 0 & 0 \\ 1 & 0 & 0 & 0 \end{bmatrix}$.

19. $\begin{bmatrix} 1 & -1 & 2 \\ -1 & 1 & 2 \\ 2 & 2 & 2 \end{bmatrix}$. 20. $\begin{bmatrix} -3 & 0 & -1 \\ 0 & -2 & 0 \\ -1 & 0 & -3 \end{bmatrix}$.

Theoretical Exercises

T.1. Show that if $\mathbf{X}$ and $\mathbf{Y}$ are vectors in R^n, then $(\mathbf{AX}) \cdot \mathbf{Y} = \mathbf{X} \cdot (\mathbf{A}^T\mathbf{Y})$.

T.2. Prove Theorem 5.6.

T.3. Show that if $\mathbf{A}$ is an orthogonal matrix, then $|\mathbf{A}| = \pm 1$.

T.4. Let $\mathbf{A}$ be an $n \times n$ orthogonal matrix and let $\mathbf{L} : R^n \to R^n$ be the linear transformation defined by $\mathbf{L}(\mathbf{X}) = \mathbf{AX}$, for $\mathbf{X}$ in R^n. Let θ be the angle between vectors $\mathbf{X}$ and $\mathbf{Y}$ in R^n. Prove that if $\mathbf{A}$ is orthogonal, then the angle between $\mathbf{L}(\mathbf{X})$ and $\mathbf{L}(\mathbf{Y})$ is also θ.

T.5. Prove Theorem 5.7 for the 2×2 case by studying the possible roots of the characteristic polynomial of $\mathbf{A}$.

T.6. Show that if $\mathbf{A}$ and $\mathbf{B}$ are orthogonal matrices, then $\mathbf{AB}$ is an orthogonal matrix.

T.7. Show that if $\mathbf{A}$ is an orthogonal matrix, then $\mathbf{A}^{-1}$ is orthogonal.

T.8. (a) Verify that the matrix $\begin{bmatrix} \cos\theta & \sin\theta \\ -\sin\theta & \cos\theta \end{bmatrix}$ is orthogonal.

(b) Prove that if $\mathbf{A}$ is an orthogonal 2×2 matrix, then there exists a real number θ such that

$$\mathbf{A} = \pm\begin{bmatrix} \cos\theta & \sin\theta \\ -\sin\theta & \cos\theta \end{bmatrix}.$$

T.9. Prove that if $\mathbf{A}^T\mathbf{AY} = \mathbf{Y}$ for any $\mathbf{Y}$ in R^n, then $\mathbf{A}^T\mathbf{A} = \mathbf{I}_n$.

6 Linear Programming

In this chapter we provide an introduction to the basic ideas and techniques of linear programming. Linear programming is a new area of applied mathematics, developed in the late 1940s to solve a number of problems for the federal government and has been applied to amazingly numerous problems in many areas. As a vital tool in management science and operations research, it has resulted in enormous savings of money. In the first section we give some examples of linear programming problems, formulate their mathematical models, and describe a geometric solution method. In the second section we present an algebraic method for solving linear programming problems.

6.1 The Linear Programming Problem; Geometric Solution

In many problems in business and industry we are interested in making decisions that will maximize or minimize some quantity. For example, a plant manager may want to determine the most economical way of shipping his goods from his factory to his markets, a hospital may want to design a diet satisfying certain nutritional requirements at a minimum cost, an investor may want to select investments that will maximize his profits, or a manufacturer may wish to blend ingredients, subject to given specifications, to maximize his profit. In this section we give several examples of linear programming problems and show how mathematical models for them can be formulated. Their geometric solution is also considered in this section.

Example 1 (A Production Problem). A small local farmer makes ice milk and ice cream each day. Suppose that each quart of ice milk requires 0.4 quart of milk and 0.2 quart of cream, while each quart of ice cream requires 0.2 quart of milk and 0.4 quart of cream. Suppose also that the profit on each quart of ice milk is 8 cents and that it is 10 cents on each quart of ice cream. If the farmer has 10 quarts of milk and 14 quarts of cream on hand each day, how many quarts of ice milk and how many quarts of ice cream should he make each day to maximize his profit?

mathematical formulation. Let x be the number of quarts of ice milk to be made and let y be the number of quarts of ice cream to be made. Since each quart of ice milk contains 0.4 quart of milk and each quart of ice cream contains 0.2 quart of milk, the total amount of milk required is

$$0.4x + 0.2y.$$

Similarly, since each quart of ice milk contains 0.2 quart of cream and each quart of ice cream contains 0.4 quart of cream, the total amount of cream required is

$$0.2x + 0.4y.$$

Since we only have 10 quarts of milk and 14 quarts of cream on hand, we must have

$$0.4x + 0.2y \leqslant 10$$

$$0.2x + 0.4y \leqslant 14.$$

Of course, x and y cannot be negative, so we must also have

$$x \geqslant 0 \quad \text{and} \quad y \geqslant 0.$$

Since the profit on each quart of ice milk is 8 cents and the profit on each quart of ice cream is 10 cents, the total profit (in cents) is

$$z = 8x + 10y.$$

Our problem can be stated in mathematical form as: Find values of x and y that will maximize

$$z = 8x + 10y$$

subject to the following restrictions that must be satisfied by x and y:

$$\begin{aligned} 0.4x + 0.2y &\leqslant 10 \\ 0.2x + 0.4y &\leqslant 14 \\ x &\geqslant 0 \\ y &\geqslant 0. \end{aligned}$$

Example 2 (The Transportation Problem). Suppose that a plastics plant manufactures 50 tons of Styrofoam per week. The weekly demand at two retail outlets, A and B, is 20 tons and 15 tons, respectively. The cost of transporting a ton of the product from the plant to the retail outlet is: $50 to A and $60 to B. Find how many tons of the product should be sent to each retail outlet to minimize the shipping cost.

mathematical formulation. Let x denote the number of tons of Styrofoam sent to outlet A and let y denote the number of tons sent to outlet B. Then the total amount sent from the plant is

$$x + y.$$

Since the plant only makes 50 tons of Styrofoam per week, we must have

$$x + y \leqslant 50.$$

Also, since the demand at outlet A is 20 tons, we need at least 20 tons of x to be delivered there; that is, we must have

$$x \geqslant 20.$$

Similarly, since the demand at outlet B is 15 tons, we must have

$$y \geqslant 15.$$

Of course, x and y cannot be negative, so we require

$$x \geqslant 0$$
$$y \geqslant 0.$$

The total transportation cost (in dollars) is

$$z = 50x + 60y,$$

which we want to minimize. Thus a mathematical statement of our problem is: Find values of x and y that will minimize

$$z = 50x + 60y$$

subject to the following restrictions that must be satisfied by x and y:

$$\begin{aligned} x + y &\leqslant 50 \\ x &\geqslant 20 \\ y &\geqslant 15 \\ x &\geqslant 0 \\ y &\geqslant 0. \end{aligned}$$

Example 3 (The Diet Problem). A nutritionist is planning a menu that includes foods A and B as its main staples. Suppose that each ounce of food A contains 2 units of protein, 1 unit of iron, and 1 unit of thiamine; each ounce of food B contains 1 unit of protein, 1 unit of iron, and 3 units of thiamine. Suppose that each ounce of A costs 30 cents, while each ounce of B costs 40 cents. The nutritionist wants the meal to provide at least 12 units of protein, at least 9 units of iron, and at least 15 units of thiamine. How many ounces of each of the foods should be used to minimize the cost of the meal?

mathematical formulation. Let x denote the number of ounces of food A to be used and let y be the number of ounces of food B. The number of units of protein supplied by the meal is

$$2x + y,$$

so we must have

$$2x + y \geqslant 12.$$

The number of units of iron supplied by the meal is

$$x + y,$$

so we must have

$$x + y \geqslant 9.$$

Since the number of units of thiamine supplied by the meal is

$$x + 3y,$$

we must have

$$x + 3y \geqslant 15.$$

Of course, we also require

$$x \geqslant 0, \qquad y \geqslant 0.$$

The cost of the meal (in cents) is

$$z = 30x + 40y,$$

which we want to minimize. Thus a mathematical formulation of our problem is: Find values of x and y that will minimize

$$z = 30x + 40y$$

subject to the restrictions

$$\begin{aligned} 2x + y &\geqslant 12 \\ x + y &\geqslant 9 \\ x + 3y &\geqslant 15 \\ x &\geqslant 0 \\ y &\geqslant 0. \end{aligned}$$

We now see that a linear programming problem has the following general form: Find values of $x_1, x_2, \ldots, x_n$ that will minimize or maximize

$$z = c_1x_1 + c_2x_2 + \cdots + c_nx_n \tag{1}$$

subject to

$$\begin{cases} a_{11}x_1 + a_{12}x_2 + \cdots + a_{1n}x_n (\leqslant)(\geqslant)(=)\, b_1 \\ a_{21}x_1 + a_{22}x_2 + \cdots + a_{2n}x_n (\leqslant)(\geqslant)(=)\, b_2 \\ \vdots \\ a_{m1}x_1 + a_{m2}x_2 + \cdots + a_{mn}x_n (\leqslant)(\geqslant)(=)\, b_m \end{cases} \tag{2}$$

$$x_j \geqslant 0 \qquad \text{for } j = 1, 2, \ldots, n, \tag{3}$$

where in (2) one and only one of the symbols $\leqslant$, $=$, $\geqslant$ occurs in each inequality. The linear function in (1) is called the **objective function**. The equalities or inequalities in (2) and (3) are called **constraints**. The term **linear** in linear programming means that the objective function (1) and each of the constraints in (2) are linear functions of the variables $x_1, x_2, \ldots, x_n$. The word "programming," *not* to be confused with its use in computer programming, refers to the applications in planning or allocation problems.

Geometric Solution

We now develop a geometric method for solving linear programming problems in two variables. This will enable us to solve the problems posed earlier. Since linear programming deals with systems of linear inequalities, we first consider these from a geometric point of view.

Consider Example 1. Find the set of points that maximize

$$z = 8x + 10y \tag{4}$$

subject to the constraints

$$\begin{aligned} 0.4x + 0.2y &\leqslant 10 \\ 0.2x + 0.4y &\leqslant 14 \\ x &\geqslant 0 \\ y &\geqslant 0. \end{aligned} \tag{5}$$

The set of points satisfying the system of four inequalities (5) consists of those points satisfying each of the inequalities. In Figure 6.1(a) we show the set of points satisfying the inequality $x \geqslant 0$, in Figure 6.1(b) the set of points satisfying the inequality $y \geqslant 0$. The set of points satisfying both inequalities $x \geqslant 0$ and $y \geqslant 0$ is the intersection of the regions in Figure 6.1(a) and (b). This set of points, the set of all points in the first quadrant, is shown in Figure 6.1(c).

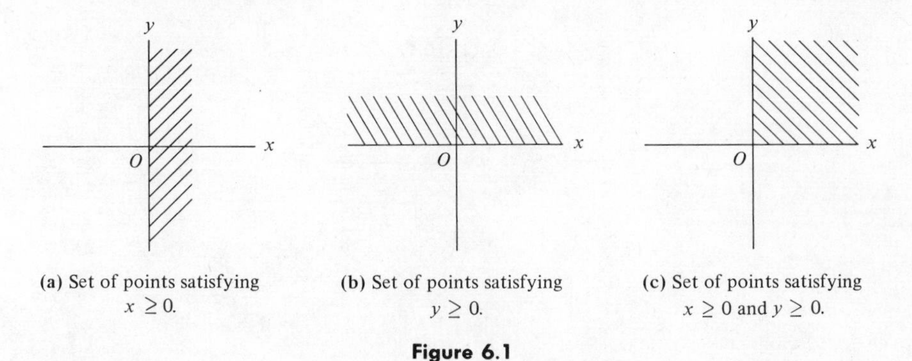

(a) Set of points satisfying $x \geq 0$.

(b) Set of points satisfying $y \geq 0$.

(c) Set of points satisfying $x \geq 0$ and $y \geq 0$.

Figure 6.1

We must now consider the set of points satisfying the inequality

$$0.4x + 0.2y \leqslant 10. \tag{6}$$

First, consider the set of points satisfying the strict inequality

$$0.4x + 0.2y < 10. \tag{7}$$

Now the straight line

$$0.4x + 0.2y = 10 \tag{8}$$

divides the set of points not on the line into two regions (Figure 6.2). The line itself, drawn in dashed form, belongs to neither region.

To determine the region determined by inequality (7), we choose any point not on the line as a test point and see on which side of the line it lies. The point (10, 20) does not lie on the line (8) and can thus serve as a test point. Since its coordinates satisfy inequality (7), the correct region determined by (7) is shown in Figure 6.3(a). Another possible test point is (20, 20), since it does not lie on the line. Its coordinates do not satisfy inequality (7), so the test point does not lie in the correct region. Thus the correct region would again be as shown in Figure 6.3(a).

The set of points satisfying inequality (6) consists of the set of points satisfying (7) as well as the points on the line (8). Thus the region, shown in Figure 6.3(b), includes the straight line, which is drawn solidly.

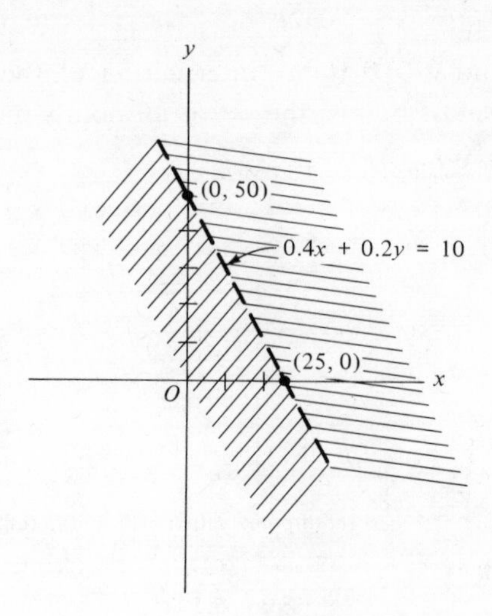

Figure 6.2

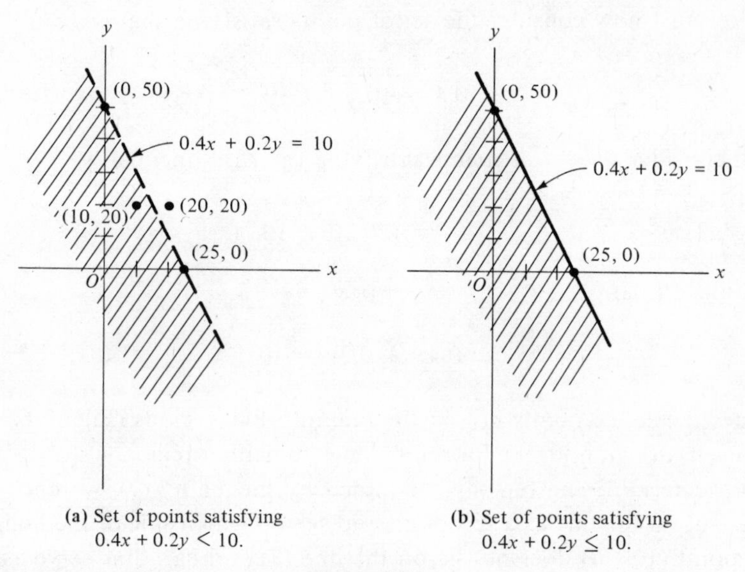

(a) Set of points satisfying $0.4x + 0.2y < 10$.

(b) Set of points satisfying $0.4x + 0.2y \leq 10$.

Figure 6.3

Next, the set of points satisfying the inequality

$$0.2x + 0.4y \leqslant 14 \tag{9}$$

is shown in Figure 6.4.

The set of points satisfying inequalities (6) and (9) is the intersection of

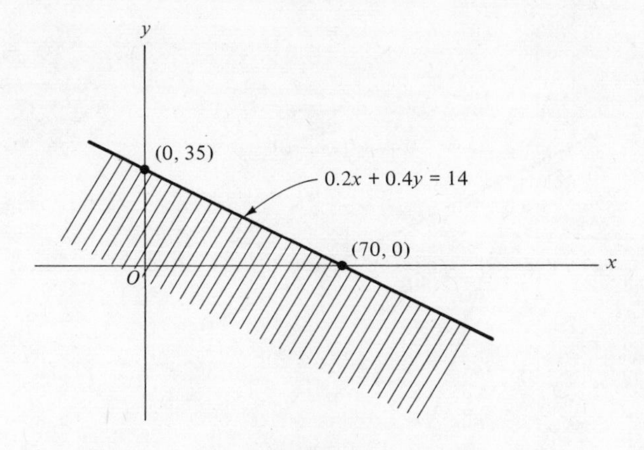

Figure 6.4

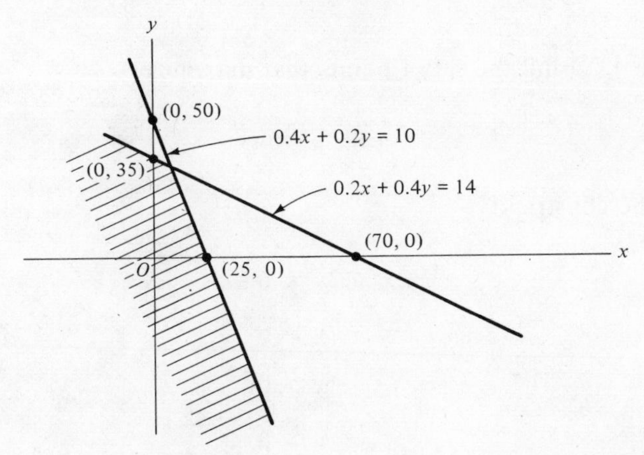

Figure 6.5 The set of points satisfying $0.4x + 0.2y \leqslant 10$ and $0.2x + 0.4y \leqslant 14$.

the regions in Figures 6.3(b) and 6.4; it is the shaded region shown in Figure 6.5.

Finally, the set of points satisfying the four inequalities (5) is the intersection of the regions in Figures 6.1(c) and 6.5. That is, it is the set of points in Figure 6.5 that lie in the first quadrant. This set of points is shown in Figure 6.6.

Thus the set of points satisfying a system of inequalities is the intersection of the sets of points satisfying each of the inequalities.

A point satisfying the constraints of a linear programming problem is called a **feasible solution**; the set of all such points is the **feasible region**. It should be noted that not every linear programming problem has a feasible region, as the next example shows.

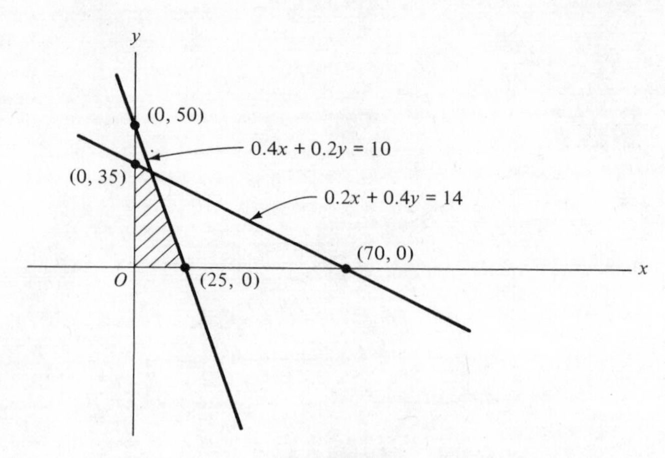

Figure 6.6

Example 4. Find the set of points that maximize

$$z = 8x + 10y \tag{10}$$

subject to the constraints

$$\begin{aligned} 2x + 3y &\leqslant 6 \\ x + 2y &\geqslant 6 \\ x &\geqslant 0 \\ y &\geqslant 0. \end{aligned} \tag{11}$$

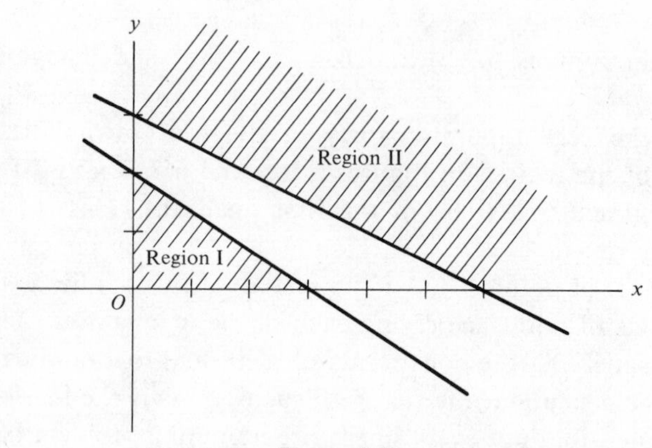

Figure 6.7

In Figure 6.7 we have labeled as region I the set of points satisfying the first, third, and fourth inequalities; that is, region I is the set of points in the first quadrant satisfying the first inequality in (11). Similarly, region II is the set of points satisfying the second, third, and fourth inequalities. It is clear that there are no points satisfying all four inequalities.

To solve our given linear programming problem, we must find a feasible solution that makes the objective function (1) as large as possible. Such a solution is called an **optimal solution**. Since there are infinitely many feasible points, it appears, at first glance, quite difficult to tell whether a feasible solution is optimal or not. In Figure 6.8 we have reproduced Figure 6.6 by showing the feasible region for our linear programming problem along with a number of indicated feasible points: O, A, B, C, D, and E. Points F and G are not in the feasible region.

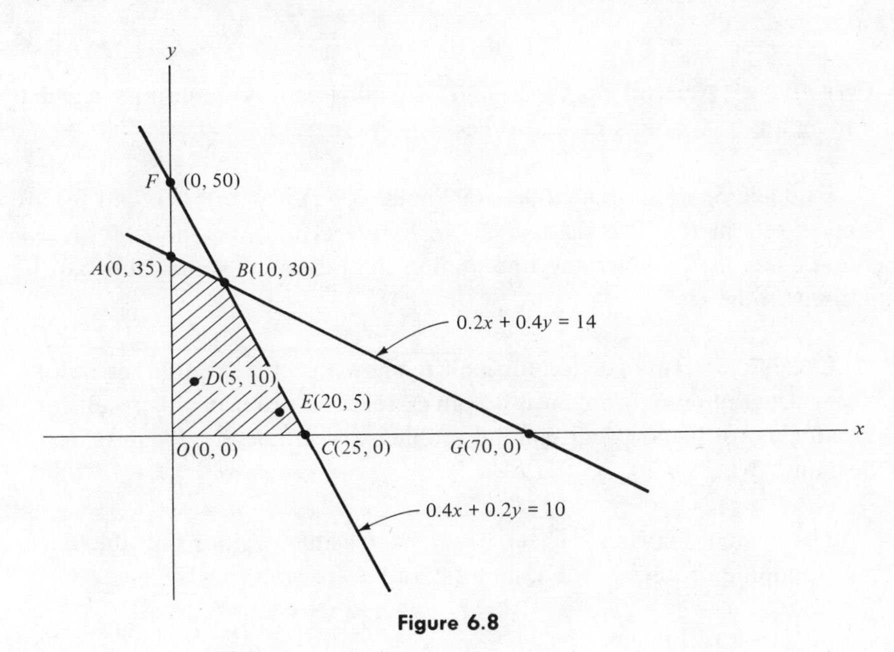

Figure 6.8

In Table 6.1 we have tabulated the value of the objective function (in cents) for each of the points O, A, B, C, D, and E. Among the feasible solutions O, A, B, C, D, and E we see that the value of the objective function is largest for the point $B(10, 30)$. However, we do not know if there is another feasible solution at which the value of the objective function will be larger than it is at B. Since there are infinitely many feasible solutions in the region, we cannot examine the value of the objective function at every one of these. However, we shall be able to find optimal solutions without examining all feasible solutions, but we first consider some auxiliary notions.

TABLE 6.1

Point	Value of $z = 8x + 10y$ (cents)
$O(0, 0)$	0
$A(0, 35)$	350
$B(10, 30)$	380
$C(25, 0)$	200
$D(5, 10)$	140
$E(20, 5)$	210

DEFINITION A set of points S in R^n is called **convex** if the line segment joining any two points in S also lies entirely in S.

Example 5. The shaded sets in Figure 6.9(a), (b), (c), (d), and (e) are convex sets in R^2. The shaded sets in Figure 6.10(a), (b), and (c) are *not* convex sets in R^2, since the line joining the indicated points does not lie entirely in the set.

Example 6. It is not too difficult to show that the feasible region of a linear programming problem is a convex set. The proof of this result for a broad class of linear programming problems is outlined in Exercise T.1 of Section 6.2.

The reader may also check that the feasible regions for the linear programming problems in Examples 2 and 3 are convex.

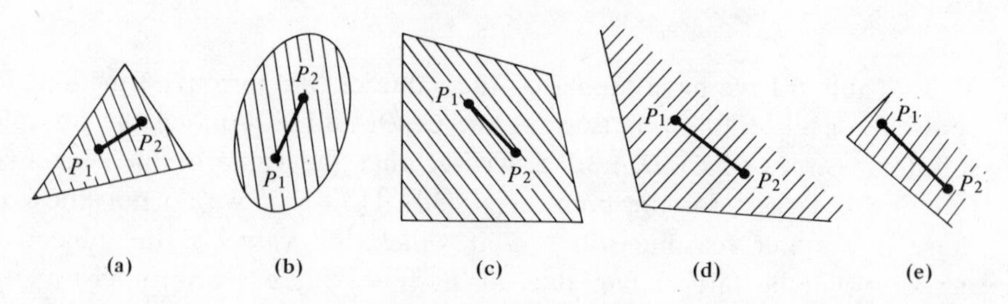

Figure 6.9 Convex sets in R^2.

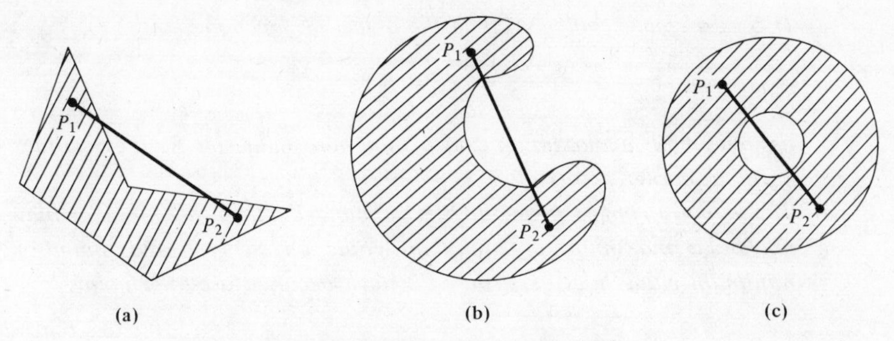

Figure 6.10 Nonconvex sets in R^2.

We shall now limit our further discussion of convex sets to such sets in R^2, although the ideas presented here can be generalized to convex sets in R^n.

Convex sets are either **bounded** or **unbounded**. A bounded convex set is one that can be enclosed by a rectangle; an unbounded convex set is one that cannot be so enclosed. The convex sets in Figure 6.9(a), (b), and (c) are bounded; the convex sets in Figure 6.9(d) and (e) are unbounded. A bounded convex set is called a **convex polygon**.

DEFINITION A **corner point** in a convex set S is a point *in* S that is the intersection of two boundary lines.

Example 7. In Figure 6.11 we have marked the corner points of the convex sets in Figure 6.9(a), (c), and (d). The convex set in Figure 6.9(b) has no corner points.

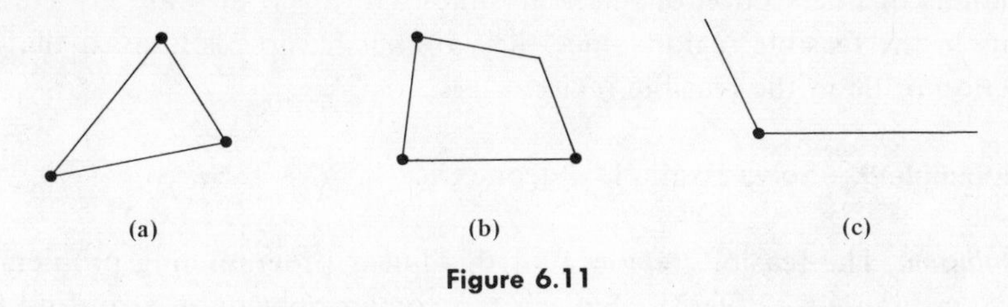

Figure 6.11

The basic result connecting convex sets, corner points, and linear programming problems is the following theorem, whose proof we omit.

THEOREM 6.1. *Let S be the feasible region of a linear programming problem.*

(a) *If S is a convex polygon, then the objective function*

$$z = ax + by$$

assumes both a maximum and a minimum value on S; these values occur at corner points of S.

(b) *If S is not a convex polygon (S is unbounded), then there may or may not be a maximum or minimum value on S. If a maximum or minimum value does exist on S, it must occur at a corner point.*

Thus, when the feasible region S of a linear programming problem is a convex polygon, a method for solving the problem consists in finding the corner points of S and evaluating the value of the objective function $z = ax + by$ at each corner point. An optimal solution is a corner point at which the value of z is a maximum.

Example 8. Consider Example 1 again. The feasible region shown in Figure 6.8 is a convex polygon and its corner points are $O(0, 0)$, $A(0, 35)$, $B(10, 30)$, and $C(25, 0)$. From Table 6.1 we see that the value of z is a maximum at the corner point $B(10, 30)$. Thus the optimal solution is

$$x = 10 \quad \text{and} \quad y = 30.$$

This means that the farmer should make 10 quarts of ice milk and 30 quarts of ice cream to maximize his profit. If he follows this course of action, his maximum profit will be $3.80 each day.

Referring back to Figure 6.8, we note that the points (70, 0) and (0, 50) are points of intersection of boundary lines. However, they are not corner points in the feasible region, since they are *not* feasible solutions; that is, they do not lie in the feasible region.

Example 9. Solve Example 2 geometrically.

solution. The feasible region S of this linear programming problem is shown in Figure 6.12 (verify). Since S is a convex polygon, we can find the minimum value of z by evaluating z at each of the corner points. In Table 6.2 we have tabulated the value of the objective function for each of the points $A(20, 15)$, $B(20, 30)$, and $C(35, 15)$. The value of z is a minimum at the corner point $A(20, 15)$. Thus the optimal solution is

$$x = 20 \quad \text{and} \quad y = 15,$$

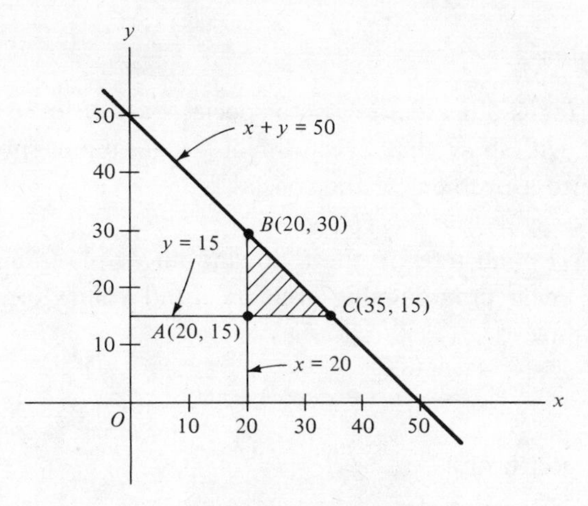

Figure 6.12

which means that the manufacturer should send 20 tons of Styrofoam to outlet A and 15 tons to outlet B to minimize his shipping cost. If he follows this course of action, his minimum shipping cost will be \$1900.

TABLE 6.2

Point	Value of $z = 50x + 60y$ (dollars)
$A(20, 15)$	1900
$B(20, 30)$	2800
$C(35, 15)$	2650

The feasible region of Example 3 is not a convex polygon (verify). We shall not deal further with such problems in this book.

In general, a linear programming problem may have

1. No feasible solution; that is, there are no points satisfying all the constraints in Equations (2) and (3);
2. A unique optimal solution;
3. More than one optimal solution (see Exercise 4); or
4. No maximum (or minimum) value in the feasible region; that is, it may be possible to choose a point in the feasible region to make the objective function as large (or as small) as we please.

Slack Variables

We shall now focus our attention on a special class of linear programming problems and will show that every linear programming problem can be transformed into a problem of the special type.

DEFINITION We shall refer to the following linear programming problem as a **standard linear programming problem**. Find values of $x_1, x_2, \ldots, x_n$ that will maximize

$$z = c_1x_1 + c_2x_2 + \cdots + c_nx_n \tag{12}$$

subject to the constraints

$$\left\{\begin{aligned} a_{11}x_1 + a_{12}x_2 + \cdots + a_{1n}x_n &\leqslant b_1 \\ a_{21}x_1 + a_{22}x_2 + \cdots + a_{2n}x_n &\leqslant b_2 \\ &\vdots \\ a_{m1}x_1 + a_{m2}x_2 + \cdots + a_{mn}x_n &\leqslant b_m. \end{aligned}\right. \tag{13}$$

$$x_j \geqslant 0 \quad \text{for} \quad j = 1, 2, \ldots, n, \tag{14}$$

Example 10. Example 1 is a standard linear programming problem.

Example 11. The linear programming problem:

$$\text{Minimize } z = 3x - 4y$$

subject to

$$\begin{aligned} 2x - 3y &\leqslant 6 \\ x + y &\leqslant 8 \\ x &\geqslant 0 \\ y &\geqslant 0 \end{aligned}$$

is *not* a standard linear programming problem, since the objective function is to be minimized and not maximized.

Example 12. The linear programming problem:

$$\text{Maximize } z = 12x - 15y$$

subject to

$$3x - y \geqslant 4$$
$$2x + 3y \leqslant 6$$
$$x \geqslant 0$$
$$y \geqslant 0$$

is *not* a standard linear programming problem, since one of the inequalities is of the form $\geqslant$; in a standard linear programming problem every inequality must be of the form $\leqslant$.

Example 13. The linear programming problem:

$$\text{Maximize } z = 8x + 10y$$

subject to

$$3x + y = 4$$
$$2x - 3y \leqslant 5$$
$$x \geqslant 0$$
$$y \geqslant 0$$

is not a standard linear programming problem, since the first constraint is an equation and not an inequality of the form $\leqslant$.

Every linear programming problem can be transformed into a standard linear programming problem.

Minimization Problem as Maximization Problem

Every maximization problem can be viewed as a minimization problem, and conversely. This follows from the observation that

$$\begin{aligned} &\text{minimum of } c_1x_1 + c_2x_2 + \cdots + c_nx_n \\ &\qquad = \text{maximum of } -(c_1x_1 + c_2x_2 + \cdots + c_nx_n). \end{aligned} \tag{15}$$

Reversing an Inequality

Consider the inequality

$$d_1x_1 + d_2x_2 + \cdots + d_nx_n \geqslant -b.$$

Multiplying both sides of this inequality by -1 reverses the inequality, yielding

$$-d_1x_1 - d_2x_2 - \cdots - d_nx_n \leqslant b.$$

Example 14. Consider the linear programming problem:

$$\text{Maximize } w = 8x + 10y$$

subject to

$$\begin{aligned} -0.4x - 0.2y &\geqslant -10 \\ 0.2x + 0.4y &\leqslant 14 \\ x &\geqslant 0 \\ y &\geqslant 0. \end{aligned} \tag{16}$$

Multiplying the first inequality in (16) by (-1), we obtain the standard linear programming problem of Example 1.

It is not too difficult to change a number of equalities into inequalities of the form $\leqslant$. Thus Example 13 can be transformed into a standard linear programming problem. In this book we shall not encounter linear programming problems of the type in Example 13, so we shall not continue to examine such problems.

It is not easy to handle, algebraically, systems of linear inequalities. However, as we have seen in Chapter 1, it is not difficult at all to deal with systems of linear equations. Accordingly, we shall change our given standard linear programming problem into a problem where we must find nonnegative variables maximizing a linear objective function and satisfying a system of linear equations. Every solution to the given problem yields a solution to the new problem, and conversely, every solution to the new problem yields a solution to the given problem.

Consider the constraint

$$d_1x_1 + d_2x_2 + \cdots + d_nx_n \leqslant b. \tag{17}$$

Since the left side of (17) is not larger than the right side, we can make (17) into an equation by adding the unknown nonnegative quantity u to its left side, to obtain

$$d_1x_1 + d_2x_2 + \cdots + d_nx_n + u = b. \tag{18}$$

The quantity u in (18) is called a **slack variable**, since it takes up the slack between the two sides of the inequality.

We now change each of the constraints in (2) into an equation by introducing a nonnegative slack variable. Thus the ith inequality

$$a_{i1}x_1 + a_{i2}x_2 + \cdots + a_{in}x_n \leqslant b_i \qquad (1 \leqslant i \leqslant m) \tag{19}$$

is converted into the equation

$$a_{i1}x_1 + a_{i2}x_2 + \cdots + a_{in}x_n + x_{n+i} = b_i \qquad (1 \leqslant i \leqslant m),$$

by introducing the nonnegative slack variable x_{n+i}. Our new problem can now be stated as follows.

New Problem

Find values of $x_1, x_2, \ldots, x_n, x_{n+1}, \ldots, x_{n+m}$ that will maximize

$$z = c_1x_1 + c_2x_2 + \cdots + c_nx_n \tag{20}$$

subject to

$$\left\{\begin{array}{llll} a_{11}x_1 + a_{12}x_2 + \cdots + a_{1n}x_n + x_{n+1} & & & = b_1 \\ a_{21}x_1 + a_{22}x_2 + \cdots + a_{2n}x_n & + x_{n+2} & & = b_2 \\ \vdots & & \ddots & \vdots \\ a_{m1}x_1 + a_{m2}x_2 + \cdots + a_{mn}x_n & & + x_{n+m} & = b_m \end{array}\right. \tag{21}$$

$$x_1 \geqslant 0, \ldots, x_n \geqslant 0, x_{n+1} \geqslant 0, x_{n+2} \geqslant 0, \ldots, x_{n+m} \geqslant 0. \tag{22}$$

Thus the new problem has $m + n$ equations in $m + n$ unknowns. Solving the original problem is equivalent to solving the new problem in the following sense. If $x_1, x_2, \ldots, x_n$ is a feasible solution to the given problem as defined by (1), (2), and (3), then

$$x_1 \geqslant 0, \quad x_2 \geqslant 0, \ldots, x_n \geqslant 0.$$

Also, $x_1, x_2, \ldots, x_n$ satisfy each of the constraints in (2). Let x_{n+i},

$1 \leqslant i \leqslant m$, be defined by

$$x_{n+i} = b_i - a_{i1}x_1 - a_{i2}x_2 - \cdots - a_{in}x_n.$$

That is, x_{n+i} is the difference between the right side of inequality (19) and its left side. Then

$$x_{n+1} \geqslant 0, \quad x_{n+2} \geqslant 0, \ldots, x_{n+m} \geqslant 0,$$

so that $x_1, x_2, \ldots, x_n, x_{n+1}, \ldots, x_{n+m}$ satisfy (21) and (22).

Conversely, suppose that $x_1, x_2, \ldots, x_n, x_{n+1}, \ldots, x_{n+m}$ satisfy (21) and (22). It is then clear that $x_1, x_2, \ldots, x_n$ satisfy (2) and (3).

Example 15. Consider the problem of Example 1.

Introducing the slack variables v and w, we formulate our new problem as: Find values of x, y, v, and w that will maximize

$$z = 8x + 10y$$

subject to

$$\begin{aligned} 0.4x + 0.2y + v \qquad &= 10 \\ 0.2x + 0.4y \qquad + w &= 14 \end{aligned}$$

$$x \geqslant 0, \quad y \geqslant 0, \quad v \geqslant 0, \quad w \geqslant 0.$$

The slack variable v is the difference between the total amount of milk on hand, 10 quarts, and the amount $0.4x + 0.2y$ of milk actually used. The slack variable w is the difference between the total amount of cream on hand, 14 quarts, and the amount $0.2x + 0.4y$ of cream actually used.

Consider the feasible solution to the given problem

$$x = 5, \qquad y = 10,$$

which represents point D in Figure 6.8. We then obtain the slack variables

$$\begin{aligned} v &= 10 - 0.4(5) - 0.2(10) \\ &= 10 - 2 - 2 = 6 \end{aligned}$$

and

$$\begin{aligned} w &= 14 - 0.2(5) - 0.4(10) \\ &= 14 - 1 - 4 = 9, \end{aligned}$$

so that

$$x = 5, \quad y = 10, \quad v = 6, \quad \text{and} \quad w = 9$$

is a feasible solution to the new problem. Of course, the solution $x = 5$, $y = 10$ is not an optimal solution, since $z = 8(5) + 10(10) = 140$, and we recall that the maximum value of z is attained for

$$x = 10 \qquad \text{and} \qquad y = 30.$$

In this case, the corresponding optimal solution to the new problem is

$$x = 10, \quad y = 30, \quad v = 0, \quad \text{and} \quad w = 0.$$

6.1 Exercises

In Exercises 1 through 9 formulate mathematically each linear programming problem.

1. A steel producer makes two types of steel: regular and special. A ton of regular steel requires 2 hours in the open-hearth furnace and 5 hours in the soaking pit; a ton of special steel requires 2 hours in the open-hearth furnace and 3 hours in the soaking pit. The open-hearth furnace is available 8 hours per day and the soaking pit is available 15 hours per day. The profit on a ton of regular steel is \$42 and it is \$50 on a ton of special steel. Determine how many tons of each type of steel should be made to maximize the profit.
2. A trust fund is planning to invest up to \$6000 in two types of bonds: A and B. Bond A is safer than bond B and carries a dividend of 8 per cent, and bond B carries a dividend of 10 per cent. Suppose that the fund's rules state that no more than \$4000 may be invested in bond B, while at least \$1500 must be invested in bond A. How much should be invested in each type of bond to maximize the fund's return?
3. Solve Exercise 2 if the fund has the following additional rule: "The amount invested in bond B cannot exceed one half the amount invested in bond A."
4. A trash-removal company carries industrial waste in sealed containers in its fleet of trucks. Suppose that each container from the Smith Corporation weighs 6 pounds and is 3 cubic feet in volume, while each container from the Johnson Corporation weighs 12 pounds and is 1 cubic foot in volume. The company charges the Smith Corporation 30 cents for each container carried on a trip, and 60 cents for each container from the Johnson Corporation. If a truck cannot carry more than 18,000 pounds

or more than 1800 cubic feet in volume, how many containers from each customer should the company carry in a truck on each trip to maximize the revenue per truckload?

5. A television producer designs a program that will include a comedian, musical interludes, and must include time for commercials. The advertiser insists on at least 2 minutes of advertising time, the station insists on no more than 4 minutes of advertising time, and the comedian insists on at least 24 minutes of the comedy program. Also, the total time allotted for the advertising and comedy portions of the program cannot exceed 30 minutes. If it has been determined that each minute of advertising (very creative) attracts 40,000 viewers and each minute of the comedy program attracts 45,000 viewers, how should the time be divided between advertising and programming to maximize the number of viewers?
6. A small generator burns two types of fuel: low sulfur (L) and high sulfur (H) to produce electricity. For each hour of use, each gallon of L emits 3 units of sulfur dioxide, generates 4 kilowatts, and costs 60 cents, while each gallon of H emits 5 units of sulfur dioxide, generates 4 kilowatts, and costs 50 cents. The environmental protection agency insists that the maximum amount of sulfur dioxide that can be emitted per hour is 15 units. Suppose that at least 16 kilowatts must be generated per hour. How many gallons of L and how many gallons of H should be used hourly to minimize the cost of the fuel used?
7. The Protein Diet Club serves a luncheon consisting of two dishes, A and B. Suppose that each unit of **A** has 1 gram of fat, 1 gram of carbohydrate, and 4 grams of protein, whereas each unit of B has 2 grams of fat, 1 gram of carbohydrate, and 6 grams of protein. If the dietician planning the luncheon wants to provide no more than 10 grams of fat or more than 7 grams of carbohydrate, how many units of A and how many units of B should be served to maximize the amount of protein consumed?
8. In designing a new transportation system, a company is considering two types of buses, types A and B. A type A bus can carry 40 passengers and requires 2 mechanics for servicing, a type B bus can carry 60 passengers and requires 3 mechanics for servicing. Suppose that the company must transport at least 300 people daily and that insurance rules for the size of the garage allow no more than 12 mechanics on the payroll. If each type A bus costs \$20,000 and each type B bus costs \$25,000, how many buses of each type should be bought to minimize the cost?
9. An animal feed producer mixes two types of grain: A and B. Each unit of grain A contains 2 grams of fat, 1 gram of protein, and 80 calories. Each unit of grain B contains 3 grams of fat, 3 grams of protein, and 60 calories. Suppose that the producer wants each unit of the final product to yield at least 18 grams of fat, at least 12 grams of protein, and at least 480 calories. If each unit of A costs 10 cents and each unit of B costs 12

cents, how many units of each type of grain should the producer use to minimize his cost?

In Exercises 10 through 13 sketch the set of points satisfying the given system of inequalities.

10.
$$\begin{aligned} x &\leqslant 6 \\ x &\geqslant 2 \\ y &\leqslant 4 \\ y &\geqslant 1 \\ x + y &\leqslant 6. \end{aligned}$$

11.
$$\begin{aligned} 2x - y &\leqslant 6 \\ 2x + y &\leqslant 10 \\ x &\geqslant 0 \\ y &\geqslant 0. \end{aligned}$$

12.
$$\begin{aligned} x + y &\leqslant 3 \\ 5x + 4y &\geqslant 20 \\ x &\geqslant 0 \\ y &\geqslant 0. \end{aligned}$$

13.
$$\begin{aligned} x + y &\geqslant 4 \\ x + 4y &\geqslant 8 \\ x &\geqslant 0 \\ y &\geqslant 0. \end{aligned}$$

In Exercises 14 and 15 solve the given linear programming problem geometrically.

14. Maximize $z = 3x + 2y$
subject to
$$\begin{aligned} 2x - 3y &\leqslant 6 \\ x + y &\leqslant 4 \\ x &\geqslant 0 \\ y &\geqslant 0. \end{aligned}$$

15. Minimize $z = 3x - y$
subject to
$$\begin{aligned} -3x + 2y &\leqslant 6 \\ 5x + 4y &\geqslant 20 \\ 8x + 3y &\leqslant 24 \\ x &\geqslant 0 \\ y &\geqslant 0. \end{aligned}$$

16. Solve the problem in Exercise 1 geometrically.
17. Solve the problem in Exercise 2 geometrically.
18. Solve the problem in Exercise 3 geometrically.
19. Solve the problem in Exercise 4 geometrically.
20. Solve the problem in Exercise 5 geometrically.
21. Solve the problem in Exercise 6 geometrically.
22. Solve the problem in Exercise 7 geometrically.
23. Solve the problem in Exercise 8 geometrically.
24. Which of the following are standard linear programming problems?

(a) Maximize $z = 2x - 3y$
subject to
$$\begin{aligned} 2x - 3y &\leqslant 4 \\ 3x + 2y &\geqslant 5. \end{aligned}$$

(b) Maximize $z = 2x + 3y$
subject to
$$\begin{aligned} 2x + 3y &\leqslant 4 \\ 3x + 2y &\leqslant 5 \\ x &\geqslant 0 \\ y &\geqslant 0. \end{aligned}$$

(c) Minimize $z = 2x_1 - 3x_2 + x_3$
subject to

$$\begin{aligned} 2x_1 + 3x_2 + 2x_3 &\leqslant 6 \\ 3x_1 \qquad - 2x_3 &\leqslant 4 \\ x_1 &\leqslant 0 \\ x_2 &\geqslant 0 \\ x_3 &\geqslant 0. \end{aligned}$$

(d) Maximize $z = 2x + 2y$
subject to

$$\begin{aligned} 2x + 3y &\leqslant 4 \\ 3x &\leqslant 5 \\ x &\geqslant 0. \end{aligned}$$

25. Which of the following are standard linear programming problems?

(a) Maximize $z = 3x_1 + 2x_2 + x_3$
subject to

$$\begin{aligned} 2x_1 + 3x_2 + x_3 &\leqslant 4 \\ 3x_1 - 2x_2 \qquad &\leqslant 5 \\ x_1 &\geqslant 0 \\ x_2 &\geqslant 0 \\ x_3 &\geqslant 0. \end{aligned}$$

(b) Minimize $z = 2x + 3y$
subject to

$$\begin{aligned} 2x + 4y &\leqslant 2 \\ 3x + 2y &\leqslant 4 \\ x &\geqslant 0 \\ y &\geqslant 0. \end{aligned}$$

(c) Maximize $z = 3x_1 + 4x_2 + x_3$
subject to

$$\begin{aligned} 2x_1 + 4x_2 + x_3 &\leqslant 2 \\ 3x_1 - 2x_2 + x_3 &\leqslant 4 \\ 2x_1 \qquad + x_3 &\leqslant 8 \\ x_1 &\geqslant 0 \\ x_2 &\geqslant 0. \end{aligned}$$

(d) Maximize $z = 2x_1 + 3x_2 + x_3$
subject to

$$\begin{aligned} 2x_1 + 3x_2 + 5x_3 &\leqslant 8 \\ 3x_1 - 2x_2 + 2x_3 &= 4 \\ 2x_1 + x_2 + 3x_3 &\leqslant 6 \\ x_1 &\geqslant 0 \\ x_2 &\geqslant 0 \\ x_3 &\geqslant 0. \end{aligned}$$

In Exercises 26 and 27 formulate each problem as a standard linear programming problem.

26. Minimize $z = -2x_1 + 3x_2 + 2x_3$
subject to

$$\begin{aligned} 2x_1 + x_2 + 2x_3 &\leqslant 12 \\ x_1 + x_2 - 3x_3 &\leqslant 8 \\ x_1 &\geqslant 0 \\ x_2 &\geqslant 0 \\ x_3 &\geqslant 0. \end{aligned}$$

27. Maximize $z = 3x_1 - x_2 + 6x_3$

subject to

$$\begin{aligned} 2x_1 + 4x_2 + x_3 &\leq 4 \\ -3x_1 + 2x_2 - 3x_3 &\geq -4 \\ 2x_1 + x_2 - x_3 &\leq 8 \\ x_1 &\geq 0 \\ x_2 &\geq 0 \\ x_3 &\geq 0. \end{aligned}$$

In Exercises 28 and 29 formulate the given linear programming problem as a new problem with slack variables.

28. Maximize $z = 2x + 8y$
subject to

$$\begin{aligned} 2x + 3y &\leq 18 \\ 3x - 2y &\leq 6 \\ x &\geq 0 \\ y &\geq 0. \end{aligned}$$

29. Maximize $z = 2x_1 + 3x_2 + 7x_3$
subject to

$$\begin{aligned} 3x_1 + x_2 - 4x_3 &\leq 3 \\ x_1 - 2x_2 + 6x_3 &\leq 21 \\ x_1 - x_2 - x_3 &\leq 9 \\ x_1 &\geq 0 \\ x_2 &\geq 0 \\ x_3 &\geq 0. \end{aligned}$$

6.2 The Simplex Method

The simplex method for solving linear programming problems was developed by George B. Dantzig, presently at Stanford University, in connection with his work on planning problems for the federal government. In this section we present the essential features of the method, illustrating them with examples. A number of proofs will be omitted, and for further details the interested reader may consult the references given at the end of this chapter. It is convenient to introduce matrix terminology into our further discussion of linear programming.

Matrix Notation

We again restrict our attention to the standard linear programming problem: Maximize

$$c_1x_1 + c_2x_2 + \cdots + c_nx_n \tag{1}$$

subject to

$$\begin{cases} a_{11}x_1 + a_{12}x_2 + \cdots + a_{1n}x_n \leqslant b_1 \\ a_{21}x_1 + a_{22}x_2 + \cdots + a_{2n}x_n \leqslant b_2 \\ \quad\vdots \qquad\qquad\qquad\qquad\qquad \vdots \\ a_{m1}x_1 + a_{m2}x_2 + \cdots + a_{mn}x_n \leqslant b_m \end{cases} \tag{2}$$

$$x_j \geqslant 0, \qquad \text{for } j = 1, 2, \ldots, n. \tag{3}$$

If we let

$$\mathbf{A} = \begin{bmatrix} a_{11} & a_{12} & \cdots & a_{1n} \\ a_{21} & a_{22} & \cdots & a_{2n} \\ \vdots & \vdots & & \vdots \\ a_{m1} & a_{m2} & \cdots & a_{mn} \end{bmatrix}, \qquad \mathbf{B} = \begin{bmatrix} b_1 \\ b_2 \\ \vdots \\ b_m \end{bmatrix}, \qquad \mathbf{X} = \begin{bmatrix} x_1 \\ x_2 \\ \vdots \\ x_n \end{bmatrix},$$

and

$$\mathbf{C} = \begin{bmatrix} c_1 & c_2 \cdots & c_n \end{bmatrix},$$

then the given problem can be stated as follows: Find a vector $\mathbf{X}$ in R^n that will maximize the objective function

$$z = \mathbf{CX} \tag{4}$$

subject to

$$\mathbf{AX} \leqslant \mathbf{B} \tag{5}$$

$$\mathbf{X} \geqslant \mathbf{0}, \tag{6}$$

where $\mathbf{X} \geqslant \mathbf{0}$ means that each entry of $\mathbf{X}$ is nonnegative and $\mathbf{AX} \leqslant \mathbf{B}$

means that each entry of **AX** is less than or equal to the corresponding entry in **B**.

A vector **X** in R^n satisfying (5) and (6) is called a **feasible solution** to the given problem, and a feasible solution maximizing the objective function (4) is called an **optimal solution**.

Example 1. We can write the problem in Example 1 of Section 6.1 in matrix form as follows: Find a vector **X** in R^2 that will maximize

$$z = \begin{bmatrix} 8 & 10 \end{bmatrix} \begin{bmatrix} x \\ y \end{bmatrix}$$

subject to

$$\begin{bmatrix} 0.4 & 0.2 \\ 0.2 & 0.4 \end{bmatrix} \begin{bmatrix} x \\ y \end{bmatrix} \leqslant \begin{bmatrix} 10 \\ 14 \end{bmatrix}$$

$$\begin{bmatrix} x \\ y \end{bmatrix} \geqslant \begin{bmatrix} 0 \\ 0 \end{bmatrix}.$$

Feasible solutions are the vectors

$$\begin{bmatrix} 0 \\ 0 \end{bmatrix}, \quad \begin{bmatrix} 0 \\ 35 \end{bmatrix}, \quad \begin{bmatrix} 10 \\ 30 \end{bmatrix}, \quad \begin{bmatrix} 25 \\ 0 \end{bmatrix}, \quad \begin{bmatrix} 5 \\ 10 \end{bmatrix}, \quad \text{and} \quad \begin{bmatrix} 20 \\ 5 \end{bmatrix}.$$

An optimal solution is the vector

$$\begin{bmatrix} 10 \\ 30 \end{bmatrix}.$$

The new problem with slack variables can also be written in matrix form as follows: Find a vector **X** that will maximize

$$z = \mathbf{CX} \tag{7}$$

subject to

$$\mathbf{AX} = \mathbf{B} \tag{8}$$

$$\mathbf{X} \geqslant \mathbf{0}, \tag{9}$$

where now

$$\mathbf{A} = \begin{bmatrix} a_{11} & a_{12} & \cdots & a_{1n} & 1 & 0 & \cdots & 0 \\ a_{21} & a_{22} & & a_{2n} & 0 & 1 & \cdots & 0 \\ \vdots & \vdots & & \vdots & & & & \vdots \\ a_{m1} & a_{m2} & & a_{mn} & 0 & & \cdots & 1 \end{bmatrix}, \qquad \mathbf{B} = \begin{bmatrix} b_1 \\ b_2 \\ \vdots \\ b_m \end{bmatrix},$$

$$\mathbf{X} = \begin{bmatrix} x_1 \\ x_2 \\ \vdots \\ x_n \\ x_{n+1} \\ \vdots \\ x_{n+m} \end{bmatrix}, \qquad \text{and} \qquad \mathbf{C} = [c_1 \ \ c_2 \cdots c_n \ \ 0 \cdots 0].$$

A vector $\mathbf{X}$ satisfying (8) and (9) is called a **feasible solution** to the new problem, and a feasible solution maximizing the objective function (7) is called an **optimal solution**. Throughout this chapter, we now make the additional assumption that in all standard linear programming problems

$$b_1 \geqslant 0, \quad b_2 \geqslant 0, \ldots, b_m \geqslant 0.$$

We shall use Example 1 of Section 6.1 as our principal illustrative example in this section. Multiplying and dividing by appropriate factors to eliminate decimals, we can state our illustrative problem as follows.

Illustrative Problem

Find values of x and y that will maximize

$$z = 8x + 10y \tag{10}$$

subject to

$$2x + y \leqslant 50 \tag{11}$$

$$x + 2y \leqslant 70$$

$$x \geqslant 0, \qquad y \geqslant 0. \tag{12}$$

The new problem with slack variables u and v is: Find values of x, y, u, and v that will maximize

$$z = 8x + 10y \tag{13}$$

subject to

$$\begin{aligned} 2x + y + u \quad &= 50 \\ x + 2y + \quad v &= 70 \end{aligned} \tag{14}$$

$$x \geqslant 0, \quad y \geqslant 0, \quad u \geqslant 0, \quad v \geqslant 0. \tag{15}$$

DEFINITION The vector $\mathbf{X}$ in R^{n+m} is called a **basic solution** of the new problem if it is obtained by setting n of the variables in (8) equal to zero and solving for the remaining m variables. The m variables that we solve for are called **basic variables**, and the n variables set equal to zero are called **nonbasic variables**. The vector $\mathbf{X}$ is called a **basic feasible solution** if it is a basic solution that also satisfies (9).

Basic feasible solutions are important because the following theorem can be established.

THEOREM 6.2. *If a linear programming problem has an optimal solution, then it has a basic feasible solution that is optimal.*

Thus to solve a linear programming problem we need only search for basic feasible solutions. In our illustrative example we can select two of the four variables x, y, u, and v as nonbasic variables by setting them equal to zero and solve for the remaining two variables; that is, we solve for the basic variables. Thus, if

$$x = y = 0,$$

then

$$u = 50 \quad \text{and} \quad v = 70.$$

The vector

$$\mathbf{X}_1 = \begin{bmatrix} 0 \\ 0 \\ 50 \\ 70 \end{bmatrix}$$

is a basic feasible solution, which gives rise to the feasible solution

$$\begin{bmatrix} 0 \\ 0 \end{bmatrix}$$

to the original problem specified by (10), (11), and (12). The variables x and y are nonbasic and the variables u and v are basic. The convex region of solutions to the original problem has been sketched in Figure 6.8. Thus the vector $\mathbf{X}_1$ corresponds to the corner point O.

If we let the variables x and u be nonbasic ($x = u = 0$), then $y = 50$ and $v = -30$. The vector

$$\mathbf{X}_2 = \begin{bmatrix} 0 \\ 50 \\ 0 \\ -30 \end{bmatrix}$$

is a basic solution that is not feasible, since v is negative. It corresponds to the point F in Figure 6.8, which is not a feasible solution to the original problem.

In Table 6.3 we have tabulated all the possible choices for basic solutions. The basic variables are shaded and the corresponding point from Figure 6.8 is indicated in the table. It can be seen in this example, and proved in general, that every basic feasible solution determines a corner point, and conversely, each corner point determines a basic feasible solution.

TABLE 6.3

x	y	u	v	Type of solution	Corresponding point in Figure 6.8
0	0	50	70	Basic feasible solution	O
0	50	0	− 30	Not a basic feasible solution	F
0	35	15	0	Basic feasible solution	A
25	0	0	45	Basic feasible solution	C
70	0	− 90	0	Not a basic feasible solution	G
10	30	0	0	Basic feasible solution	B

One method of solving our linear programming problem would be to obtain all the basic solutions, discard those which are not feasible, and evaluate the objective function at each basic feasible solution, selecting that one, or ones, for which we get a maximum value of the objective function. The number of possible basic solutions is

$$\binom{n}{m} = \frac{n!}{m!\,(n-m)!}.$$

That is, it is the number of ways of selecting m objects out of n given objects.

The simplex method is a procedure that enables us to go from a given corner point (basic feasible solution) to an adjacent corner point in such a way that the value of the objective function increases as we move from corner point to corner point until we either obtain an optimal solution or find that the given problem has no finite optimal solution. The simplex method thus consists of two steps: (1) a way of checking whether a given basic feasible solution is an optimal solution, and (2) a way of obtaining another basic feasible solution with a larger value of the objective function. In actual practice the simplex method does not consider every basic feasible solution; rather, it works with only a small number of these. We shall now turn to a detailed discussion of this powerful method, using our illustrative example as a guide.

Selecting an Initial Basic Feasible Solution

We can take all the original (nonslack) variables as our nonbasic variables and set them equal to zero. We then solve for the slack variables, our basic variables. In our example, we set

$$x = y = 0$$

and solve for u and v:

$$u = 50 \qquad \text{and} \qquad v = 70.$$

Thus the initial basic feasible solution is the vector

$$\begin{bmatrix} 0 \\ 0 \\ 50 \\ 70 \end{bmatrix},$$

which yields the origin as a corner point.

It is convenient to develop a tabular method for displaying the given problem and the initial basic feasible solution. First, we write (13) as the equation

$$-8x - 10y + z = 0, \tag{13'}$$

with z being considered as another variable. We now form the **initial tableau** (Tableau 1). The variables x, y, u, v, and z are written in the top row as labels on the corresponding columns. Constraints (14) are entered in the top two rows followed by Equation (13′) in the bottom row. The bottom row of the tableau is called the **objective row**. Along the left-hand side of the tableau we indicate which variable is a basic variable in the corresponding equation. Thus u is a basic variable in the first equation and v is a basic variable in the second equation. A basic variable can also be described as a variable, other than z, which is present in exactly one equation, and there it appears with a coefficient of $+1$. In the tableau, the value of the basic variable is explicitly given in the rightmost column. In general, for the problem given by (7), (8), and (9), the initial tableau is Tableau 2.

TABLEAU 1

	x	y	u	v	z	
u	2	1	1	0	0	50
v	1	2	0	1	0	70
	-8	-10	0	0	1	0

TABLEAU 2

	x_1	x_2		x_n	x_{n+1}	x_{n+2}	$\cdots$	x_{n+m}	z	
x_{n+1}	a_{11}	a_{12}	$\cdots$	a_{1n}	1	0	$\cdots$	0	0	b_1
x_{n+2}	a_{21}	a_{22}	$\cdots$	a_{2n}	0	1	$\cdots$	0	0	b_2
$\vdots$	$\vdots$						$\vdots$		0	$\vdots$
x_{n+m}	a_{m1}	a_{m2}	$\cdots$	a_{mn}	0	0	$\cdots$	1	0	b_m
	$-c_1$	$-c_2$		$-c_n$	0	0	$\cdots$	0	1	0

The initial tableau (Tableau 1) shows the values of the basic variables u and v, and therefore the nonbasic variables have values

$$x = 0, \qquad y = 0.$$

The value of the objective function for this initial basic feasible solution is

$$c_1x + c_2y + 0(u) + 0(v) = 8(0) + 10(0) + 0(0) + 0(v) = 0,$$

which is the entry in the objective row and rightmost column. It is clear that this solution is not optimal, for using the bottom row of the initial tableau, we can write

$$z = 0 + 8x + 10y - 0u - 0v. \tag{16}$$

Now the value of z can be increased by increasing either x or y, since both of these variables appear in (16) with positive coefficients. Since (16) contains terms with positive coefficients if and only if the objective row of our initial tableau has negative entries under the columns labeled with variables, we see that we can increase z by increasing any variable with a negative entry in the objective row. Thus we obtain the following optimality criterion for determining whether the feasible solution indicated in a tableau is an optimal solution yielding a maximum value for the objective function z.

Optimality criterion. If the objective row of a tableau has no negative entries in the columns labeled with variables, then the indicated solution is optimal and we can stop our computation.

Selecting the Entering Variable

If the objective row of a tableau has negative entries in the columns labeled with variables, then the indicated solution is not optimal and we must continue our search for an optimal solution.

The simplex method moves from one corner point to an adjacent corner point in such a way that the value of the objective function increases. This is done by increasing *one* variable at a time. The largest increase in z per unit increase in a variable occurs for the variable with the most negative entry in the objective row. In Tableau 1, the most negative entry in the objective row is -10, and since it occurs in the y-column, this is the variable to be increased. The variable to be increased is called the

entering variable, because in the next iteration it will become a basic variable, thereby *entering* the set of basic variables. If there are several candidates for entering variables, choose one. An increase in y must be accompanied by a decrease in some of the other variables. This can be seen if we solve the equations in (14) for u and v:

$$\begin{aligned} u &= 50 - 2x - y \\ v &= 70 - x - 2y. \end{aligned}$$

Since we only increase y, we keep $x = 0$, and obtain

$$\begin{aligned} u &= 50 - y \\ v &= 70 - 2y, \end{aligned} \tag{17}$$

so that as y increases, both u and v decrease. Equations (17) also show by how much we can increase y. That is, since u and v must be nonnegative, we must have

$$\begin{aligned} y &\leqslant \tfrac{50}{1} = 50 \\ y &\leqslant \tfrac{70}{2} = 35. \end{aligned}$$

We now see that the allowable increase in y can be no larger than the smaller of the ratios $\frac{50}{1}$ and $\frac{70}{2}$. Taking y as 35, we obtain the new basic feasible solution,

$$x = 0, \quad y = 35, \quad u = 15, \quad v = 0.$$

The basic variables are y and u; the variables x and v are nonbasic. The objective function for this solution now has the value

$$z = 8(0) + 10(35) + 0(15) + 0(0),$$

which is much better than the earlier value of 0. This solution yields the corner point A in Figure 6.8, which is adjacent to O.

Choosing the Departing Variable

Since the variable $v = 0$, it is not basic and is called the **departing variable**, for it has *departed* from the set of basic variables. The column of the entering variable is called the **pivotal column**; the row that is labeled by the departing variable is called the **pivotal row**.

Let us look more closely at the selection of the departing variable. The choice of this variable was closely related to the determination of how far we could increase the entering variable (y in our example). To obtain this number, we formed the ratios (called **θ-ratios**) of the entries above the objective row in the rightmost column of the tableau by the corresponding entries in the pivotal column. The smallest of these ratios tells how far the entering variable can be increased. The basic variable labeling the row for which this smallest ratio occurs (the pivotal row) is then the departing variable. In our example, the ratios, formed by using the rightmost column and the y-column, are $\frac{50}{1}$ and $\frac{70}{2}$. The smallest of these ratios, 35, occurs for the second row, which means that the second row is the pivotal row and the basic variable, v, labeling it becomes the departing variable and is no longer basic. If the smallest of the ratios is not selected, then one of the variables in the new solution becomes negative and the new solution is no longer feasible (Exercise T.2). What happens if there are entries in the pivotal column that are either zero or negative? If any entry in the pivotal column is negative, then the corresponding ratio is also negative; in this case the equation associated with the negative entry imposes no restriction on how far the entering variable can be increased. Suppose, for example, that the y-column in our initial tableau is

$$\begin{bmatrix} -3 \\ 2 \end{bmatrix} \quad \text{instead of} \quad \begin{bmatrix} 1 \\ 2 \end{bmatrix}.$$

Then instead of (17) we have

$$u = 50 + 3y$$
$$v = 70 - 2y,$$

and since u must be nonnegative, we have

$$y \geqslant -\tfrac{50}{3},$$

which puts no limitation at all on how far y can be increased. If an entry in the pivotal column is zero, then the corresponding ratio cannot be formed (we cannot divide by zero), and again the associated equation puts no limitation on how far the entering variable can be increased. Thus, in forming the ratios, we only use the positive entries above the objective row in the pivotal column.

If all the entries above the objective row in the pivotal column are either zero or negative, then the entering variable can be made as large as we please. This means that the problem has no finite optimal solution.

Obtaining a New Tableau

We must now obtain a new tableau indicating the new basic variables and the new basic feasible solution. Solving the second equation in (14) for y, we obtain

$$y = 35 - \tfrac{1}{2}x - \tfrac{1}{2}v, \tag{18}$$

and substituting this expression for y in the first equation in (14), we have

$$\tfrac{3}{2}x + u - \tfrac{1}{2}v = 15. \tag{19}$$

The second equation in (14) can be written, upon dividing by 2 (the coefficient of y), as

$$\tfrac{1}{2}x + y + \tfrac{1}{2}v = 35. \tag{20}$$

Substituting (18) for y in (13′), we have

$$-3x + 5v + z = 350. \tag{21}$$

Since $x = 0$, $v = 0$, we obtain the value of z for the current basic feasible solution as

$$z = 350.$$

Equations (19), (20), and (21) yield our new tableau (Tableau 3). Observe in the tableau that we have labeled the basic variables in each row. Comparing Tableau 1 with Tableau 3, we observe that we can transform the former to the latter by elementary row operations as follows.

TABLEAU 3

	x	y	u	v	z	
u	$\frac{3}{2}$	0	1	$-\frac{1}{2}$	0	15
y	$\frac{1}{2}$	1	0	$\frac{1}{2}$	0	35
	-3	0	0	5	1	350

STEP 1. Locate and circle the entry in the pivotal row and pivotal column. This entry is called the **pivot**. Mark the pivotal column by placing

an arrow ↓ above the entering variable and mark the pivotal row by placing an arrow ← to the left of the departing variable.

STEP 2. If the pivot is k, multiply the pivotal row by $1/k$, making the entry that was the pivot a 1.

STEP 3. Add appropriate multiples of the pivotal row to all other rows (including the objective row) so that all elements in the pivotal column except for the 1 where the pivot was located become zero.

STEP 4. In the new tableau replace the label on the pivotal row by the entering variable.

These four steps form a process called **pivotal elimination**. It is one of the iterations of the procedure described in Section 1.3 for transforming a matrix to reduced row echelon form.

We now repeat Tableau 1, with the arrows placed next to the entering and departing variables and with the pivot circled.

TABLEAU 1

	x	↓ y	u	v	z	
u	2	1	1	0	0	50
← v	1	(2)	0	1	0	70
	−8	−10	0	0	0	10

Performing pivotal elimination on Tableau 1 yields Tableau 3. We now repeat the entire procedure, using Tableau 3 as our initial tableau. Since the most negative entry in the objective row of Tableau 3, -3, occurs in the first column, x is the entering variable and the first column is the pivotal column. To determine the departing variable, we form the ratios of the entries in the rightmost column (except for the objective row) by the corresponding entries of the pivotal column for those entries in the pivotal column which are positive and select the smallest of these ratios. Since both entries in the pivotal column are positive, the ratios are $15/\frac{3}{2}$ and $35/\frac{1}{2}$. The smallest of these, $15/\frac{3}{2} = 10$, occurs for the first row, so the departing variable is u and the pivotal row is the first row. Tableau 3, with the pivotal column, pivotal row, and circled pivot, is shown again here.

TABLEAU 3

	x ↓	y	u	v	z	
← u	($\frac{3}{2}$)	0	1	$-\frac{1}{2}$	0	15
y	$\frac{1}{2}$	1	0	$\frac{1}{2}$	0	35
	-3	0	0	5	1	350

Performing pivotal elimination on Tableau 3 yields Tableau 4.

TABLEAU 4

	x	y	u	v	z	
x	1	0	$\frac{2}{3}$	$-\frac{1}{3}$	0	10
y	0	1	$-\frac{1}{3}$	$\frac{2}{3}$	0	30
	0	0	2	4	1	380

Since the objective row of Tableau 4 has no negative entries in the columns labeled with variables, we conclude, by the optimality criterion, that we are finished and that the indicated solution is optimal. Thus the optimal solution is

$$x = 10, \quad y = 30, \quad u = 0, \quad \text{and} \quad v = 0,$$

which corresponds to the corner point $B(10, 30)$. Thus the simplex method started from the corner point $O(0, 0)$ moved to the adjacent corner point $A(0, 35)$ and then to the corner point $B(10, 30)$, which is adjacent to A. The value of the objective function increased from 0 to 350 to 380, respectively.

Summary of the Simplex Method

STEP 1. Set up the initial tableau.

STEP 2. Apply the optimality test. If the objective row has no negative entries in the columns labeled with variables, then the indicated solution is optimal; we stop our computations.

STEP 3. Choose a pivotal column by determining the column with the most negative entry in the objective row. If there are several candidates for a pivotal column, choose any one.

STEP 4. Choose a pivotal row: form the ratios of the rightmost column (except for the objective row) by the corresponding entries of the pivotal column for those entries in the pivotal column which are positive. The pivotal row is the row for which the smallest of these ratios occurs. If there is a tie, so that the smallest ratio occurs at more than one row, choose any one of the qualifying rows. If none of the entries in the pivotal column above the objective row is positive, the problem has no finite optimum. We stop our computation.

STEP 5. Perform pivotal elimination to construct a new tableau and return to step 2.

Figure 6.13 gives a flow chart for the simplex algorithm.

We have restricted *our* discussion of the simplex method to standard linear programming problems in which all the right-hand entries are nonnegative. It should be noted that the method applies to the general linear programming problem. For additional details we refer the reader to the references at the end of this chapter.

Example 2. Maximize

$$z = 4x_1 + 8x_2 + 5x_3$$

subject to

$$\begin{aligned} x_1 + 2x_2 + 3x_3 &\leqslant 18 \\ x_1 + 4x_2 + x_3 &\leqslant 6 \\ 2x_1 + 6x_2 + 4x_3 &\leqslant 15 \\ x_1 \geqslant 0, \quad x_2 \geqslant 0, \quad x_3 &\geqslant 0. \end{aligned}$$

The new problem with slack variables is

$$\text{Maximize } z = 4x_1 + 8x_2 + 5x_3$$

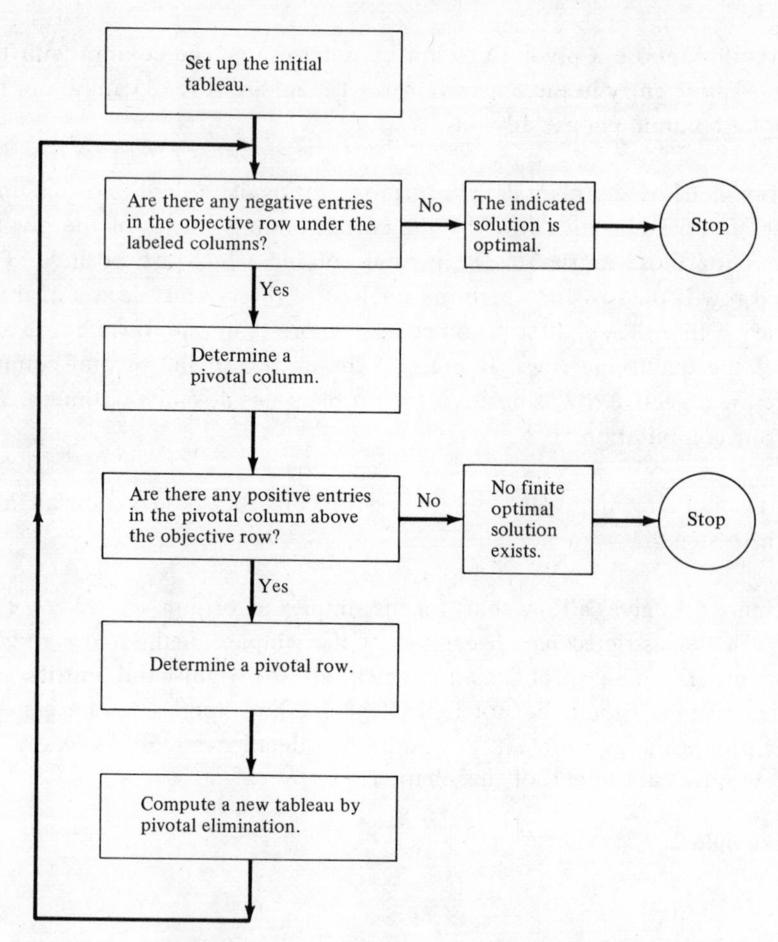

Figure 6.13 Flow chart for the simplex method.

subject to

$$x_1 + 2x_2 + 3x_3 + x_4 + \qquad = 18$$

$$x_1 + 4x_2 + x_3 \qquad + x_5 \qquad = 6$$

$$2x_1 + 6x_2 + 4x_3 \qquad + x_6 = 15$$

$$x_1 \geqslant 0, \quad x_2 \geqslant 0, \quad x_3 \geqslant 0, \quad x_4 \geqslant 0, \quad x_5 \geqslant 0, \quad x_6 \geqslant 0.$$

The initial tableau and the succeeding tableaus are

	x_1	↓ x_2	x_3	x_4	x_5	x_6	z	
x_4	1	2	3	1	0	0	0	18
← x_5	1	4	1	0	1	0	0	6
x_6	2	6	4	0	0	1	0	15
	-4	-8	-5	0	0	0	1	0

	x_1	x_2	↓ x_3	x_4	x_5	x_6	z	
x_4	$\frac{1}{2}$	0	$\frac{5}{2}$	1	$-\frac{1}{2}$	0	0	15
x_2	$\frac{1}{4}$	1	$\frac{1}{4}$	0	$\frac{1}{4}$	0	0	$\frac{3}{2}$
← x_6	$\frac{1}{2}$	0	$\frac{5}{2}$	0	$-\frac{3}{2}$	1	0	6
	-2	0	-3	0	2	0	1	12

	↓ x_1	x_2	x_3	x_4	x_5	x_6	z	
x_4	0	0	0	1	1	-1	0	9
← x_2	$\frac{1}{5}$	1	0	0	$\frac{2}{5}$	$-\frac{1}{10}$	0	$\frac{9}{10}$
x_3	$\frac{1}{5}$	0	1	0	$-\frac{3}{5}$	$\frac{2}{5}$	0	$\frac{12}{5}$
	$-\frac{7}{5}$	0	0	0	$\frac{1}{5}$	$\frac{6}{5}$	1	$\frac{96}{5}$

	x_1	x_2	x_3	x_4	x_5	x_6	z	
x_4	0	0	0	1	1	-1	0	9
x_1	1	5	0	0	2	$-\frac{1}{2}$	0	$\frac{9}{2}$
x_3	0	-1	1	0	-1	$\frac{1}{2}$	0	$\frac{3}{2}$
	0	7	0	0	3	$\frac{1}{2}$	1	$\frac{51}{2}$

Hence an optimal solution is

$$x_1 = \tfrac{9}{2}, \qquad x_2 = 0, \qquad x_3 = \tfrac{3}{2}.$$

The slack variables are

$$x_4 = 9, \qquad x_5 = 0, \qquad x_6 = 0,$$

and the optimal value of z is $\frac{51}{2}$.

A number of difficulties can arise in using the simplex method. We shall briefly describe one of these and we again refer the reader to the references at the end of this chapter for an extended discussion of computational considerations.

Degeneracy

Suppose that a basic variable becomes zero in one of the tableaus in the simplex method. Then one of the ratios used to determine the next pivotal row may be zero, in which case the pivotal row is the one labeled with the basic variable that is zero (in this case, zero is the smallest of the ratios). Recall that the entering variable is increased from zero to the smallest ratio. Since the smallest ratio is zero, the entering variable remains at the value zero. In the new tableau all the old basic variables have the same values that they had in the old tableau; the value of the objective function has not been increased. The new tableau looks like the old one, except that a basic variable with value zero has become nonbasic, and its place has been taken by a nonbasic variable also entering with value zero. A basic feasible solution in which one or more of the basic variables are zero is called a **degenerate basic feasible solution**. It can be shown that when a tie occurs for the smallest ratio in determining the pivotal row, a degenerate solution will arise and there are several optimal solutions. When no degenerate solution occurs, the value of the objective function improves as we move from one basic feasible solution to another. The procedure stops after a finite number of steps, since the number of basic feasible solutions is finite. However, if we have a degenerate basic feasible solution, we may return to a basic feasible solution that had already been found at an earlier iteration and thus enter an infinite cycle. Fortunately, cycling has never occurred in a practical linear programming problem, although several examples have been carefully constructed to show that cycling can occur. In actual practice, when degeneracy occurs, it is handled by merely ignoring it. The value of the objective function may remain constant for a few iterations and will then start to increase. Moreover, a technique called **perturbation** has been developed for handling degeneracy. This technique calls for making slight changes in the rightmost column of the tableau so that the troublesome ties in the ratios will no longer occur.

There are many computer programs implementing the simplex method and other algorithms in the area of mathematical programming (which includes integer programming and nonlinear programming). Some of these programs can do extensive data manipulation to prepare the input for the problem, solve the problem, and then prepare elaborate reports that can be used by management in decision making. Moreover, some programs can handle large problems having as many as 8200 inequalities and 100,000 unknowns.

FURTHER READING

Garvin, W. W. *Introduction to Linear Programming*. New York: McGraw-Hill Book Company, 1960.

Gass, Saul I. *Linear Programming*, 4th ed. New York: McGraw-Hill Book Company, 1975.

Kuester, James L. and Joe H. Mize. *Optimization Techniques with FORTRAN*. New York: McGraw-Hill Book Company, 1974.

Wolfe, Carvel S. *Linear Programming with FORTRAN*. Glenview, Ill.: Scott, Foresman and Co., 1970.

6.2 Exercises

In Exercises 1 through 4 write the initial simplex tableau for each given linear programming problem.

1. Maximize $z = 3x + 7y$
subject to

$$\begin{aligned} 3x - 2y &\leqslant 7 \\ 2x + 5y &\leqslant 6 \\ 2x + 3y &\leqslant 8 \\ x \geqslant 0, \quad y &\geqslant 0. \end{aligned}$$

2. Maximize $z = 2x_1 + 3x_2 - 4x_3$
subject to

$$\begin{aligned} 3x_1 - 2x_2 + x_3 &\leqslant 4 \\ 2x_1 + 4x_2 + 5x_3 &\leqslant 6 \\ x_1 \geqslant 0, \quad x_2 \geqslant 0, \quad x_3 &\geqslant 0. \end{aligned}$$

3. Maximize $z = 2x_1 + 2x_2 + 3x_3 + x_4$
subject to

$$3x_1 - 2x_2 + x_3 + x_4 \leq 6$$
$$x_1 + x_2 + x_3 + x_4 \leq 8$$
$$2x_1 - 3x_2 - x_3 + 2x_4 \leq 10$$
$$x_1 \geq 0, \quad x_2 \geq 0, \quad x_3 \geq 0, \quad x_4 \geq 0.$$

4. Maximize $z = 2x_1 - 3x_2 + x_3$
subject to

$$x_1 - 2x_2 + 4x_3 \leq 5$$
$$2x_1 + 2x_2 + 4x_3 \leq 5$$
$$3x_1 + x_2 - x_3 \leq 7$$
$$x_1 \geq 0, \quad x_2 \geq 0, \quad x_3 \geq 0.$$

In Exercises 5 through 11 solve each given linear programming by the simplex method. Some of these problems may have no finite optimal solution.

5. Maximize $z = 2x + 3y$
subject to

$$3x + 5y \leq 6$$
$$2x + 3y \leq 7$$
$$x \geq 0, \quad y \geq 0.$$

6. Maximize $z = 2x + 5y$
subject to

$$3x + 7y \leq 6$$
$$2x + 6y \leq 7$$
$$3x + 2y \leq 5$$
$$x \geq 0, \quad y \geq 0.$$

7. Maximize $z = 2x + 5y$
subject to

$$2x - 3y \leq 4$$
$$x - 2y \leq 6$$
$$x \geq 0, \quad y \geq 0.$$

8. Maximize $z = 3x_1 + 2x_2 + 4x_3$
subject to

$$\begin{aligned} x_1 - x_2 - x_3 &\leq 6 \\ -2x_1 + x_2 - 2x_3 &\leq 7 \\ 3x_1 + x_2 - 4x_3 &\leq 8 \\ x_1 \geq 0, \quad x_2 \geq 0, \quad x_3 &\geq 0. \end{aligned}$$

9. Maximize $z = 2x_1 - 4x_2 + 5x_3$
subject to

$$\begin{aligned} 3x_1 + 2x_2 + x_3 &\leq 6 \\ 3x_1 - 6x_2 + 7x_3 &\leq 9 \\ x_1 \geq 0, \quad x_2 \geq 0, \quad x_3 &\geq 0. \end{aligned}$$

10. Maximize $z = 2x_1 + 4x_2 - 3x_3$
subject to

$$\begin{aligned} 5x_1 + 2x_2 + x_3 &\leq 5 \\ 3x_1 - 2x_2 + 3x_3 &\leq 10 \\ 4x_1 + 5x_2 - x_3 &\leq 20 \\ x_1 \geq 0, \quad x_2 \geq 0, \quad x_3 &\geq 0. \end{aligned}$$

11. Maximize $z = x_1 + 2x_2 - x_3 + 5x_4$
subject to

$$\begin{aligned} 2x_1 + 3x_2 + x_3 - x_4 &\leq 8 \\ 3x_1 + x_2 - 4x_3 + 5x_4 &\leq 9 \\ x_1 \geq 0, \quad x_2 \geq 0, \quad x_3 \geq 0, \quad x_4 &\geq 0. \end{aligned}$$

12. Solve Exercise 1 in Section 6.1 by the simplex method.

13. Solve Exercise 4 in Section 6.1 by the simplex method.

14. Solve Exercise 7 in Section 6.1 by the simplex method.

15. A power plant burns coal, oil, and gas to generate electricity. Suppose that each ton of coal generates 600 kilowatts, emits 20 units of sulfur dioxide and 15 units of particulate matter, and costs \$200; each ton of oil generates 550 kilowatts, emits 18 units of sulfur dioxide and 12 units of particulate matter, and costs \$220; each ton of gas generates 500 kilowatts, emits 15 units of sulfur dioxide and 10 units of particulate matter, and costs \$250. The environmental protection agency restricts the daily emission of sulfur dioxide to no more than 60 units and no more than 75 units of particulate matter. If the power plant wants to spend no more than \$2000 per day on fuel, how much fuel of each type should be bought to maximize the amount of power generated?

Theoretical Exercises

T.1. Consider the standard linear programming problem:

Maximize $z = \mathbf{CX}$

subject to

$$\mathbf{AX} \leqslant \mathbf{B}$$
$$\mathbf{X} \geqslant \mathbf{0},$$

where $\mathbf{A}$ is an $m \times n$ matrix. Show that the feasible region (the set of all feasible solutions) of this problem is a convex set.

[*Hint*: If $\mathbf{X}$ and $\mathbf{Y}$ are in R^n, the line segment joining them is the set of points $r\mathbf{X} + (1 - r)\mathbf{Y}$, where $0 \leqslant r \leqslant 1$.]

T.2. Suppose that in selecting the departing variable, the minimum is not chosen. Show that the resulting solution is not feasible.

7 Applications

7.1 Lines and Planes

Lines

We may recall that in R^2 a line is determined by specifying its slope and one of its points. In R^3 a line is determined by specifying its direction and one of its points. Let $\mathbf{U} = (u, v, w)$ be a nonzero vector in R^3, and let $P_0 = (x_0, y_0, z_0)$ be a point in R^3. Let $\mathbf{X}_0$ be the position vector of P_0. Then the line L through P_0 and parallel to $\mathbf{U}$ consists of the points $P(x, y, z)$ whose position vector $\mathbf{X}$ satisfies (Figure 7.1)

$$\mathbf{X} = \mathbf{X}_0 + t\mathbf{U} \qquad \text{where } -\infty < t < \infty. \tag{1}$$

Equation (1) is called the **parametric equation** of L, since it contains the parameter t, which can be assigned any real number. Equation (1) can also

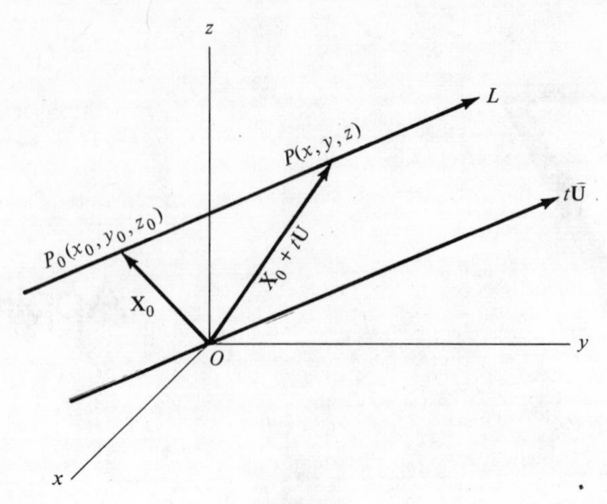

Figure 7.1

be written in terms of the components as

$$\begin{aligned} x &= x_0 + tu \\ y &= y_0 + tv \qquad -\infty < t < \infty \\ z &= z_0 + tw, \end{aligned} \tag{2}$$

which are called the **parametric equations** of L.

Example 1. The parametric equations of the line through the point $P_0(-3, 2, 1)$, which is parallel to the vector $\mathbf{U} = (2, -3, 4)$, are

$$\begin{aligned} x &= -3 + 2t \\ y &= 2 - 3t \\ z &= 1 + 4t. \end{aligned}$$

Example 2. Find parametric equations for the line L through the points $P_0(2, 3, -4)$ and $P_1(3, -2, 5)$.

solution. The desired line is parallel to the vector $\mathbf{U} = \overrightarrow{P_0P_1}$. Now

$$\mathbf{U} = (3 - 2, -2 - 3, 5 - (-4)) = (1, -5, 9).$$

Since P_0 lies on the line, we can write the parametric equations of L as

$$\begin{aligned} x &= 2 + t \\ y &= 3 - 5t \qquad -\infty < t < \infty \\ z &= -4 + 9t. \end{aligned}$$

Of course, we could have used the point P_2 instead of P_1. In fact, we could use any point on the line in the parametric equation for L. Thus a line can be represented in infinitely many ways in parametric form. If u, v, and w are all nonzero in (2), we can solve each equation for t and equate the results to obtain the equations in **symmetric form** for the line through P_0 and parallel to $\mathbf{U}$:

$$\frac{x - x_0}{u} = \frac{y - y_0}{v} = \frac{z - z_0}{w}.$$

The equations in symmetric form for a line are useful in some analytic geometry applications.

Example 3. The equations in symmetric form for the line in Example 2 are

$$\frac{x - 2}{1} = \frac{y - 3}{-5} = \frac{z + 4}{9}.$$

Planes

A plane in R^3 can be determined by specifying a point in the plane and a vector perpendicular to it. A vector perpendicular to a plane is called a **normal** to the plane.

To obtain an equation of the plane passing through the point $P_0(x_0, y_0, z_0)$ and having the nonzero vector $\mathbf{N} = (a, b, c)$ as a normal, we proceed as follows. A point $P(x, y, z)$ lies in the plane if and only if the vector $\overrightarrow{P_0P}$ is perpendicular to $\mathbf{N}$ (Figure 7.2). Thus $P(x, y, z)$ lies in the plane if and only if

$$\mathbf{N} \cdot \overrightarrow{P_0P} = 0. \tag{3}$$

Since

$$\overrightarrow{P_0P} = (x - x_0, y - y_0, z - z_0),$$

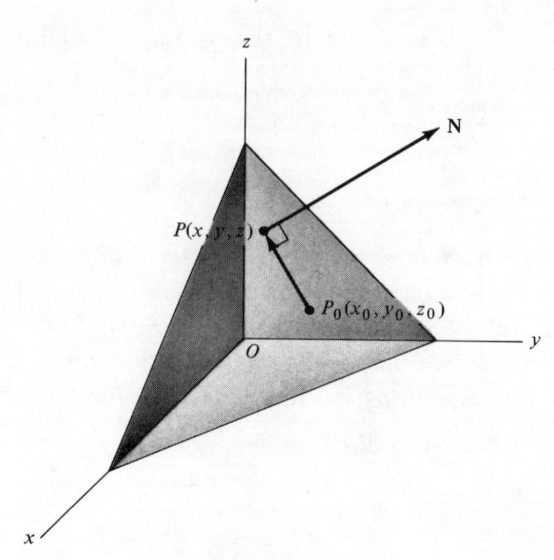

Figure 7.2

we can write (3) as

$$a(x - x_0) + b(y - y_0) + c(z - z_0) = 0. \tag{4}$$

Example 4. Find an equation of the plane passing through the point (3, 4, − 3) and perpendicular to the vector $\mathbf{N} = (5, -2, 4)$.

solution. Substituting in (4), we obtain the equation of the plane as

$$5(x - 3) - 2(y - 4) + 4(z + 3) = 0. \tag{5}$$

A plane is also determined by three noncollinear points, as we show in the following example.

Example 5. Find an equation of the plane passing through the points $P_1(2, -2, 1)$, $P_2(-1, 0, 3)$, and $P_3(5, -3, 4)$.

solution. The nonparallel vectors $\overrightarrow{P_1P_2} = (-3, 2, 2)$ and $\overrightarrow{P_1P_3} = (3, -1, 3)$ lie in the plane, since the points P_1, P_2, and P_3 lie in the plane. The vector

$$\mathbf{N} = \overrightarrow{P_1P_2} \times \overrightarrow{P_1P_3} = (8, 15, -3)$$

is then perpendicular to both $\overrightarrow{P_1P_2}$ and $\overrightarrow{P_1P_3}$ and is thus a normal to the plane. Using the vector **N** and the point $P_1(2, -2, 1)$ in (4), we obtain an equation of the plane as

$$8(x - 2) + 15(y + 2) - 3(z - 1) = 0.$$

If we multiply out and simplify, (4) can be rewritten as

$$ax + by + cz + d = 0. \tag{6}$$

Example 6. Equation (5) for the plane in Example 5 can be rewritten as

$$8x + 15y - 3z + 17 = 0. \tag{7}$$

It is not difficult to show (Exercise T.1) that the graph of an equation of the form given in (6), where a, b, c, and d are constants, is a plane with normal $\mathbf{N} = (a, b, c)$.

Example 7. An alternative solution to Example 5 is as follows. Let the equation of the desired plane be

$$ax + by + cz + d = 0, \tag{8}$$

where a, b, c, and d are to be determined. Since P_1, P_2, and P_3 lie in the plane, their coordinates satisfy (8). Thus we obtain the linear system

$$\begin{aligned} 2a - 2b + c + d &= 0 \\ -a \phantom{{}-3b} + 3c + d &= 0 \\ 5a - 3b + 4c + d &= 0. \end{aligned}$$

Solving this system, we have

$$a = \tfrac{8}{17}r, \qquad b = \tfrac{15}{17}r, \qquad c = -\tfrac{3}{17}r, \qquad \text{and} \qquad d = r,$$

where r is any real number. Letting $r = 17$, we obtain

$$a = 8, \qquad b = 15, \qquad c = -3, \qquad \text{and} \qquad d = 17,$$

which yields (7) as in the first solution.

The equations of the line in symmetric form can be used to determine two planes whose intersection is the given line.

Example 8. Find two planes whose intersection is the line

$$\begin{aligned} x &= -2 + 3t \\ y &= 3 - 2t \qquad -\infty < t < \infty \\ z &= 5 + 4t. \end{aligned}$$

solution. First, find the equations of the line in symmetric form as

$$\frac{x+2}{3} = \frac{y-3}{-2} = \frac{z-5}{4}.$$

The given line is then the intersection of the planes

$$\frac{x+2}{3} = \frac{y-3}{-2} \qquad \text{and} \qquad \frac{x+2}{3} = \frac{z-5}{4}.$$

Thus the given line is the intersection of the planes

$$2x + 3y - 5 = 0 \qquad \text{and} \qquad 4x - 3z + 23 = 0.$$

Two planes are either parallel or they intersect in a straight line. They are parallel if their normals are parallel. In the following example we determine the line of intersection of two planes.

Example 9. Find the parametric equations of the line of intersection of the planes

$$\pi_1 : 2x + 3y - 2z + 4 = 0 \qquad \text{and} \qquad \pi_2 : x - y + 2z + 3 = 0.$$

solution. Solving the linear system consisting of the equations for π_1 and π_2, we obtain (verify)

$$\begin{aligned} x &= -\tfrac{13}{5} - \tfrac{4}{5}t \\ y &= \tfrac{2}{5} + \tfrac{6}{5}t \qquad -\infty < t < \infty \\ z &= 0 + \phantom{\tfrac{6}{5}}t \end{aligned}$$

as the parametric equations for the line of intersection of the planes (see Figure 7.3).

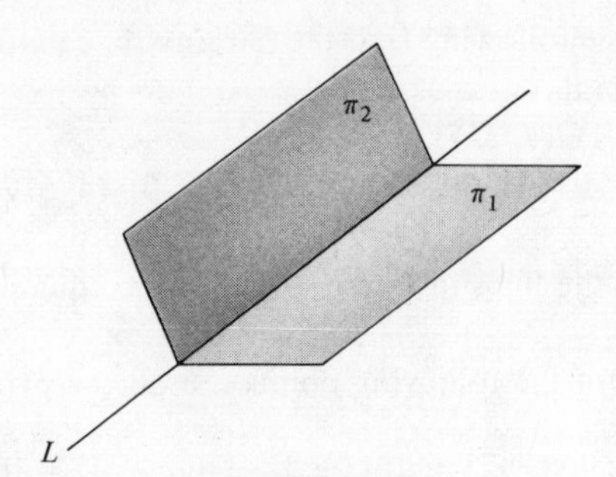

Figure 7.3

Consider now the subspaces of R^2. First, we have $\{0\}$ and R^2, the trivial subspaces of dimensions 0 and 2, respectively. Now the subspace V of R^2 spanned by a nonzero vector $\mathbf{X}$ is a one-dimensional subspace of R^2, and V is a line through the origin. We have now classified all the subspaces of R^2: R^2, the origin, and all the lines through the origin. In Exercise T.5 the reader is asked to carry out a similar discussion for the classification of all the subspaces of R^3.

7.1 Exercises

1. State which of the following points are on the line

$$\begin{aligned} x &= 3 + 2t \\ y &= -2 + 3t \qquad -\infty < t < \infty \\ z &= 4 - 3t. \end{aligned}$$

(a) $(1, 1, 1)$. (b) $(1, -1, 0)$.
(c) $(1, 0, -2)$. (d) $(4, -\frac{1}{2}, \frac{5}{2})$.

2. State which of the following points are on the line

$$\frac{x - 4}{-2} = \frac{y + 3}{2} = \frac{z - 4}{-5}.$$

(a) $(0, 1, -6)$. (b) $(1, 2, 3)$. (c) $(4, -3, 4)$. (d) $(0, 1, -1)$.

3. In each of the following find the parametric equations of the line through the point $P_0 = (x_0, y_0, z_0)$, which is parallel to the vector $\mathbf{U}$.
(a) $P_0 = (3, 4, -2)$, $\mathbf{U} = (4, -5, 2)$.
(b) $P_0 = (3, 2, 4)$, $\mathbf{U} = (-2, 5, 1)$.
(c) $P_0 = (0, 0, 0)$, $\mathbf{U} = (2, 2, 2)$.
(d) $P_0 = (-2, -3, 1)$, $\mathbf{U} = (2, 3, 4)$.

4. In each of the following find the parametric equations of the line through the given points.
(a) $(2, -3, 1), (4, 2, 5)$. (b) $(-3, -2, -2), (5, 5, 4)$.
(c) $(-2, 3, 4), (2, -3, 5)$. (d) $(0, 0, 0), (4, 5, 2)$.

5. For each of the lines in Exercise 4 find the equations in symmetric form.

6. State which of the following points are on the plane

$$3(x - 2) + 2(y + 3) - 4(z - 4) = 0.$$

(a) $(0, -2, 3)$. (b) $(1, -2, 3)$. (c) $(1, -1, 3)$. (d) $(0, 0, 4)$.

7. In each of the following find an equation of the plane passing through the given point and perpendicular to the given vector $\mathbf{N}$.
(a) $(0, 2, -3)$, $\mathbf{N} = (3, -2, 4)$. (b) $(-1, 3, 2)$, $\mathbf{N} = (0, 1, -3)$.
(c) $(-2, 3, 4)$, $\mathbf{N} = (0, 0, -4)$. (d) $(5, 2, 3)$, $\mathbf{N} = (-1, -2, 4)$.

8. In each of the following find an equation of the plane passing through the given three points.
(a) $(0, 1, 2), (3, -2, 5), (2, 3, 4)$.
(b) $(2, 3, 4), (-1, -2, 3), (-5, -4, 2)$.
(c) $(1, 2, 3), (0, 0, 0), (-2, 3, 4)$.
(d) $(1, 1, 1), (2, 3, 4), (-5, 3, 2)$.

9. In each of the following find parametric equations of the line of intersection of the given planes.
(a) $2x + 3y - 4z + 5 = 0$ and $-3x + 2y + 5z + 6 = 0$.
(b) $3x - 2y - 5z + 4 = 0$ and $2x + 3y + 4z + 8 = 0$.
(c) $-x + 2y + z = 0$ and $2x - y + 2z + 8 = 0$.

10. In each of the following find a pair of planes whose intersection is the given line.

(a) $$\begin{aligned} x &= 2 - 3t \\ y &= 3 + t \\ 2 &= 2 - 4t. \end{aligned}$$

(b) $$\frac{x - 2}{-2} = \frac{y - 3}{4} = \frac{z + 4}{3}.$$

(c) $$\begin{aligned} x &= 4t \\ y &= 1 + 5t \\ z &= 2 - t. \end{aligned}$$

11. Are the points $(2, 3, -2)$, $(4, -2, -3)$, and $(0, 8, -1)$ on the same line?

12. Are the points $(-2, 4, 2)$, $(3, 5, 1)$, and $(4, 2, -1)$ on the same line?

13. Find the point of intersection of the lines

$$\begin{aligned} x &= 2 - 3s \\ y &= 3 + 2s \\ z &= 4 + 2s \end{aligned} \quad \text{and} \quad \begin{aligned} x &= 5 + 2t \\ y &= 1 - 3t \\ z &= 2 + t. \end{aligned}$$

14. Which of the following pairs of lines are perpendicular?

(a) $$\begin{aligned} x &= 2 + 2t \\ y &= -3 - 3t \\ z &= 4 + 4t \end{aligned} \quad \text{and} \quad \begin{aligned} x &= 2 + t \\ y &= 4 - t \\ z &= 5 - t. \end{aligned}$$

(b) $$\begin{aligned} x &= 3 - t \\ y &= 4 + t \\ z &= 2 + 2t \end{aligned} \quad \text{and} \quad \begin{aligned} x &= 2t \\ y &= 3 - 2t \\ z &= 4 + 2t. \end{aligned}$$

15. Show that the following parametric equations define the same line.

$$\begin{aligned} x &= 2 + 3t \\ y &= 3 - 2t \\ z &= -1 + 4t \end{aligned} \quad \text{and} \quad \begin{aligned} x &= -1 - 9t \\ y &= 5 + 6t \\ z &= -5 - 12t. \end{aligned}$$

16. Find the parametric equations for the line passing through the point $(3, -1, -3)$ and perpendicular to the line passing through the points $(3, -2, 4)$ and $(0, 3, 5)$.

17. Find an equation for the plane passing through the point $(-2, 3, 4)$ and perpendicular to the line passing through the points $(4, -2, 5)$ and $(0, 2, 4)$.

18. Find the point of intersection of the line

$$\begin{aligned} x &= 2 - 3t \\ y &= 4 + 2t \\ z &= 3 - 5t \end{aligned}$$

and the plane $2x + 3y + 4z + 8 = 0$.

19. Find a plane containing the lines

$$\begin{aligned} x &= 3 + 2t \\ y &= 4 - 3t \\ z &= 5 + 4t \end{aligned} \quad \text{and} \quad \begin{aligned} x &= 1 - 2t \\ y &= 7 + 4t \\ z &= 1 - 3t. \end{aligned}$$

20. Find a plane passing through the point $(2, 4, -3)$ and parallel to the plane $-2x + 4y - 5z + 6 = 0$.

21. Find a line passing through the point $(-2, 5, -3)$ and perpendicular to the plane $2x - 3y + 4z + 7 = 0$.

Theoretical Exercises

T.1. Show that the equation $ax + by + cz + d = 0$, where a, b, c, and d are constants, is a plane with normal $\mathbf{N} = (a, b, c)$.

T.2. Let the lines L_1 and L_2 be given parametrically by

$$L_1 : \mathbf{X} = \mathbf{X}_0 + s\mathbf{U} \quad \text{and} \quad L_2 : \mathbf{X} = \mathbf{X}_1 + t\mathbf{V}.$$

Prove that

(a) L_1 and L_2 are parallel if and only if $\mathbf{U} = a\mathbf{V}$ for some scalar a.

(b) L_1 and L_2 are identical if and only if $\mathbf{X}_1 - \mathbf{X}_0$ and $\mathbf{U}$ are both parallel to $\mathbf{V}$.

(c) L_1 and L_2 are perpendicular if and only if $\mathbf{U} \cdot \mathbf{V} = 0$.

(d) L_1 and L_2 intersect if and only if $\mathbf{X}_1 - \mathbf{X}_0$ is a linear combination of $\mathbf{U}$ and $\mathbf{V}$.

T.3. The lines L_1 and L_2 are said to be **skew** if they are not parallel and do not intersect. Give an example of skew lines L_1 and L_2.

T.4. Consider the planes $a_1x + b_1y + c_1z + d_1 = 0$ and $a_2x + b_2y + c_2z + d_2 = 0$ with normals $\mathbf{N}_1$ and $\mathbf{N}_2$, respectively. Prove that if the planes are identical, then $d_1\mathbf{N}_2 = d_2\mathbf{N}_1$.

T.5. Classify all the subspaces of R^3.

T.6. (a) Show that the set of all points on the plane $ax + by + cz = 0$ is a subspace of R^3.

(b) Find a basis for the plane $2x - 3y + 4z = 0$.

7.2 Quadratic Forms

Let

$$ax^2 + 2bxy + cy^2 = d, \tag{1}$$

where a, b, c, and d are real numbers, be the equation of a conic centered at the origin of a rectangular Cartesian coordinate system in two-dimensional space. In many applications we are interested in identifying the type of conic represented by the equation; that is, we have to decide whether the equation represents an ellipse, a hyperbola, a parabola, or any degenerate forms of the first two of these, such as circles and straight lines. The approach frequently used is to rotate the x- and y-axes (or change coordinates) so as to obtain a new set of axes, x' and y', in which the $x'y'$ term is no longer present; we can then identify the type of conic rather easily. The analogous problem can be studied for quadric surfaces in three-dimensional space, that is, surfaces whose equation is

$$ax^2 + 2dxy + 2exz + by^2 + 2fyz + cz^2 = g, \tag{2}$$

where a, b, c, d, e, f, and g are real numbers. The solution is to again change the x-, y-, and z-axes to x'-, y'-, and z'-axes so as to eliminate the $x'y'$, $x'z'$, and $y'z'$ terms. The expressions on the left side of (1) and (2) are quadratic forms. Quadratic forms arise in statistics, mechanics, and in other problems in physics; in quadratic programming; in the study of maxima and minima of functions of several variables; and in other applied problems. In this section we apply our results on eigenvalues and eigenvectors of matrices to give a brief treatment of real quadratic forms in n variables, generalizing the techniques used for quadratic forms in two and three variables.

DEFINITION If $\mathbf{A}$ is a symmetric matrix ($\mathbf{A} = \mathbf{A}^T$), then the function $\mathbf{Q} : R^n \to R^1$ (a real-valued function defined on R^n) defined by

$$\mathbf{Q}(\mathbf{X}) = \mathbf{X}^T\mathbf{A}\mathbf{X},$$

where

$$\mathbf{X} = \begin{bmatrix} x_1 \\ x_2 \\ \cdot \\ \cdot \\ \cdot \\ x_n \end{bmatrix}$$

is called a **quadratic form in the n variables** $x_1, x_2, \ldots, x_n$. The matrix $\mathbf{A}$ is called the **matrix of the quadratic** form $\mathbf{Q}$. We shall also denote the quadratic form by $\mathbf{Q}(\mathbf{X})$.

Example 1. The left side of Equation (1) is the quadratic form in the variables x and y:

$$\mathbf{Q}(\mathbf{X}) = \mathbf{X}^T\mathbf{A}\mathbf{X},$$

where

$$\mathbf{X} = \begin{bmatrix} x \\ y \end{bmatrix} \quad \text{and} \quad \mathbf{A} = \begin{bmatrix} a & b \\ b & c \end{bmatrix}.$$

Example 2. The left side of Equation (2) is the quadratic form

$$\mathbf{Q(X)} = \mathbf{X}^T\mathbf{AX},$$

where

$$\mathbf{X} = \begin{bmatrix} x \\ y \\ z \end{bmatrix} \quad \text{and} \quad \mathbf{A} = \begin{bmatrix} a & d & e \\ d & b & f \\ e & f & c \end{bmatrix}.$$

Example 3. The following expressions are quadratic forms:

(a)
$$3x^2 - 5xy - 7y^2 = \begin{bmatrix} x & y \end{bmatrix} \begin{bmatrix} 3 & -\frac{5}{2} \\ -\frac{5}{2} & -7 \end{bmatrix} \begin{bmatrix} x \\ y \end{bmatrix}.$$

(b)
$$3x^2 - 7xy + 5xz + 4y^2 - 4yz - 3z^2$$

$$= \begin{bmatrix} x & y & z \end{bmatrix} \begin{bmatrix} 3 & -\frac{7}{2} & \frac{5}{2} \\ -\frac{7}{2} & 4 & -2 \\ \frac{5}{2} & -2 & -3 \end{bmatrix} \begin{bmatrix} x \\ y \\ z \end{bmatrix}.$$

Suppose now that $\mathbf{Q(X)} = \mathbf{X}^T\mathbf{AX}$ is a quadratic form. To simplify the quadratic form, we change from the variables $x_1, x_2, \ldots, x_n$ to the variables $y_1, y_2, \ldots, y_n$. Suppose that the old variables are related to the new variables by $\mathbf{X} = \mathbf{PY}$ for some orthogonal matrix $\mathbf{P}$. Then

$$\mathbf{Q(X)} = \mathbf{X}^T\mathbf{AX} = (\mathbf{PY})^T\mathbf{A}(\mathbf{PY}) = \mathbf{Y}^T(\mathbf{P}^T\mathbf{AP})\mathbf{Y} = \mathbf{Y}^T\mathbf{BY},$$

where $\mathbf{B} = \mathbf{P}^T\mathbf{AP}$. We shall let the reader verify the fact that if $\mathbf{A}$ is a symmetric matrix, then $\mathbf{P}^T\mathbf{AP}$ is also symmetric (Exercise T.1). Thus

$$\mathbf{Q}'(\mathbf{Y}) = \mathbf{Y}^T\mathbf{BY}$$

is another quadratic form and $\mathbf{Q(X)} = \mathbf{Q}'(\mathbf{Y})$.

This situation is important enough to formulate the following definitions.

DEFINITION Two $n \times n$ matrices $\mathbf{A}$ and $\mathbf{B}$ are said to be **congruent** if $\mathbf{B} = \mathbf{P}^T\mathbf{AP}$ for a nonsingular matrix $\mathbf{P}$.

DEFINITION Two quadratic forms $\mathbf{Q}$ and $\mathbf{Q}'$ with matrices $\mathbf{A}$ and $\mathbf{B}$, respectively, are said to be **equivalent** if $\mathbf{A}$ and $\mathbf{B}$ are congruent.

Example 4. Consider the quadratic form in the variables x and y defined by

$$\mathbf{Q}(\mathbf{X}) = 2x^2 + 2xy + 2y^2 = \begin{bmatrix} x & y \end{bmatrix} \begin{bmatrix} 2 & 1 \\ 1 & 2 \end{bmatrix} \begin{bmatrix} x \\ y \end{bmatrix}. \tag{3}$$

We now change from the variables x and y to the variables x' and y'. Suppose that the old variables are related to the new variables by the equations

$$x = \frac{1}{\sqrt{2}} x' + \frac{1}{\sqrt{2}} y' \quad \text{and} \quad y = \frac{1}{\sqrt{2}} x' - \frac{1}{\sqrt{2}} y', \tag{4}$$

which can be written in matrix form as

$$\mathbf{X} = \begin{bmatrix} x \\ y \end{bmatrix} = \begin{bmatrix} \frac{1}{\sqrt{2}} & \frac{1}{\sqrt{2}} \\ \frac{1}{\sqrt{2}} & -\frac{1}{\sqrt{2}} \end{bmatrix} \begin{bmatrix} x' \\ y' \end{bmatrix} = \mathbf{PY},$$

where the orthogonal (hence nonsingular) matrix

$$\mathbf{P} = \begin{bmatrix} \frac{1}{\sqrt{2}} & \frac{1}{\sqrt{2}} \\ \frac{1}{\sqrt{2}} & -\frac{1}{\sqrt{2}} \end{bmatrix} \quad \text{and} \quad \mathbf{Y} = \begin{bmatrix} x' \\ y' \end{bmatrix}.$$

We shall soon see why and how this particular matrix $\mathbf{P}$ was selected. Substituting in (3), we obtain

$$\mathbf{Q}(\mathbf{X}) = \mathbf{X}^T\mathbf{A}\mathbf{X} = (\mathbf{PY})^T\mathbf{A}(\mathbf{PY}) = \mathbf{Y}^T\mathbf{P}^T\mathbf{A}\mathbf{P}\mathbf{Y}$$

$$= \begin{bmatrix} x' & y' \end{bmatrix} \begin{bmatrix} \frac{1}{\sqrt{2}} & \frac{1}{\sqrt{2}} \\ \frac{1}{\sqrt{2}} & -\frac{1}{\sqrt{2}} \end{bmatrix}^T \begin{bmatrix} 2 & 1 \\ 1 & 2 \end{bmatrix} \begin{bmatrix} \frac{1}{\sqrt{2}} & \frac{1}{\sqrt{2}} \\ \frac{1}{\sqrt{2}} & -\frac{1}{\sqrt{2}} \end{bmatrix} \begin{bmatrix} x' \\ y' \end{bmatrix}$$

$$= \begin{bmatrix} x' & y' \end{bmatrix} \begin{bmatrix} 3 & 0 \\ 0 & 1 \end{bmatrix} \begin{bmatrix} x' \\ y' \end{bmatrix} = \mathbf{Q}'(\mathbf{Y}).$$

Thus the matrices

$$\begin{bmatrix} 2 & 1 \\ 1 & 2 \end{bmatrix} \quad \text{and} \quad \begin{bmatrix} 3 & 0 \\ 0 & 1 \end{bmatrix}$$

are congruent and the quadratic forms $\mathbf{Q}$ and $\mathbf{Q}'$ are equivalent.

The equation

$$\mathbf{Q}(\mathbf{X}) = 2x^2 + 2xy + 2y^2 = 9 \tag{5}$$

represents a conic section. Since $\mathbf{Q}$ is the quadratic form defined in Example 4, it is equivalent to the quadratic form

$$\mathbf{Q}'(\mathbf{Y}) = 3x'^2 + y'^2.$$

Now the equation

$$\mathbf{Q}'(\mathbf{Y}) = 3x'^2 + y'^2 = 9 \tag{6}$$

is the equation of an ellipse. We can then conclude that the conic section whose equation is (6) is the same ellipse as that represented by (5).

We now turn to the question of how to select the matrix $\mathbf{P}$.

THEOREM 7.1 (The Principal Axis Theorem). *Any quadratic form in n variables $\mathbf{Q}(\mathbf{X}) = \mathbf{X}^T\mathbf{A}\mathbf{X}$ is equivalent to a quadratic form, $\mathbf{Q}'(\mathbf{Y}) = \lambda_1 y_1^2 + \lambda_2 y_2^2 + \cdots + \lambda_n y_n^2$, where*

$$\mathbf{Y} = \begin{bmatrix} y_1 \\ y_2 \\ \cdot \\ \cdot \\ \cdot \\ y_n \end{bmatrix}$$

and $\lambda_1, \lambda_2, \ldots, \lambda_n$ are the eigenvalues of the matrix $\mathbf{A}$ of $\mathbf{Q}$.

proof. If $\mathbf{A}$ is the matrix of $\mathbf{Q}$, then, since $\mathbf{A}$ is symmetric, we know by Theorem 5.7 that $\mathbf{A}$ can be diagonalized by an orthogonal matrix. This means that there exists an orthogonal matrix $\mathbf{P}$ such that $\mathbf{B} = \mathbf{P}^{-1}\mathbf{A}\mathbf{P}$ is a diagonal matrix. Since $\mathbf{P}$ is orthogonal, $\mathbf{P}^{-1} = \mathbf{P}^T$, so $\mathbf{B} = \mathbf{P}^T\mathbf{A}\mathbf{P}$. Moreover, the elements on the main diagonal of $\mathbf{B}$ are the eigenvalues

$\lambda_1, \lambda_2, \ldots, \lambda_n$ of **A**. The quadratic form **Q**′ with matrix **B** is given by

$$\mathbf{Q}'(\mathbf{Y}) = \lambda_1 y_1^2 + \lambda_2 y_2^2 + \cdots + \lambda_n y_n^2;$$

Q and **Q**′ are equivalent.

Example 5. Consider the quadratic form **Q** in the variables x, y, and z, defined by

$$\mathbf{Q}(\mathbf{X}) = 2x^2 + 4y^2 + 6yz - 4z^2.$$

The matrix of **Q** is

$$\mathbf{A} = \begin{bmatrix} 2 & 0 & 0 \\ 0 & 4 & 3 \\ 0 & 3 & -4 \end{bmatrix}$$

and the eigenvalues of **A** are

$$\lambda_1 = 2, \qquad \lambda_2 = 5, \qquad \text{and} \qquad \lambda_3 = -5.$$

Let **Q**′ be the quadratic form in the variables x', y', and z' defined by

$$\mathbf{Q}'(\mathbf{Y}) = 2x'^2 + 5y'^2 - 5z'^2.$$

Then **Q** and **Q**′ are equivalent. Now

$$\mathbf{D} = \begin{bmatrix} 2 & 0 & 0 \\ 0 & 5 & 0 \\ 0 & 0 & -5 \end{bmatrix}$$

is the matrix of **Q**′; **A** and **D** are congruent matrices.

Note that to apply Theorem 7.1 to diagonalize a given quadratic form, as shown in Example 5, we do not need to know the eigenvectors of **A** (nor the matrix **P**); we only require the eigenvalues of **A**.

We now return to quadratic forms in two and three variables. The equation $\mathbf{X}^T\mathbf{A}\mathbf{X} = 1$, where **X** is a vector in R^2 is the locus of a central conic in the plane. From Theorem 7.1 it follows that there is a Cartesian coordinate system in the plane with respect to which the equation of the central conic is $ax^2 + by^2 = 1$, where a and b are real numbers. Similarly, the equation $\mathbf{X}^T\mathbf{A}\mathbf{X} = 1$, where **X** is a vector in R^3, is the locus of a central quadric surface in R^3. From Theorem 7.1 it follows that there is a Cartesian coordinate system in 3-space with respect to which the equation of the quadric surface is $ax^2 + by^2 + cz^2 = 1$, where a, b, and c are real numbers. The principal axes of the conic or surface lie along the coordinate axes and this is the reason for calling Theorem 7.1 the **principal axis theorem**.

Example 6. Consider the conic section whose equation is (5),

$$\mathbf{Q}(\mathbf{X}) = 2x^2 + 2xy + 2y^2 = 9.$$

This conic section can also be described by the equation

$$\mathbf{Q}'(\mathbf{Y}) = 3x'^2 + y'^2 = 9,$$

which can be rewritten as

$$\frac{x'^2}{3} + \frac{y'^2}{9} = 1.$$

This is an ellipse (Figure 7.4) whose major axis is along the y'-axis. The

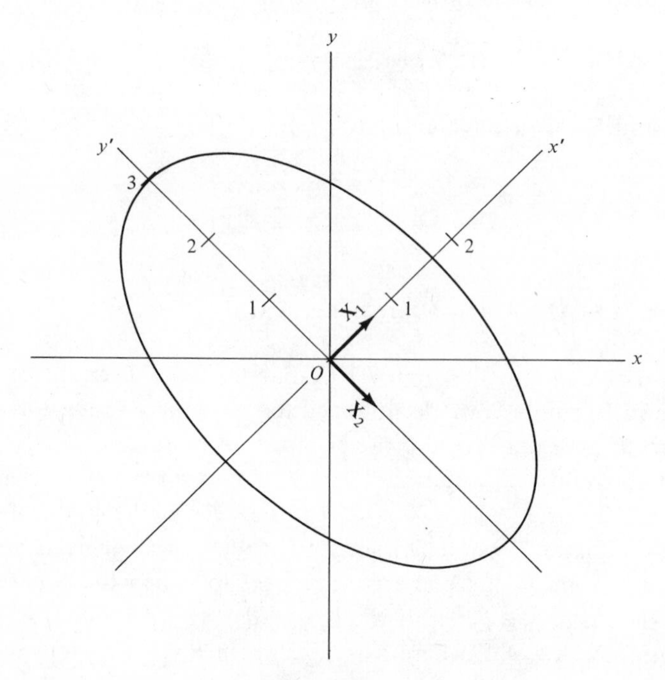

Figure 7.4

major axis is of length 3; the minor axis is of length $\sqrt{3}$. We now note that there is a very close connection between the eigenvectors of the matrix of (5) and the location of the x'- and y'-axes.

Since $\mathbf{X} = \mathbf{PY}$, we have $\mathbf{Y} = \mathbf{P}^{-1}\mathbf{X} = \mathbf{P}^T\mathbf{X} = \mathbf{PX}$ ($\mathbf{P}$ is orthogonal and symmetric). Thus

$$x' = \frac{1}{\sqrt{2}}x + \frac{1}{\sqrt{2}}y \quad \text{and} \quad y' = \frac{1}{\sqrt{2}}x - \frac{1}{\sqrt{2}}y.$$

This means that, in terms of the x- and y-axes, the x'-axis lies along the vector

$$\mathbf{X}_1 = \begin{bmatrix} \dfrac{1}{\sqrt{2}} \\ \dfrac{1}{\sqrt{2}} \end{bmatrix}$$

and the y'-axis lies along the vector

$$\mathbf{X}_2 = \begin{bmatrix} \dfrac{1}{\sqrt{2}} \\ -\dfrac{1}{\sqrt{2}} \end{bmatrix}.$$

Now $\mathbf{X}_1$ and $\mathbf{X}_2$ are the columns of the matrix

$$\mathbf{P} = \begin{bmatrix} \dfrac{1}{\sqrt{2}} & \dfrac{1}{\sqrt{2}} \\ \dfrac{1}{\sqrt{2}} & -\dfrac{1}{\sqrt{2}} \end{bmatrix},$$

which in turn are eigenvectors of the matrix of (5). Thus the x'- and y'-axes lie along the eigenvectors of the matrix of (5) (see Figure 7.4).

The situation described in Example 6 is true in general. Thus the principal axes of a conic or surface lie along the eigenvectors of the matrix of the quadratic form.

Let $\mathbf{Q}(\mathbf{X}) = \mathbf{X}^T\mathbf{AX}$ be a quadratic form in n variables. Then we know that $\mathbf{Q}$ is equivalent to the quadratic form $\mathbf{Q}'(\mathbf{Y}) = \lambda_1 y_1^2 + \lambda_2 y_2^2 + \cdots + \lambda_n y_n^2$, where $\lambda_1, \lambda_2, \ldots, \lambda_n$ are the eigenvalues of the matrix $\mathbf{A}$ of $\mathbf{Q}$. We can label the eigenvalues $\lambda_1, \lambda_2, \ldots, \lambda_n$, so that all the positive eigenvalues of **A**, if any, are listed first, followed by all the negative eigenvalues of **A**, if any, followed by the zero eigenvalues, if any. Thus, let $\lambda_1, \lambda_2, \ldots, \lambda_p$ be positive, $\lambda_{p+1}, \lambda_{p+2}, \ldots, \lambda_r$ be negative, and $\lambda_{r+1}, \lambda_{r+2}, \ldots, \lambda_n$ be zero. We now define the diagonal matrix **H** whose entries on the main diagonal are

$$\frac{1}{\sqrt{\lambda_1}}, \frac{1}{\sqrt{\lambda_2}}, \ldots, \frac{1}{\sqrt{\lambda_p}}, \frac{1}{\sqrt{-\lambda_{p+1}}}, \frac{1}{\sqrt{-\lambda_{p+2}}},$$

$$\ldots, \frac{1}{\sqrt{-\lambda_r}}, 1, 1, \ldots, 1,$$

with $n - r$ ones. Let **D** be the diagonal matrix whose entries on the main diagonal are $\lambda_1, \lambda_2, \ldots, \lambda_p, \lambda_{p+1}, \ldots, \lambda_r, \lambda_{r+1}, \ldots, \lambda_n$; **A** and **D** are congruent. Let $\mathbf{D}_1 = \mathbf{H}^T\mathbf{D}\mathbf{H}$ be the matrix whose diagonal elements are 1, $1, \ldots, 1, -1, \ldots, -1, 0, 0, \ldots, 0$ (p ones, $n - r$ zeros); **D** and $\mathbf{D}_1$ are then congruent. From Exercise T.2 it follows that **A** and $\mathbf{D}_1$ are congruent. In terms of quadratic forms, we have established Theorem 7.2.

THEOREM 7.2. *A quadratic form* $\mathbf{Q}(\mathbf{X}) = \mathbf{X}^T\mathbf{A}\mathbf{X}$ *in n variables is equivalent to a quadratic form* $\mathbf{Q}'(\mathbf{Y}) = y_1^2 + y_2^2 + \cdots + y_p^2 - y_{p+1}^2 - y_{p+2}^2 - \cdots - y_r^2$.

It is clear that the rank of the matrix $\mathbf{D}_1$ is r, the number of nonzero entries on its main diagonal. Now it can be shown that congruent matrices have equal ranks. Since the rank of $\mathbf{D}_1$ is r, the rank of **A** is also r. We also refer to r as the **rank** of the quadratic form **Q** whose matrix is **A**. It can be shown that the number p of positive terms in the quadratic form $\mathbf{Q}'$ of Theorem 7.2 is unique; that is, no matter how we simplify the given quadratic form **Q** to obtain an equivalent quadratic form, the latter will always have p positive terms. Hence the quadratic form $\mathbf{Q}'$ in Theorem 7.2 is unique; it is often called the **canonical form** of a quadratic form in n variables. The difference between the number of positive eigenvalues and the number of negative eigenvalues is $s = p - (r - p) = 2p - r$ and is called the **signature** of the quadratic form. Thus, if **Q** and $\mathbf{Q}'$ are equivalent quadratic forms, then they have equal ranks and signatures. However, it can also be shown that if **Q** and $\mathbf{Q}'$ have equal ranks and signatures, then they are equivalent.

Example 7. Consider the quadratic form

$$Q(X) = 3x_2^2 + 8x_2x_3 - 3x_3^2 = X^TAX$$

$$= \begin{bmatrix} x_1 & x_2 & x_3 \end{bmatrix} \begin{bmatrix} 0 & 0 & 0 \\ 0 & 3 & 4 \\ 0 & 4 & -3 \end{bmatrix} \begin{bmatrix} x_1 \\ x_2 \\ x_3 \end{bmatrix}.$$

The eigenvalues of **A** are

$$\lambda_1 = 5, \qquad \lambda_2 = -5, \qquad \text{and} \qquad \lambda_3 = 0.$$

In this case **A** is congruent to

$$D = \begin{bmatrix} 5 & 0 & 0 \\ 0 & -5 & 0 \\ 0 & 0 & 0 \end{bmatrix}.$$

If we let

$$H = \begin{bmatrix} \frac{1}{\sqrt{5}} & 0 & 0 \\ 0 & \frac{1}{\sqrt{5}} & 0 \\ 0 & 0 & 1 \end{bmatrix},$$

then

$$D_1 = H^TDH = \begin{bmatrix} 1 & 0 & 0 \\ 0 & -1 & 0 \\ 0 & 0 & 0 \end{bmatrix}$$

and **A** are congruent, and the given quadratic form is equivalent to the canonical form

$$Q'(Y) = y_1^2 - y_2^2.$$

The rank of **Q** is 2, and since $p = 1$, the signature $s = 2p - r = 0$.

As a final application of quadratic forms we consider positive-definite symmetric matrices.

DEFINITION A symmetric $n \times n$ matrix $\mathbf{A}$ is called **positive definite** if $\mathbf{X}^T\mathbf{A}\mathbf{X} > 0$ for every nonzero vector $\mathbf{X}$ in R^n.

If $\mathbf{A}$ is a symmetric matrix, then $\mathbf{X}^T\mathbf{A}\mathbf{X}$ is a quadratic form $\mathbf{Q}(\mathbf{X}) = \mathbf{X}^T\mathbf{A}\mathbf{X}$ and, by Theorem 7.1, $\mathbf{Q}$ is equivalent to $\mathbf{Q}'$, where $\mathbf{Q}'(\mathbf{Y}) = \lambda_1 y_1^2 + \lambda_2 y_2^2 + \cdots + \lambda_p y_p^2 + \lambda_{p+1} y_{p+1}^2 + \lambda_{p+2} y_{p+2}^2 + \cdots + \lambda_r y_r^2$. Now $\mathbf{A}$ is positive definite if and only if $\mathbf{Q}'(\mathbf{Y}) > 0$ for each $\mathbf{Y} \neq \mathbf{0}$. However, this can happen if and only if all summands in $\mathbf{Q}'(\mathbf{Y})$ are positive. These remarks have established the following theorem.

THEOREM 7.3. *A symmetric matrix* $\mathbf{A}$ *is positive definite if and only if all the eigenvalues of* $\mathbf{A}$ *are positive.*

A quadratic form is then called **positive definite** if its matrix is positive definite.

7.2 Exercises

1. Write each of the following quadratic forms as $\mathbf{X}^T\mathbf{A}\mathbf{X}$.
(a) $-3x^2 + 5xy - 2y^2$.
(b) $2x_1^2 + 3x_1x_2 - 5x_1x_3 + 7x_2x_3$.
(c) $3x_1^2 + x_1x_2 - 2x_1x_3 + x_2^2 - 4x_2x_3 - 2x_3^2$.

2. Write each of the following quadratic forms as $\mathbf{X}^T\mathbf{A}\mathbf{X}$.
(a) $x^2 - 4xy - 3y^2 + 6yz + 4z^2$.
(b) $4x^2 - 6xy + 2y^2$.
(c) $-2x_1x_2 + 4x_1x_3 + 6x_2x_3$.

3. For each of the following symmetric matrices $\mathbf{A}$ find a diagonal matrix $\mathbf{D}$ that is congruent to $\mathbf{A}$.

(a) $\mathbf{A} = \begin{bmatrix} -1 & 0 & 0 \\ 0 & 1 & 1 \\ 0 & 1 & 1 \end{bmatrix}$. (b) $\mathbf{A} = \begin{bmatrix} 1 & 1 & 1 \\ 1 & 1 & 1 \\ 1 & 1 & 1 \end{bmatrix}$.

In Exercises 4 through 8 find a quadratic form of the type in Theorem 7.1 that is equivalent to the given quadratic form.

4. $2x^2 - 4xy - y^2$.

5. $x_1^2 + x_2^2 + 2x_2x_3 + x_3^2$.

6. $2xz$.

7. $2x_2^2 + 4x_2x_3 + 2x_3^2$.

8. $-2x_1^2 - 4x_2^2 - 6x_2x_3 + 4x_3^2$.

In Exercises 9 through 13 find a quadratic form of the type in Theorem 7.2 that is equivalent to the given quadratic form.

9. $2x^2 + 4xy + 2y^2$.
10. $x_1^2 + 2x_1x_2 + x_2^2 + x_3^2$.
11. $2x_1^2 + 4x_2^2 + 10x_2x_3 + 4x_3^2$.
12. $2x^2 + 3y^2 + 4yz + 3z^2$.
13. $-3x_1^2 + 2x_2^2 + 4x_2x_3 + 2x_3^2$.

14. Let $\mathbf{Q}(\mathbf{X}) = 4x^2 - 10xy + 4y^2 = 1$ be the equation of a central conic. Identify the conic by finding a quadratic form of the type in Theorem 7.2 that is equivalent to $\mathbf{Q}$.
15. Let $\mathbf{Q}(\mathbf{X}) = x^2 + 2y^2 + 2yz + 2z^2 = 1$ be the equation of a central quadric surface. Identify the quadric surface by finding a quadratic form of the type in Theorem 7.2 that is equivalent to $\mathbf{Q}$.
16. Let $\mathbf{Q}(\mathbf{X}) = 4x_2^2 - 10x_2x_3 + 4x_3^2$ be a quadratic form in three variables. Find a quadratic form of the type in Theorem 7.2 that is equivalent to $\mathbf{Q}$. What is the rank of $\mathbf{Q}$? What is the signature of $\mathbf{Q}$?
17. Which of the following quadratic forms in three variables are equivalent?

$$\mathbf{Q}_1(\mathbf{X}) = x_1^2 + 2x_1x_2 + x_2^2 + x_3^2$$

$$\mathbf{Q}_2(\mathbf{X}) = 2x_2^2 + 2x_2x_3 + 2x_3^2$$

$$\mathbf{Q}_3(\mathbf{X}) = 3x_2^2 + 8x_2x_3 - 3x_3^2$$

$$\mathbf{Q}_4(\mathbf{X}) = 3x_2^2 - 4x_2x_3 + 3x_3^2.$$

18. Which of the following matrices are positive definite?

(a) $\begin{bmatrix} 2 & -1 \\ -1 & 2 \end{bmatrix}$. (b) $\begin{bmatrix} 2 & 1 \\ 1 & 2 \end{bmatrix}$. (c) $\begin{bmatrix} 0 & -1 \\ -1 & 0 \end{bmatrix}$.

(d) $\begin{bmatrix} 1 & 1 \\ 1 & 1 \end{bmatrix}$. (e) $\begin{bmatrix} 2 & 2 \\ 2 & 2 \end{bmatrix}$.

Theoretical Exercises

T.1. Prove that if $\mathbf{A}$ is a symmetric matrix, then $\mathbf{P}^T\mathbf{A}\mathbf{P}$ is also symmetric.

T.2. If $\mathbf{A}$, $\mathbf{B}$, and $\mathbf{C}$ are $n \times n$ symmetric matrices, prove that

(a) $\mathbf{A}$ and $\mathbf{A}$ are congruent.

(b) If $\mathbf{A}$ and $\mathbf{B}$ are congruent, then $\mathbf{B}$ and $\mathbf{A}$ are congruent.

(c) If $\mathbf{A}$ and $\mathbf{B}$ are congruent and if $\mathbf{B}$ and $\mathbf{C}$ are congruent, then $\mathbf{A}$ and $\mathbf{C}$ are congruent.

T.3. Prove that if $\mathbf{A}$ is symmetric, then $\mathbf{A}$ is congruent to a diagonal matrix $\mathbf{D}$.

T.4. Find all quadratic forms $\mathbf{Q}(\mathbf{X}) = \mathbf{X}^T\mathbf{A}\mathbf{X}$ in two variables of the type described in Theorem 7.2. What conics do the equations $\mathbf{X}^T\mathbf{A}\mathbf{X} = 1$ represent?

T.5. Find all quadratic forms $\mathbf{Q}(\mathbf{X}) = \mathbf{X}^T\mathbf{A}\mathbf{X}$ in two variables of rank 1 of the type described in Theorem 7.2. What conics do the equations $\mathbf{X}^T\mathbf{A}\mathbf{X} = 1$ represent?

T.6. Let $\mathbf{A} = \begin{bmatrix} a & b \\ b & d \end{bmatrix}$ be a 2×2 symmetric matrix. Prove that $\mathbf{A}$ is positive definite if and only if $|\mathbf{A}| > 0$ and $a > 0$.

T.7. Prove that a symmetric matrix $\mathbf{A}$ is positive definite if and only if $\mathbf{A} = \mathbf{P}^T\mathbf{P}$ for a nonsingular matrix $\mathbf{P}$.

7.3 Graph Theory

Graph theory is a new area of applied mathematics which is being widely used in formulating models in many problems in business, the social sciences, and the physical sciences. These applications include communications problems and the study of organizations and social structures. In this section we present a very brief introduction to the subject as it relates to matrices and show how these elementary and simple notions can be used in formulating models of some important problems.

Graphs

DEFINITION A **graph** G is a finite set of points, called **vertices**, together with a finite set of **edges**, each of which joins a pair of vertices.

The vertices of a graph are represented by dots and the edges by straight lines or by arcs.

Example 1. Figure 7.5 shows examples of graphs. The graph in Figure 7.5(a) has four vertices, P_1, P_2, P_3, and P_4, and six edges, P_1P_2, P_1P_3, P_1P_4, P_2P_3, P_2P_4, and P_3P_4. The edges P_1P_4 and P_2P_3 intersect at a point other than a vertex, so they are drawn as shown. The graph in Figure 7.5(b) has four vertices, P_1, P_2, P_3, and P_4, and two edges, P_1P_2 and P_1P_3. Vertices P_2 and P_3 are not joined by an edge. Moreover, vertex P_4 is not joined to any vertex. The graph in Figure 7.5(c) has three vertices, P_1, P_2, and P_3, and three edges, P_1P_2, P_1P_3, and P_3P_3. An edge of the form P_3P_3, joining a vertex to itself, is called a **loop**.

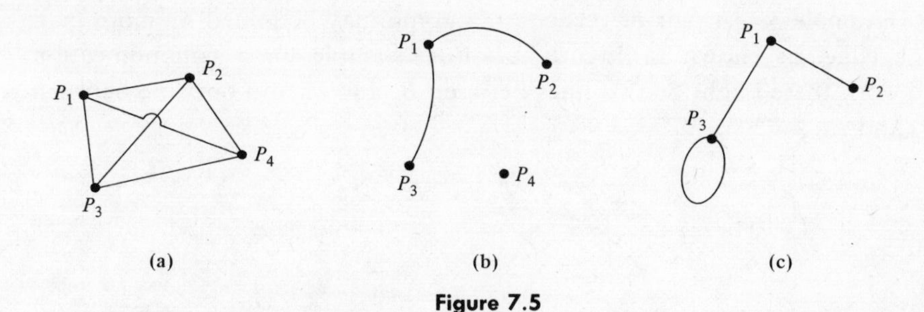

Figure 7.5

A graph is called **complete** if every pair of vertices is joined by an edge. The complete graph with n vertices is denoted by K_n. The graph in Figure 7.5(a) is the complete graph K_4.

Example 2. A bowling league consists of seven teams: T_1, T_2, T_3, T_4, T_5, T_6, and T_7. Suppose that after a number of games have been played we have the following situation:

T_1 has played T_2, T_3, and T_5
T_2 has played T_1, T_3, and T_5
T_3 has played T_1, T_2, and T_4
T_4 has played T_3 and T_7
T_5 has played T_1 and T_2
T_6 has not played anyone
T_7 has played T_4.

This situation can be described by a graph in which the teams are represented by the vertices, and an edge joining two vertices means that the corresponding teams have played each other. We thus obtain the graph in Figure 7.6.

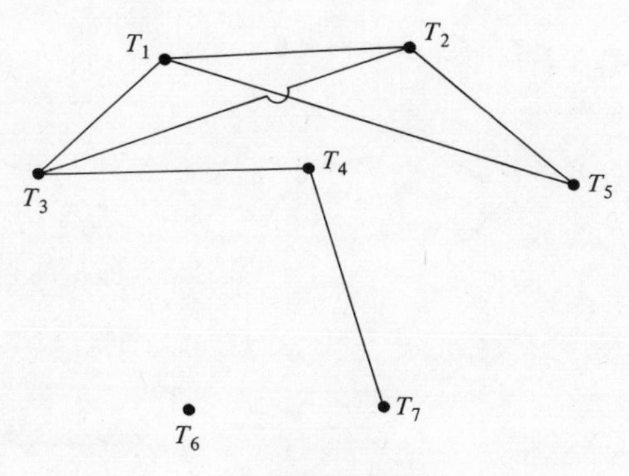

Figure 7.6

Example 3. A pair of vertices in a graph may be joined by more than one edge, as shown in Figure 7.7. For example, in a communication network there might be two lines between P_1 and P_2 and one line between P_2 and P_3.

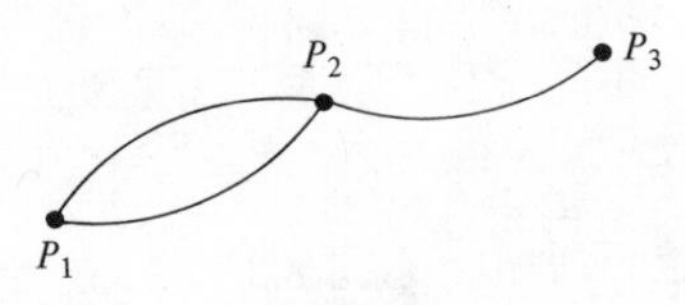

Figure 7.7

A graph is determined by the vertices and edges joining the vertices, not by the particular appearance of the configuration. Thus two graphs G and G' are said to be **equal** if they have the same number of vertices P_1, $P_2, \ldots, P_n$ and if in each case they can be relabeled in such a way that the number of edges between P_i and P_j is the same in both G and G'.

Example 4. The graph in Figure 7.8 is equal to the graph in Figure 7.5(a) (verify).

Matrices provide a convenient way of describing graphs, and since matrices lend themselves well to computer use, they make it possible to use the computer for extensive computational work in graph theory.

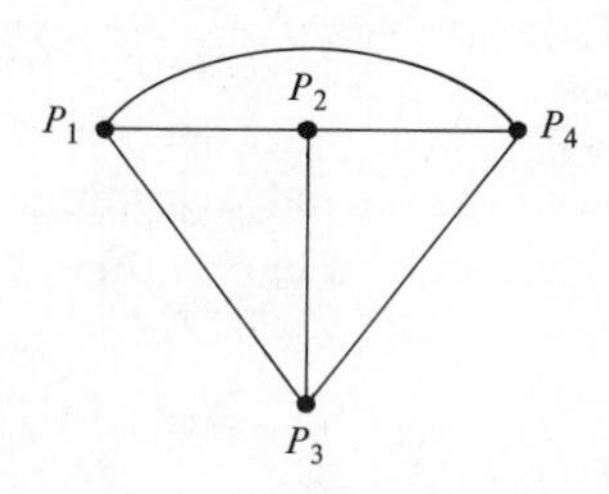

Figure 7.8

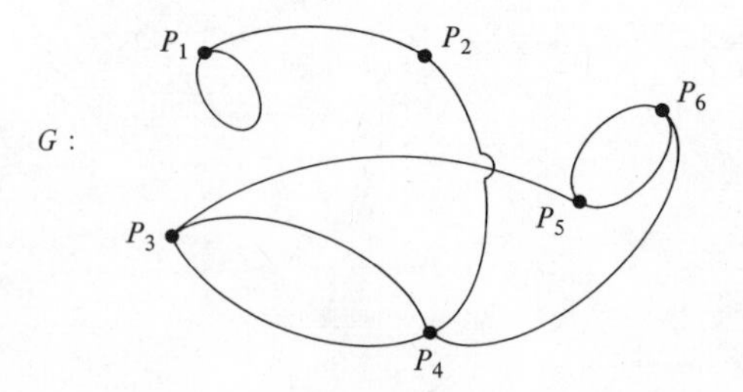

Figure 7.9

If we are given a graph G with n vertices, we can associate a matrix $\mathbf{M}(G)$ with it by letting the i, jth element be the number of edges from vertex P_i to vertex P_j. The matrix $\mathbf{M}(G)$ is called the **matrix representing** G.

Example 5. The matrix of the graph in Figure 7.9 is $\mathbf{M}(G)$.

$$\mathbf{M}(G) = \begin{matrix} & P_1 & P_2 & P_3 & P_4 & P_5 & P_6 \\ P_1 & 1 & 1 & 0 & 0 & 0 & 0 \\ P_2 & 1 & 0 & 0 & 1 & 0 & 0 \\ P_3 & 0 & 0 & 0 & 2 & 1 & 0 \\ P_4 & 0 & 1 & 2 & 0 & 0 & 1 \\ P_5 & 0 & 0 & 1 & 0 & 0 & 2 \\ P_6 & 0 & 0 & 0 & 1 & 2 & 0 \end{matrix}.$$

Conversely, a given $n \times n$ matrix gives rise to a graph G with the given matrix representing G.

Example 6. The matrix

$$\begin{bmatrix} 0 & 2 & 1 & 0 & 0 \\ 2 & 0 & 0 & 1 & 0 \\ 1 & 0 & 0 & 1 & 1 \\ 0 & 1 & 1 & 0 & 2 \\ 0 & 0 & 1 & 2 & 0 \end{bmatrix}$$

determines the graph G, shown in Figure 7.10.

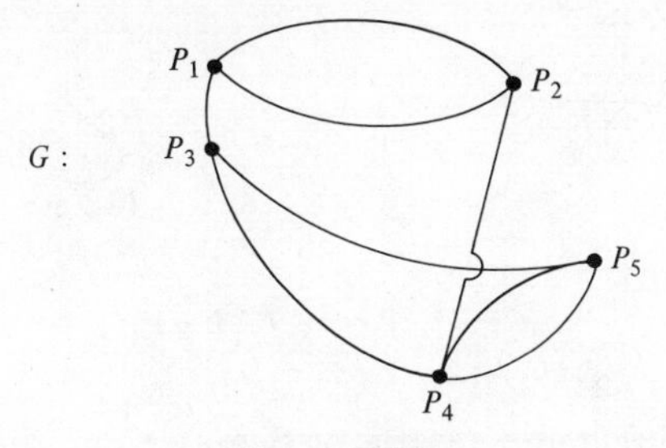

Figure 7.10

Of course, it is easy to see that graphs G and G' are equal if and only if there is a labeling of the vertices for which the matrices representing them are equal.

There is another matrix associated with a graph, which we now define.

DEFINITION If G is a graph that has n vertices, then the matrix $\mathbf{I}(G)$, whose i, jth element is 1 if there is at least one edge between P_i and P_j and zero otherwise, is called the **incidence matrix of** G. Note that $\mathbf{I}(G)$ is a symmetric matrix.

Example 7. The incidence matrix of the graph in Figure 7.9 is

$$\mathbf{I}(G) = \begin{bmatrix} 1 & 1 & 0 & 0 & 0 & 0 \\ 1 & 0 & 0 & 1 & 0 & 0 \\ 0 & 0 & 0 & 1 & 1 & 0 \\ 0 & 1 & 1 & 0 & 0 & 1 \\ 0 & 0 & 1 & 0 & 0 & 1 \\ 0 & 0 & 0 & 1 & 1 & 0 \end{bmatrix}.$$

Of course, from a matrix whose entries are zeros and ones we can obtain a graph whose incidence matrix is the given matrix. However, this graph need not be unique.

Example 8. The matrix

$$\begin{bmatrix} 0 & 1 & 0 & 1 \\ 1 & 0 & 1 & 0 \\ 0 & 1 & 1 & 1 \\ 1 & 0 & 1 & 0 \end{bmatrix}$$

gives rise to the graphs in Figure 7.11.

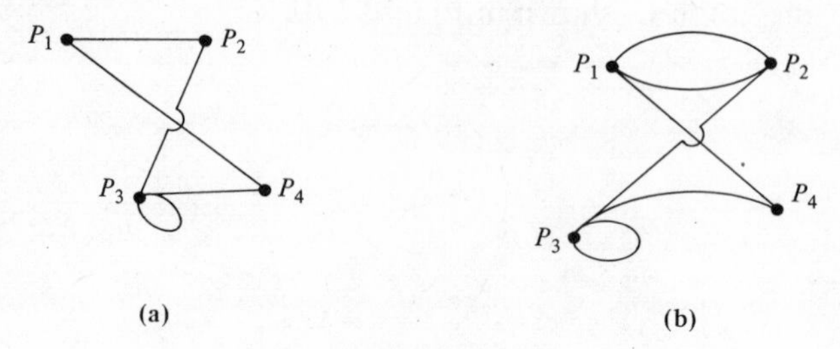

Figure 7.11

Sometimes the incidence matrix of G may equal the matrix representing G. This will occur if no vertex in G has multiple edges.

The graph of Example 2 indicated the teams that played each other. However, if we wanted to indicate the winner in each competition, we need to develop the notion of a directed graph.

Digraphs

DEFINITION A **directed graph**, or a **digraph**, is a finite set of points, called **vertices**, together with a finite set of **directed edges**, each of which joins a pair of distinct vertices. Thus a digraph contains no loops. Moreover, the directed edge P_iP_j is now different from the directed edge P_jP_i. The matrix $\mathbf{I}(G)$, whose i, jth element is 1 if there is a directed edge from P_i to P_j and zero otherwise, is called the **incidence matrix of** G. The incidence matrix of a digraph need not be symmetric.

Example 9. In Figure 7.12 are shown four examples of digraphs. The digraph in Figure 7.12(a) has vertices P_1, P_2, and P_3 and directed edges P_1P_2 and P_2P_3; the digraph in Figure 7.12(b) has vertices P_1, P_2, P_3, and P_4 and directed edges P_1P_2 and P_1P_3; the digraph in Figure 7.12(c) has vertices P_1, P_2, and P_3 and directed edges P_1P_2, P_1P_3, and P_3P_1; the digraph in Figure 7.12(d) has vertices P_1, P_2, and P_3 and directed edges P_2P_1, P_2P_3, P_1P_3, and P_3P_1. The directed edges P_1P_3 and P_3P_1 are indicated as $P_1 \leftrightarrow P_3$.

Example 10. Consider Example 2 again. Suppose we now know that

T_1 has defeated T_2 and T_5 and lost to T_3
T_2 has defeated T_5 and lost to T_1 and T_3
T_3 has defeated T_1 and T_2 and lost to T_4
T_4 has defeated T_3 and lost to T_7
T_5 has lost to T_1 and T_2
T_6 has not played anyone
T_7 has defeated T_4.

We now obtain the directed graph in Figure 7.13, where $T_i \to T_j$ means that T_i defeated T_j.

Of course, digraphs can be used in a great many situations including communications problems, family relationships, social structures, street maps, and ecological chains. We shall deal with some of these below and in the exercises.

Models in Sociology and in Communications

Suppose that we have n individuals P_1, P_2, . . . , P_n, some of whom are related to each other. We assume that no one is related to himself. Examples of such relations are:

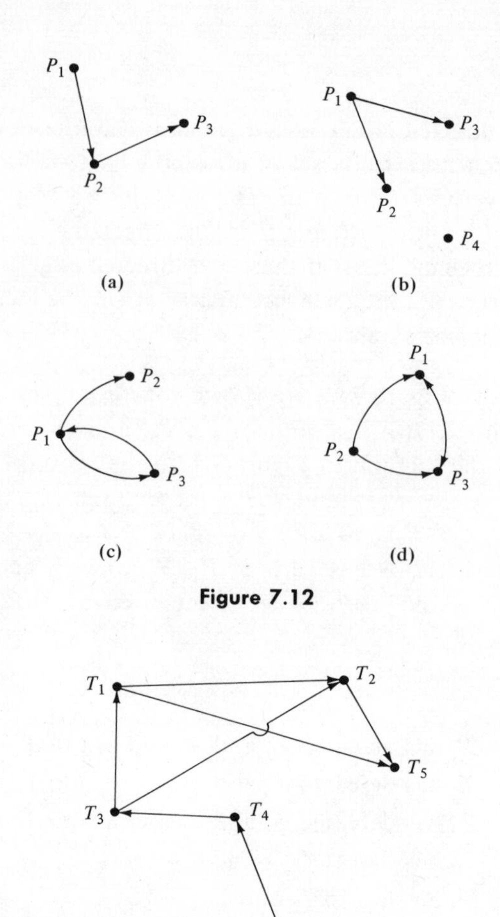

Figure 7.12

Figure 7.13

1. P_i has access to P_j. In this case it may or may not be the case that if P_i has access to P_j, then P_j has access to P_i. For example, many emergency telephones on turnpikes allow a distressed traveler to contact a nearby emergency station but make no provision for the station to contact the traveler. This model can thus be represented by a digraph G as follows. Let $P_1, P_2, \ldots, P_n$ be the vertices of G and draw a directed edge from P_i to P_j if P_i has access to P_j. It is important to observe that this relation need not be transitive. That is, P_i may have access to P_j and P_j may have access to P_k, but P_i need not have access to P_k.

2. P_i influences P_j. This situation is identical to that in (a): if P_i influences P_j, then it may or may not happen that P_j influences P_i.

3. For every pair of individuals P_i, P_j, either P_i dominates P_j or P_j dominates P_i. The graph representing this situation is the complete directed graph with n vertices. Such graphs are often called **dominance digraphs**.

Example 11. Suppose that six individuals have been meeting in group therapy for a long time and their leader, who is not part of the group, has drawn the digraph G in Figure 7.14 to describe the influence relations among the various individuals. The incidence matrix for G is

$$\mathbf{I}(G) = \begin{array}{c} \\ P_1 \\ P_2 \\ P_3 \\ P_4 \\ P_5 \\ P_6 \end{array} \begin{array}{c} \begin{array}{cccccc} P_1 & P_2 & P_3 & P_4 & P_5 & P_6 \end{array} \\ \begin{bmatrix} 0 & 0 & 0 & 0 & 1 & 0 \\ 0 & 0 & 0 & 0 & 1 & 0 \\ 1 & 1 & 0 & 0 & 1 & 0 \\ 0 & 1 & 0 & 0 & 1 & 0 \\ 0 & 0 & 0 & 0 & 0 & 0 \\ 0 & 1 & 0 & 1 & 0 & 0 \end{bmatrix} \end{array}.$$

Looking at the rows of $\mathbf{I}(G)$, we see that P_3 has three 1s in its row, so that P_3 influences three people—more than any other individual. Thus P_3 would be declared as the leader of the group. On the other hand, P_5 influences no one.

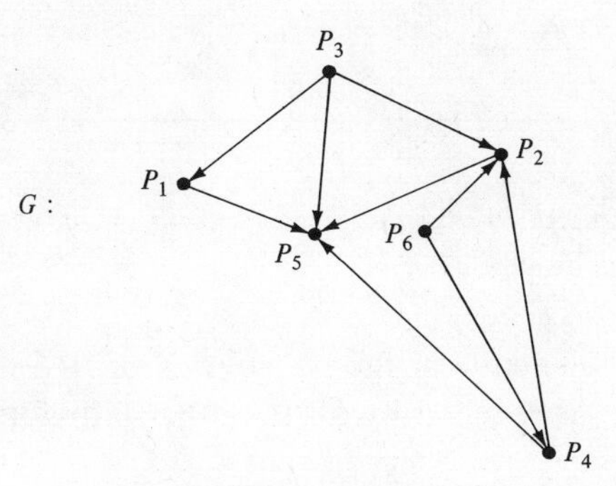

Figure 7.14

Example 12. Consider a communication network whose digraph G is shown in Figure 7.15. The incidence matrix for G is

$$\mathbf{I}(G) = \begin{array}{c} \\ P_1 \\ P_2 \\ P_3 \\ P_4 \\ P_5 \\ P_6 \end{array} \begin{array}{c} \begin{array}{cccccc} P_1 & P_2 & P_3 & P_4 & P_5 & P_6 \end{array} \\ \begin{bmatrix} 0 & 0 & 0 & 0 & 1 & 0 \\ 0 & 0 & 0 & 0 & 1 & 1 \\ 1 & 1 & 0 & 0 & 1 & 0 \\ 0 & 1 & 0 & 0 & 1 & 1 \\ 0 & 1 & 0 & 0 & 0 & 1 \\ 1 & 0 & 1 & 0 & 0 & 0 \end{bmatrix} \end{array}.$$

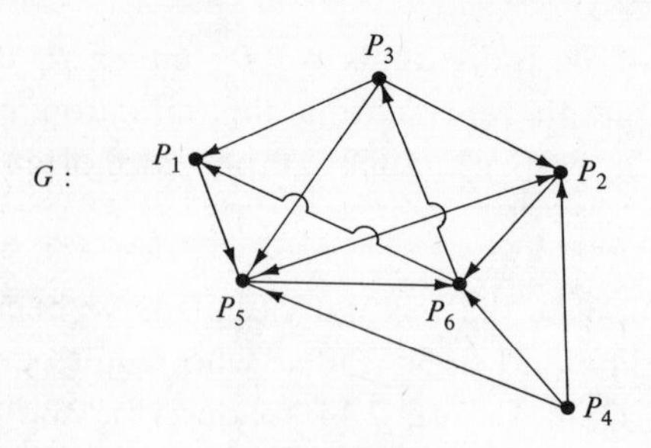

Figure 7.15

Although the relation "P_i has access to P_j" need not be transitive, we can speak of two-stage access. We say that P_i has **two-stage access** to P_k if we can find an individual P_j such that P_i has access to P_j and P_j has access to P_k. Similarly, P_i has r-stage access to P_k if we can find $r - 1$ individuals $P_{j_1}, \ldots, P_{j_{r-1}}$, such that P_i has access to P_{j_1}, P_{j_1} has access to $P_{j_2}, \ldots, P_{j_{r-2}}$ has access to $P_{j_{r-1}}$ and $P_{j_{r-1}}$ has access to P_k. Some of the r-individuals $P_{j_1}, \ldots, P_{j_{r-1}}$ may be the same.

The following theorem, whose proof we omit, can be established.

THEOREM 7.4. *Let* $\mathbf{I}(G)$ *be the incidence matrix of a digraph G and let the* kth *power of* $\mathbf{I}(G)$ *be* $\mathbf{A}$:

$$[\mathbf{I}(G)]^r = \mathbf{A} = [a_{ij}^{(r)}].$$

Then the i, jth *element in* $\mathbf{A}$, $a_{ij}^{(r)}$, *is the number of ways in which* P_i *has access to* P_j *in* r *stages.*

The sum of the elements in the jth column of $[\mathbf{I}(G)]^r$ gives the number of ways in which P_j is reached by all the other individuals in r stages.

If we let

$$\mathbf{I}(G) + [\mathbf{I}(G)]^2 + \cdots + [\mathbf{I}(G)]^r = \mathbf{B} = [b_{ij}],$$

then b_{ij} is the number of ways in which P_i has access to P_j in one, two, . . . , or r stages.

Similarly, we speak of r-stage dominance, r-stage influence, and so forth. We can also use this model to study the spread of a rumor. Thus b_{ij} in (1) is the number of ways in which P_i has spread the rumor to P_j in one, two, . . . , or r stages. In the influence relation, r-stage influence shows the effect of indirect influence.

Example 13. If G is the digraph in Example 12, then we find that

$$[\mathbf{I}(G)]^2 = \begin{array}{c} \\ P_1 \\ P_2 \\ P_3 \\ P_4 \\ P_5 \\ P_6 \end{array} \begin{array}{c} \begin{array}{cccccc} P_1 & P_2 & P_3 & P_4 & P_5 & P_6 \end{array} \\ \begin{bmatrix} 0 & 1 & 0 & 0 & 0 & 1 \\ 1 & 1 & 1 & 0 & 0 & 1 \\ 0 & 1 & 0 & 0 & 2 & 2 \\ 1 & 1 & 1 & 0 & 1 & 2 \\ 1 & 0 & 1 & 0 & 1 & 1 \\ 1 & 1 & 0 & 0 & 2 & 0 \end{bmatrix} \end{array}$$

and

$$\mathbf{I}(G) + [\mathbf{I}(G)]^2 = \mathbf{B} = \begin{array}{c} \\ P_1 \\ P_2 \\ P_3 \\ P_4 \\ P_5 \\ P_6 \end{array} \begin{array}{c} \begin{array}{cccccc} P_1 & P_2 & P_3 & P_4 & P_5 & P_6 \end{array} \\ \begin{bmatrix} 0 & 1 & 0 & 0 & 1 & 1 \\ 1 & 1 & 1 & 0 & 1 & 2 \\ 1 & 2 & 0 & 0 & 3 & 2 \\ 1 & 2 & 1 & 0 & 2 & 3 \\ 1 & 1 & 1 & 0 & 1 & 2 \\ 2 & 1 & 1 & 0 & 2 & 0 \end{bmatrix} \end{array}.$$

Since $b_{35} = 3$, there are three ways in which P_3 has access to P_5 in one or two stages: $P_3 \to P_5$, $P_3 \to P_2 \to P_5$, and $P_3 \to P_1 \to P_5$.

In studying organizational structures we often find subsets of people in which any pair of individuals are related. This is an example of a clique, which we now define.

DEFINITION A **clique** is the largest set of three or more persons such that any two individuals in the set communicate with each other (of course, we used the relation "communicate with each other" merely to be specific; other relations will do as well).

Example 14. Consider a digraph whose incidence matrix is

$$\begin{array}{c} \\ P_1 \\ P_2 \\ P_3 \\ P_4 \\ P_5 \end{array} \begin{array}{c} \begin{array}{ccccc} P_1 & P_2 & P_3 & P_4 & P_5 \end{array} \\ \begin{bmatrix} 0 & 1 & 1 & 1 & 0 \\ 1 & 0 & 1 & 1 & 0 \\ 1 & 1 & 0 & 1 & 1 \\ 1 & 1 & 1 & 0 & 0 \\ 0 & 0 & 0 & 1 & 0 \end{bmatrix} \end{array}.$$

In this case we find that the only clique is $\{P_1, P_2, P_3, P_4\}$. Observe that $\{P_1, P_2, P_3\}$ is not a clique, since it is not the largest set of individuals with the required properties.

For large groups, it is difficult to determine cliques. The following approach provides a useful method for detecting cliques that can easily be implemented on a computer. If $\mathbf{I}(G) = [a_{ij}]$ is the given incidence matrix, form a new matrix $\mathbf{S} = [s_{ij}]$:

$$s_{ij} = s_{ji} \quad \text{if} \quad a_{ij} = a_{ji} = 1;$$

otherwise, let $s_{ij} = s_{ji} = 0$. Thus $s_{ij} = 1$ if P_i and P_j communicate with each other; otherwise, $s_{ij} = 0$. It should be noted that $\mathbf{S}$ is a symmetric matrix ($\mathbf{S} = \mathbf{S}^T$).

Example 15. Consider a digraph with incidence matrix

$$\mathbf{I}(G) = \begin{array}{c} \\ P_1 \\ P_2 \\ P_3 \\ P_4 \\ P_5 \\ P_6 \end{array} \begin{array}{c} \begin{array}{cccccc} P_1 & P_2 & P_3 & P_4 & P_5 & P_6 \end{array} \\ \begin{bmatrix} 0 & 0 & 1 & 1 & 1 & 0 \\ 1 & 1 & 1 & 1 & 1 & 1 \\ 0 & 1 & 0 & 1 & 1 & 1 \\ 1 & 0 & 1 & 0 & 0 & 1 \\ 1 & 1 & 0 & 1 & 1 & 1 \\ 0 & 1 & 1 & 1 & 1 & 0 \end{bmatrix} \end{array}.$$

Then

$$\mathbf{S} = \begin{array}{c} \\ P_1 \\ P_2 \\ P_3 \\ P_4 \\ P_5 \\ P_6 \end{array} \begin{array}{c} \begin{array}{cccccc} P_1 & P_2 & P_3 & P_4 & P_5 & P_6 \end{array} \\ \begin{bmatrix} 0 & 0 & 0 & 1 & 1 & 0 \\ 0 & 1 & 1 & 0 & 1 & 1 \\ 0 & 1 & 0 & 1 & 0 & 1 \\ 1 & 0 & 1 & 0 & 0 & 1 \\ 1 & 1 & 0 & 0 & 1 & 1 \\ 0 & 1 & 1 & 1 & 1 & 0 \end{bmatrix} \end{array}.$$

The following theorem can be proved.

THEOREM 7.5. *Let $\mathbf{I}(G)$ be the incidence matrix of a digraph and let $\mathbf{S} = [s_{ij}]$ be the symmetric matrix defined above, with $\mathbf{S}^3 = [s_{ij}^{(3)}]$, where $s_{ij}^{(3)}$ is the i, jth element in $\mathbf{S}^3$. Then P_i belongs to a clique if and only if $s_{ii}^{(3)}$ is positive.*

Let us briefly consider why we examine the diagonal entries of $\mathbf{S}^3$ in Theorem 7.5. First, note that the diagonal entry $s_{ii}^{(3)}$ of $\mathbf{S}^3$ gives the number of ways in which P_i communicates with himself in three stages. If $s_{ii}^{(3)} > 0$, then there is at least one way in which P_i communicates with himself. Since a digraph has no loops, this communication must occur through two individuals: $P_i \to P_j \to P_k \to P_i$. Thus $s_{ij} \neq 0$. But $s_{ij} \neq 0$ implies that $s_{ji} \neq 0$; so $P_j \to P_i$. Similarly, any two of the individuals in $\{P_i, P_j, P_k\}$ communicate with each other. This means that P_i, P_j, and P_k all belong to the same clique.

Example 16. If

$$\mathbf{I}(G) = \begin{array}{c} \\ P_1 \\ P_2 \\ P_3 \\ P_4 \\ P_5 \end{array} \begin{array}{c} \begin{array}{ccccc} P_1 & P_2 & P_3 & P_4 & P_5 \end{array} \\ \begin{bmatrix} 0 & 1 & 1 & 0 & 1 \\ 1 & 0 & 0 & 0 & 1 \\ 1 & 0 & 0 & 1 & 0 \\ 0 & 1 & 1 & 0 & 0 \\ 1 & 0 & 0 & 0 & 0 \end{bmatrix} \end{array},$$

then

$$\mathbf{S} = \begin{array}{c} \\ P_1 \\ P_2 \\ P_3 \\ P_4 \\ P_5 \end{array} \begin{array}{c} \begin{array}{ccccc} P_1 & P_2 & P_3 & P_4 & P_5 \end{array} \\ \begin{bmatrix} 0 & 1 & 1 & 0 & 1 \\ 1 & 0 & 0 & 0 & 0 \\ 1 & 0 & 0 & 1 & 0 \\ 0 & 0 & 1 & 0 & 0 \\ 1 & 0 & 0 & 0 & 0 \end{bmatrix} \end{array}$$

and

$$\mathbf{S}^3 = \begin{array}{c} \\ P_1 \\ P_2 \\ P_3 \\ P_4 \\ P_5 \end{array} \begin{array}{c} \begin{array}{ccccc} P_1 & P_2 & P_3 & P_4 & P_5 \end{array} \\ \begin{bmatrix} 0 & 3 & 4 & 0 & 3 \\ 3 & 0 & 0 & 1 & 0 \\ 4 & 0 & 0 & 2 & 0 \\ 0 & 1 & 2 & 0 & 1 \\ 3 & 0 & 0 & 1 & 0 \end{bmatrix} \end{array}.$$

Since every diagonal entry in $\mathbf{S}^3$ is zero, we conclude that there are no cliques.

Example 17. If

$$\mathbf{I}(G) = \begin{array}{c} \\ P_1 \\ P_2 \\ P_3 \\ P_4 \\ P_5 \end{array} \begin{array}{c} \begin{array}{ccccc} P_1 & P_2 & P_3 & P_4 & P_5 \end{array} \\ \begin{bmatrix} 0 & 0 & 1 & 1 & 1 \\ 0 & 0 & 1 & 0 & 1 \\ 1 & 0 & 0 & 0 & 1 \\ 0 & 0 & 1 & 0 & 1 \\ 1 & 1 & 1 & 0 & 0 \end{bmatrix} \end{array},$$

then

$$\mathbf{S} = \begin{array}{c} \\ P_1 \\ P_2 \\ P_3 \\ P_4 \\ P_5 \end{array} \begin{array}{c} \begin{array}{ccccc} P_1 & P_2 & P_3 & P_4 & P_5 \end{array} \\ \begin{bmatrix} 0 & 0 & 1 & 0 & 1 \\ 0 & 0 & 0 & 0 & 1 \\ 1 & 0 & 0 & 0 & 1 \\ 0 & 0 & 0 & 0 & 0 \\ 1 & 1 & 1 & 0 & 0 \end{bmatrix} \end{array}$$

and

$$\mathbf{S}^3 = \begin{array}{c} \\ P_1 \\ P_2 \\ P_3 \\ P_4 \\ P_5 \end{array} \begin{array}{c} \begin{array}{ccccc} P_1 & P_2 & P_3 & P_4 & P_5 \end{array} \\ \begin{bmatrix} 2 & 1 & 3 & 0 & 4 \\ 1 & 0 & 1 & 0 & 3 \\ 3 & 1 & 2 & 0 & 4 \\ 0 & 0 & 0 & 0 & 0 \\ 4 & 3 & 4 & 0 & 2 \end{bmatrix} \end{array}.$$

Since s_{11}, s_{33}, and s_{55} are positive, we conclude that P_1, P_3, and P_5 form a clique.

We now consider the notion of a connected graph.

DEFINITION A **path** joining two individuals P_i and P_k in a digraph is a collection of distinct vertices $P_i, P_a, P_b, P_c, \ldots, P_r, P_k$ and directed edges $P_iP_a, P_aP_b, \ldots, P_rP_k$.

Example 18. Consider the digraph in Figure 7.16. The collection

$$P_2 \to P_3 \to P_4 \to P_5$$

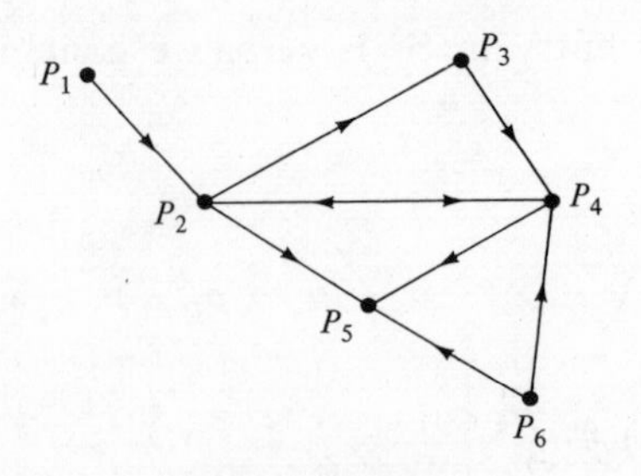

Figure 7.16

is a path. The collection

$$P_2 \to P_3 \to P_4 \to P_2 \to P_5$$

is not a path, since the vertex P_2 is repeated.

DEFINITION The digraph G is said to be **connected** if there is a path between every two distinct vertices P_i and P_j. Otherwise, G is said to be **disconnected**. That is, G is disconnected if there is a pair of vertices that cannot be joined by a path.

For many digraphs it is a tedious task to determine whether they are connected or disconnected. First, observe that if our digraph has n vertices, then the number of edges in a path from P_i to P_j cannot exceed n, since all the vertices of a path are distinct and we cannot go through more than the n vertices in the digraph. If $[\mathbf{I}(G)]^r = [a_{ij}^{(r)}]$, then $a_{ij}^{(r)}$ is the number of ways of getting from P_i to P_j in r stages. This collection of vertices and edges need not be a path, since it may contain repeated vertices. If these repeated vertices and all edges between repeated vertices are deleted, we do obtain a path between P_i and P_j with at most r edges. For example, if we have $P_1 \to P_2 \to P_4 \to P_3 \to P_2 \to P_5$, we can eliminate one P_2 and all edges between the P_2's and obtain the path $P_1 \to P_2 \to P_5$. Hence if the i, jth element in

$$[\mathbf{I}(G)] + [\mathbf{I}(G)]^2 + \cdots + [\mathbf{I}(G)]^{n-1}$$

is zero, then there is no path between P_i and P_j. Thus the following theorem, half of whose proof we have sketched here, provides a test for connected digraphs.

THEOREM 7.6. *A digraph with n vertices is connected if and only if its incidence matrix $\mathbf{I}(G)$ has the property that*

$$\mathbf{I}(G) + [\mathbf{I}(G)]^2 + \cdots + [\mathbf{I}(G)]^{n-1} = \mathbf{C}$$

has no zero entries.

Example 19. Consider the digraph in Figure 7.17. The incidence matrix is

$$\mathbf{I}(G) = \begin{matrix} & P_1 & P_2 & P_3 & P_4 & P_5 \\ P_1 & 0 & 1 & 0 & 1 & 0 \\ P_2 & 0 & 0 & 1 & 1 & 0 \\ P_3 & 0 & 0 & 0 & 1 & 1 \\ P_4 & 1 & 0 & 1 & 0 & 0 \\ P_5 & 0 & 1 & 0 & 1 & 0 \end{matrix}.$$

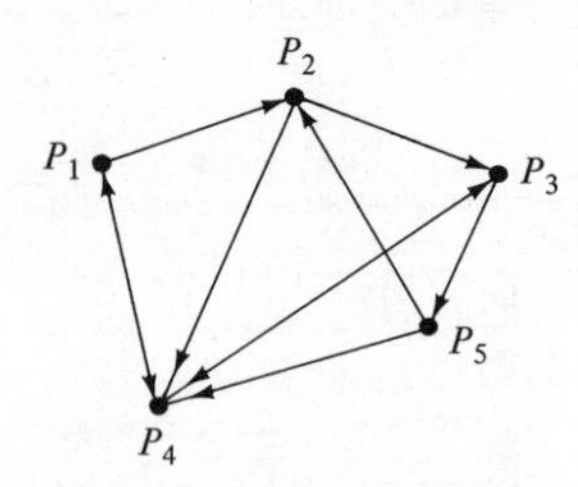

Figure 7.17

Then

$$\mathbf{C} = \mathbf{I}(G) + [\mathbf{I}(G)]^2 + [\mathbf{I}(G)]^3 + [\mathbf{I}(G)]^4$$

$$= \begin{array}{c} \\ P_1 \\ P_2 \\ P_3 \\ P_4 \\ P_5 \end{array} \begin{array}{c} \begin{array}{ccccc} P_1 & P_2 & P_3 & P_4 & P_5 \end{array} \\ \begin{bmatrix} 5 & 5 & 7 & 10 & 3 \\ 5 & 4 & 8 & 10 & 3 \\ 5 & 4 & 7 & 10 & 4 \\ 5 & 4 & 7 & 10 & 4 \\ 5 & 5 & 7 & 10 & 3 \end{bmatrix} \end{array}.$$

Since all the entries in **C** are positive, the given digraph is connected.

This approach can be used to trace the spread of a contaminant in a group of individuals; if there is a path from P_i to P_j, then the contaminant can spread from P_i to P_j.

FURTHER READING

Busacker, R. G. and T. L. Saaty. *Finite Graphs and Networks: An Introduction with Applications*. New York: McGraw-Hill Book Company, 1965.

Johnston, J. B., G. Price, and F. S. van Vleck. *Linear Equations and Matrices*. Reading, Mass.: Addison-Wesley Publishing Co., Inc., 1966.

Ore, O. *Graphs and Their Uses*. New York: Random House, Inc., 1963.

7.3 Exercises

1. Draw the complete graph on five vertices.

2. Write the matrices representing digraphs 2a and 2b on p. 319.

3. Which of the digraphs (3a, 3b, and 3c) on p. 319 are equal?

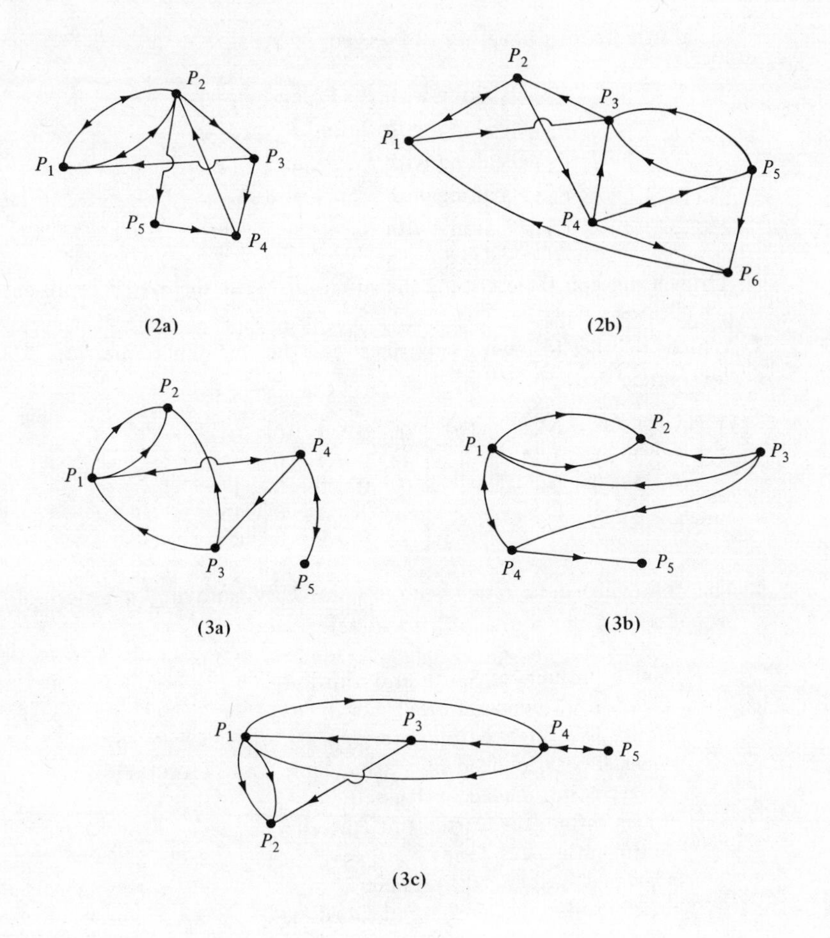

(2a) (2b)

(3a) (3b)

(3c)

4. Draw a digraph determined by the given matrix.

(a) $\begin{bmatrix} 0 & 2 & 0 & 0 & 0 \\ 0 & 0 & 1 & 1 & 0 \\ 1 & 0 & 0 & 1 & 1 \\ 1 & 1 & 0 & 0 & 2 \\ 1 & 0 & 1 & 2 & 0 \end{bmatrix}$. (b) $\begin{bmatrix} 0 & 1 & 2 & 1 \\ 1 & 0 & 0 & 1 \\ 2 & 1 & 0 & 0 \\ 0 & 1 & 0 & 0 \end{bmatrix}$.

5. Write the incidence matrices of the digraphs in Exercise 2.

6. Consider a group of five people, P_1, P_2, P_3, P_4, and P_5, who have been stationed on a remote island to operate a weather station. The following social interactions have been observed:

P_1 gets along with P_2, P_3, and P_4
P_2 gets along with P_1 and P_5
P_3 gets along with P_1, P_2, and P_4
P_4 gets along with P_2, P_3, and P_5
P_5 gets along with P_4.

Draw a digraph G describing the situation. Write the matrix representing G.

7. Which of the following matrices can be incidence matrices for a dominance digraph?

(a) $\begin{bmatrix} 0 & 1 & 1 & 0 \\ 0 & 0 & 1 & 1 \\ 0 & 0 & 0 & 0 \\ 0 & 0 & 1 & 0 \end{bmatrix}$. (b) $\begin{bmatrix} 0 & 0 & 1 & 0 & 0 \\ 1 & 0 & 1 & 0 & 1 \\ 0 & 0 & 0 & 0 & 1 \\ 1 & 1 & 1 & 0 & 1 \\ 1 & 0 & 0 & 0 & 0 \end{bmatrix}$.

8. The following data have been obtained by studying a group of six individuals in the course of a sociological study:

Carter influences Smith and Gordon
Gordon influences Jones
Smith is influenced by Peters
Russell is influenced by Carter, Smith, and Gordon
Peters is influenced by Russell
Jones influences Carter and Russell
Smith influences Jones
Carter is influenced by Peters
Peters influences Jones and Gordon.

(a) Who influences the most people?
(b) Who is influenced by the most people?

9. Consider a communication network among five individuals with incidence matrix

$$\begin{array}{c} \\ P_1 \\ P_2 \\ P_3 \\ P_4 \\ P_5 \end{array} \begin{array}{c} \begin{array}{ccccc} P_1 & P_2 & P_3 & P_4 & P_5 \end{array} \\ \begin{bmatrix} 0 & 1 & 1 & 1 & 1 \\ 0 & 0 & 0 & 1 & 1 \\ 0 & 1 & 0 & 1 & 0 \\ 0 & 0 & 0 & 0 & 1 \\ 1 & 0 & 1 & 0 & 0 \end{bmatrix} \end{array}.$$

(a) In how many ways does P_2 communicate with P_1 through one city?
(b) What is the smallest number of cities through which P_2 communicates?

10. Consider the following influence relation among five individuals:

P_1 influences P_2, P_4, and P_5
P_2 influences P_3 and P_4
P_3 influences P_1 and P_4
P_4 influences P_5
P_5 influences P_2 and P_3.

(a) Can P_4 influence P_1 in at most two stages?
(b) In how many ways does P_1 influence P_4 in three stages?
(c) In how many ways does P_1 influence P_4 in one, two, or three stages?

In Exercises 11 through 13 determine a clique, if there is one, for the digraph with given incidence matrix.

11.

$$\begin{bmatrix} 0 & 0 & 0 & 0 & 0 \\ 1 & 0 & 1 & 1 & 1 \\ 0 & 1 & 0 & 1 & 0 \\ 1 & 1 & 1 & 0 & 0 \\ 0 & 0 & 1 & 1 & 0 \end{bmatrix}.$$

12.

$$\begin{bmatrix} 0 & 1 & 1 & 0 & 1 & 1 \\ 1 & 0 & 0 & 0 & 0 & 0 \\ 0 & 0 & 0 & 1 & 0 & 0 \\ 1 & 0 & 0 & 0 & 1 & 1 \\ 1 & 0 & 1 & 1 & 0 & 1 \\ 1 & 0 & 1 & 1 & 1 & 0 \end{bmatrix}.$$

13.

$$\begin{bmatrix} 0 & 1 & 1 & 0 & 1 \\ 1 & 0 & 1 & 1 & 1 \\ 0 & 0 & 0 & 1 & 0 \\ 1 & 0 & 0 & 0 & 1 \\ 0 & 1 & 0 & 1 & 0 \end{bmatrix}.$$

14. Consider a communication network among five individuals with incidence matrix

$$\begin{bmatrix} 0 & 1 & 0 & 0 & 0 \\ 0 & 0 & 1 & 0 & 1 \\ 0 & 0 & 0 & 1 & 0 \\ 1 & 1 & 0 & 0 & 0 \\ 0 & 0 & 1 & 1 & 0 \end{bmatrix}.$$

(a) Can P_3 get a message to P_5 in at most two stages?
(b) What is the minimum number of stages that will guarantee that every person can get a message to any other (different) person?
(c) What is the minimum number of stages that will guarantee that every person can get a message to any person (including himself)?

15. Determine whether the digraph with given incidence matrix is connected or disconnected.

(a) $\begin{bmatrix} 0 & 1 & 1 & 1 \\ 0 & 0 & 1 & 1 \\ 1 & 0 & 0 & 1 \\ 0 & 0 & 1 & 0 \end{bmatrix}$. (b) $\begin{bmatrix} 0 & 1 & 0 & 0 & 0 \\ 0 & 0 & 1 & 0 & 0 \\ 1 & 0 & 0 & 0 & 0 \\ 0 & 0 & 0 & 0 & 1 \\ 0 & 0 & 1 & 1 & 0 \end{bmatrix}$.

16. Determine whether the digraph with given incidence matrix is connected or disconnected.

(a) $\begin{bmatrix} 0 & 0 & 1 & 1 & 1 \\ 1 & 0 & 1 & 1 & 0 \\ 0 & 1 & 0 & 0 & 0 \\ 0 & 1 & 0 & 0 & 1 \\ 1 & 1 & 0 & 0 & 0 \end{bmatrix}$. (b) $\begin{bmatrix} 0 & 0 & 0 & 0 & 1 \\ 0 & 0 & 1 & 1 & 0 \\ 0 & 1 & 0 & 0 & 1 \\ 1 & 0 & 0 & 0 & 0 \\ 0 & 0 & 0 & 1 & 0 \end{bmatrix}$.

17. A group of five acrobats performs a pyramiding act in which there must be a path between any two persons or the pyramid collapses. In the following incidence matrix, $a_{ij} = 1$ means that P_i supports P_j.

$$\begin{bmatrix} 0 & 0 & 1 & 1 & 0 \\ 1 & 0 & 0 & 0 & 1 \\ 1 & 0 & 0 & 1 & 0 \\ 1 & 0 & 0 & 0 & 1 \\ 0 & 1 & 1 & 0 & 0 \end{bmatrix}.$$

Can anyone be left out of the pyramid without having a collapse occur?

Theoretical Exercise

T.1. Show that a digraph which has the directed edges P_iP_j and P_jP_i cannot be a dominance graph.

7.4 The Theory of Games

There are many problems in economics, politics, warfare, business, and so on, which require decisions to be made in conflicting or competitive situations. The theory of games is a rather new area of applied mathematics that attempts to analyze conflict situations and provides a basis for rational decision making.

The theory of games was developed in the 1920s by John von Neumann and E. Borel, but the subject did not come to fruition until the publication, in 1944, of the decisive book *The Theory of Games and Economic Behavior* by John von Neumann and Oskar Morgenstern.

A **game** is a competitive situation in which each of a number of players is pursing his objective in direct conflict with the other players. Each player is doing everything he can to gain as much as possible for himself. Essentially, games are of two types. First, there are **games of chance**, such as roulette, which require no skill on the part of the players; the outcomes and winnings are determined solely by the laws of probability and can in no way be affected by any actions of the players. Second, there are **games of strategy**, such as chess, checkers, bridge, and poker, which require skill on the part of the players; the outcomes and winnings are determined by the skills of the players. By **game**, we mean a game of strategy. In addition to such parlor games there are many games in economic competition, warfare, geological exploration, farming, administration of justice, and so on, in which each player (competitor) can choose one of a set of possible moves and the outcomes depend upon the player's (competitor's) skill. The theory of games attempts to determine the best course of action for each player. The subject is still being developed and many new theoretical and applied results are needed for handling the complex games that occur in everyday problems.

We shall limit our treatment to games played by two players, usually denoted by R and C. We shall also assume that R has m possible moves (or courses of action) and that C has n moves. We now form an $m \times n$ matrix by labeling its rows, from top to bottom, with the moves of R, and labeling its columns, from left to right, with the moves of C. Entry a_{ij}, in row i and column j, indicates the amount (money or some other valuable item) paid by C to R if R makes his ith move and C makes his jth move. If a_{ij} is negative, then R pays C the amount, $-a_{ij}$. The entry a_{ij} is called a **payoff** and the matrix $\mathbf{A} = [a_{ij}]$ is called the **payoff matrix**. Such games are called **two-person games**. We also refer to them as **matrix games**. Games in which the sum of the payoffs received by R and C is a constant are called **constant-sum games**. Constant-sum games in which the constant is zero are called **zero-sum games**. Thus in a zero-sum game the amount won by one player is exactly the amount lost by the other player.

In the study of matrix games, it is always assumed that both players are equally capable, that each is playing as well as he possibly can play, and that each player makes his move without knowing what his opponent's move will be.

Example 1. Consider the game of matching pennies, consisting of two players, R and C, each of whom has a penny in his hand. Each player shows one side of the coin without knowing his opponents choice. If both players are showing the same side of the coin, then R wins \$1 from C; otherwise, C wins \$1 from R.

In this two-person, zero-sum game each player has two possible moves: he can show a tail or he can show a head. The payoff matrix is thus

$$\begin{array}{cc} & C \\ & \begin{array}{cc} H & T \end{array} \\ R & \begin{array}{c} H \\ T \end{array} \begin{bmatrix} 1 & -1 \\ -1 & 1 \end{bmatrix}. \end{array}$$

Example 2. There are two suppliers, firms R and C, of a new specialized type of tire that has 100,000 customers. Each company can advertise its product on TV or in the newspapers. A marketing firm determines that if both firms advertise on TV, then firm R gets 40,000 customers (and firm C gets 60,000 customers). If they both use newspapers, then each gets 50,000 customers. If R uses newspapers and C uses TV, then R gets 60,000 customers. If R uses TV and C uses newspapers, they each get 50,000 customers.

We can consider this situation as a game between firms R and C, with payoff matrix

		Firm C	
		T.V.	Newspapers
Firm R	T.V.	40,000	50,000
	Newspapers	60,000	50,000

The entries in the matrix indicate the number of customers secured by firm R. We think of C as initially having all 100,000 customers. Then a_{ij} represents the number of customers given up by C to R if R chooses his ith move and C chooses his jth move.

Consider now a two-person, constant-sum game with the $m \times n$ payoff matrix $\mathbf{A} = [a_{ij}]$, so that player R has m moves and player C has n moves. If player R plays his ith move, he is assured of winning at least the smallest entry in the ith row of $\mathbf{A}$, no matter what C does. Thus R's best course of action is to choose that move which will maximize his assured winnings in spite of C's best countermove. Player R will get his largest payoff by maximizing his smallest gain. Player C's goals are in direct conflict with those of player R: he is trying to keep R's winnings to a minimum. If C plays his jth move, he is assured of losing no more than the largest entry in the jth column of $\mathbf{A}$, no matter what R does. Thus C's best course of action is to choose that move which will minimize his assured losses in spite of R's best countermove. Player C will do his best by minimizing his largest loss.

DEFINITION If the payoff matrix of a matrix game contains an entry a_{rs}, which is at the same time the minimum of row r and the maximum of column s, then a_{rs} is called a **saddle point**. Also, a_{rs} is called the **value** of the game, and if the value is zero, the game is said to be **fair**.

DEFINITION A matrix game is said to be **strictly determined** if its payoff matrix has a saddle point.

If a_{rs} is a saddle point for a matrix game, then player R will be assured of winning at least a_{rs} by playing his rth move and player C will be guaranteed that he will lose no more than a_{rs} by playing his sth move. This is the best that each player can do.

Example 3. Consider a game with payoff matrix

$$\begin{array}{c} \\ R \end{array}\overset{\displaystyle C}{\begin{bmatrix} 0 & -3 & -1 & 3 \\ 3 & -2 & 2 & 4 \\ 1 & 4 & 0 & 6 \end{bmatrix}}.$$

To determine whether this game has a saddle point, we write the minimum of each row to the right of the row and the maximum of each column at the bottom of each column. Thus we have

$$\begin{array}{rccccc} & & & C & & \textbf{Row minima} \\ & 0 & -3 & -1 & 3 & -3 \\ R & 3 & -2 & 2 & 4 & 2 \\ & 1 & 4 & 0 & 6 & 0. \\ \textbf{Column maxima} & 3 & 4 & 2 & 6 & \end{array}$$

Entry $a_{23} = 2$ is both the least entry in the second row and the largest entry in the third column. Hence it is a saddle point for the game, which is then a strictly determined game. The value of the game is 2 and player R has an advantage. The best course of action for R is to play his second move; he will win 2 units from C, no matter what C does. The best course of action for C is to play his third move; he will limit his loss to 2 units, no matter what R does.

Example 4. Consider the advertising game of Example 2. The payoff matrix is

		Firm C		
		T.V.	Newspapers	Row minima
Firm R	T.V.	40,000	50,000	40,000
	Newspapers	60,000	50,000	50,000
	Column maxima	60,000	50,000	

Thus entry $a_{22} = 50{,}000$ is a saddle point. The best course of action for both firms is to advertise in newspapers. The game is strictly determined with value 50,000.

There are many games that are not strictly determined.

Example 5. Consider the game with payoff matrix

$$\begin{bmatrix} 1 & 6 & -1 \\ 3 & -2 & 4 \\ 4 & 5 & -3 \end{bmatrix} \qquad \begin{matrix} \textbf{Row minima} \\ -1 \\ -2 \\ -3 \end{matrix}$$

$$\textbf{Column maxima} \quad 4 \quad 6 \quad 4$$

It is clear that there is no saddle point.

On the other hand, a game may have more than one saddle point. However, it can be proved that all saddle points must have the same value.

Example 6. Consider the game with payoff matrix

$$\begin{bmatrix} 5 & 4 & 5 & 4 \\ 6 & -1 & 3 & 2 \\ 4 & 4 & 6 & 4 \end{bmatrix} \qquad \begin{matrix} \textbf{Row minima} \\ 4 \\ -1 \\ 6. \end{matrix}$$

$$\textbf{Column maxima} \quad 6 \quad 4 \quad 6 \quad 4$$

Entries a_{12}, a_{14}, a_{32}, and a_{34} are all saddle points and have the same value, 4. They appear shaded in the payoff matrix. The value of the game is also 4.

Consider now the penny-matching game of Example 1 with payoff matrix

$$\begin{matrix} & & C & & \\ & & H & T & \textbf{Row minima} \\ R & \begin{matrix} H \\ T \end{matrix} & \begin{bmatrix} 1 & -1 \\ -1 & 1 \end{bmatrix} & & \begin{matrix} -1 \\ -1. \end{matrix} \\ \textbf{Column maxima} & & 1 \quad\; 1 & & \end{matrix}$$

It is clear that this game is not strictly determined; that is, it has no saddle point.

To analyze this type of situation, we assume that a game is played repeatedly and that each player is trying to determine his best course of action. Thus player R tries to maximize his winnings while player C tries to minimize his losses. A **strategy** for a player is a decision for choosing his moves.

Consider now the above penny-matching game. Suppose that, in the repeated play of the game, player R always chooses the first row (he chooses to show heads), in the hope that player C will always choose the first column (play heads), thereby ensuring a win of \$1 for himself. However, as player C begins to notice that player R always chooses his first row, then player C will choose his second column, resulting in a loss of \$1 for R. Similarly, if R always chooses the second row, then C will choose the first column, resulting in a loss of \$1 for R. We can thus conclude that each player must somehow keep the other player from anticipating his choice of moves. This situation is in marked contrast with the case in strictly determined games. In a strictly determined game each player will make the same move whether or not he has advanced knowledge of his opponent's move. Thus in a nonstrictly determined game each player will make each move with a certain relative frequency.

DEFINITION Suppose that we have a matrix game with an $m \times n$ payoff matrix **A**. Let p_i, $1 \leqslant i \leqslant m$, be the probability that R chooses the ith row of **A** (that is, chooses his ith move). Let q_j, $1 \leqslant j \leqslant n$, be the probability that C chooses the jth column of **A**. The vector $P = [p_1 \quad p_2 \cdots p_m]$ is called a **strategy** for player R; the vector

$$\mathbf{Q} = \begin{bmatrix} q_1 \\ q_2 \\ \vdots \\ q_n \end{bmatrix}$$

is called a **strategy** for player C.

Of course, the probabilities p_i and q_j in this definition satisfy

$$p_1 + p_2 + \cdots + p_m = 1$$

$$q_1 + q_2 + \cdots + q_n = 1.$$

If a matrix game is strictly determined, then optimal strategies for R and C are strategies having 1 as one component and zero for all other components. Such strategies are called **pure strategies**. A strategy that is not pure is called a **mixed strategy**. Thus, in Example 3, the pure strategy for R is

$$\mathbf{p} = [0 \quad 1 \quad 0],$$

and the pure strategy for C is

$$\mathbf{Q} = \begin{bmatrix} 0 \\ 0 \\ 1 \\ 0 \end{bmatrix}.$$

Consider now a matrix game with payoff matrix

$$\mathbf{A} = \begin{bmatrix} a_{11} & a_{12} \\ a_{21} & a_{22} \end{bmatrix}. \tag{1}$$

Suppose that $\mathbf{P} = [p_1 \quad p_2]$ and $\mathbf{Q} = \begin{bmatrix} q_1 \\ q_2 \end{bmatrix}$ are strategies for R and C, respectively. Then if R plays his first row with probability p_1 and if C plays his first column with probability q_1, then R's expected payoff is $p_1 q_1 a_{11}$.

Similarly, we can examine the remaining three possibilities, obtaining Table 7.1. The expected payoff $E(\mathbf{P}, \mathbf{Q})$ of the game to R is then the sum of the four quantities in the rightmost column. We obtain

$$E(\mathbf{P}, \mathbf{Q}) = p_1q_1a_{11} + p_1q_2a_{12} + p_2q_1a_{21} + p_2q_2a_{22},$$

which can be written in matrix form (verify) as

$$E(\mathbf{P}, \mathbf{Q}) = \mathbf{PAQ}. \tag{2}$$

TABLE 7.1

Moves				
Player R	Player C	Probability	Payoff to player R	Expected payoff to player R
Row 1	Column 1	p_1q_1	a_{11}	$p_1q_1a_{11}$
Row 1	Column 2	p_1q_2	a_{12}	$p_1q_2a_{12}$
Row 2	Column 1	p_2q_1	a_{21}	$p_2q_1a_{21}$
Row 2	Column 2	p_2q_2	a_{22}	$p_2q_2a_{22}$

The same analysis applies to a matrix game with an $m \times n$ payoff matrix $\mathbf{A}$. Thus, if

$$\mathbf{P} = \begin{bmatrix} p_1 & p_2 & \cdots & p_m \end{bmatrix} \quad \text{and} \quad \mathbf{Q} = \begin{bmatrix} q_1 \\ q_2 \\ \vdots \\ q_n \end{bmatrix}$$

are strategies for R and C, respectively, then the expected payoff to player R is given by (2).

Example 7. Consider a matrix game with payoff matrix

$$\mathbf{A} = \begin{bmatrix} 2 & -2 & 3 \\ 4 & 0 & -3 \end{bmatrix}.$$

If

$$\mathbf{P} = \begin{bmatrix} \frac{1}{4} & \frac{3}{4} \end{bmatrix} \quad \text{and} \quad \mathbf{Q} = \begin{bmatrix} \frac{1}{3} \\ \frac{1}{3} \\ \frac{1}{3} \end{bmatrix}$$

are strategies for R and C, respectively, then the expected payoff to R is

$$E(\mathbf{P}, \mathbf{Q}) = \mathbf{PAQ}$$

$$= \begin{bmatrix} \frac{1}{4} & \frac{3}{4} \end{bmatrix} \begin{bmatrix} 2 & -2 & 3 \\ 4 & 0 & -3 \end{bmatrix} \begin{bmatrix} \frac{1}{3} \\ \frac{1}{3} \\ \frac{1}{3} \end{bmatrix} = \frac{1}{2}.$$

If

$$\mathbf{P} = \begin{bmatrix} \frac{3}{4} & \frac{1}{4} \end{bmatrix} \quad \text{and} \quad \mathbf{Q} = \begin{bmatrix} \frac{1}{3} \\ \frac{2}{3} \\ 0 \end{bmatrix}$$

are strategies for R and C, respectively, then the expected payoff to R is $-\frac{1}{6}$. Thus, in the first case R gains $\frac{1}{2}$ from C, whereas in the second case R loses $\frac{1}{6}$ to C.

A strategy for player R is said to be **optimal** if it guarantees R the largest possible payoff no matter what his opponent may do. Similarly, a strategy for player C is said to be **optimal** if it guarantees the smallest possible payments to R no matter what R may do.

If $\mathbf{P}$ and $\mathbf{Q}$ are optimal strategies for R and C, respectively, then the expected payoff to R, $v = E(\mathbf{P}, \mathbf{Q})$, is called the **value** of the game. Although $E(\mathbf{P}, \mathbf{Q})$ is a 1×1 matrix, we think of it merely as a number v. If the value is zero, the game is said to be **fair**. The principal task of the theory of games is the determination of optimal strategies for each player.

Consider again a matrix game with the 2×2 payoff matrix (1) and suppose that the game is not strictly determined. It can then be shown that

$$a_{11} + a_{22} - a_{12} - a_{21} \neq 0.$$

To determine an optimal strategy for R, we proceed as follows. Suppose that R's strategy is $[p_1 \quad p_2]$. Then if C plays the first column, the expected payoff to R is

$$a_{11}p_1 + a_{21}p_2. \tag{3}$$

If C plays the second column, the expected payoff to R is

$$a_{12}p_1 + a_{22}p_2. \tag{4}$$

If v is the minimum of the expected payoffs (3) and (4), then R expects to gain at least v units from C no matter what C does. Thus we have

$$a_{11}p_1 + a_{21}p_2 \geqslant v \tag{5}$$

$$a_{12}p_1 + a_{22}p_2 \geqslant v. \tag{6}$$

Moreover, player R seeks to make v as large as possible. Thus player R seeks to find p_1, p_2, and v such that

$$v \text{ is a maximum}$$

and

$$\begin{aligned} a_{11}p_1 + a_{21}p_2 - v &\geqslant 0 \\ a_{12}p_1 + a_{22}p_2 - v &\geqslant 0 \\ p_1 + p_2 &= 1 \\ p_1 \geqslant 0, \quad p_2 \geqslant 0, \quad v &\geqslant 0. \end{aligned} \tag{7}$$

Of course, (7) is a linear programming problem. It can be shown that a solution to (7), giving an optimal strategy for R, is

$$p_1 = \frac{a_{22} - a_{21}}{a_{11} + a_{22} - a_{12} - a_{21}}, \qquad p_2 = \frac{a_{11} - a_{12}}{a_{11} + a_{22} - a_{12} - a_{21}} \tag{8}$$

and

$$v = \frac{a_{11}a_{22} - a_{12}a_{21}}{a_{11} + a_{22} - a_{12} - a_{21}}. \tag{9}$$

We now find an optimal strategy for C. Suppose that C's strategy is $\begin{bmatrix} q_1 \\ q_2 \end{bmatrix}$. If R plays the first row, then the expected payoff to R is

$$a_{11}q_1 + a_{12}q_2, \tag{10}$$

while if R plays the second row, the expected payoff to R is

$$a_{21}q_1 + a_{22}q_2. \tag{11}$$

If v' is the maximum of the expected payoffs (10) and (11), then

$$a_{11}q_1 + a_{12}q_2 \leqslant v'$$

$$a_{21}q_1 + a_{22}q_2 \leqslant v'.$$

Since player C wishes to lose as little as possible, he seeks to make v' as small as possible. Thus C wants to find q_1, q_2, and v such that

$$v' \text{ is a minimum}$$

and

$$\begin{aligned} a_{11}q_1 + a_{12}q_2 - v' &\leqslant 0 \\ a_{21}q_1 + a_{22}q_2 - v' &\leqslant 0 \\ q_1 + q_2 &= 1 \\ q_1 \geqslant 0, \quad q_2 \geqslant 0, \quad v' &\geqslant 0. \end{aligned} \tag{12}$$

Problem (12) is also a linear programming problem. It can be shown that a solution to (8), giving an optimal solution for C, is

$$q_1 = \frac{a_{22} - a_{12}}{a_{11} + a_{22} - a_{12} - a_{21}}, \qquad q_2 = \frac{a_{11} - a_{21}}{a_{11} + a_{22} - a_{12} - a_{21}} \tag{13}$$

and

$$v' = \frac{a_{11}a_{22} - a_{12}a_{21}}{a_{11} + a_{22} - a_{12} - a_{21}}. \tag{14}$$

Thus $v = v'$ when both players use their optimal strategies.

Example 8. For the penny-matching game of Example 1, we have upon substituting in (8), (9), and (13),

$$p_1 = q_1 = \tfrac{1}{2} \qquad \text{and} \qquad q_1 = q_2 = \tfrac{1}{2}, \quad v = 0,$$

so that optimal strategies for R and C are

$$\begin{bmatrix} \frac{1}{2} & \frac{1}{2} \end{bmatrix} \qquad \text{and} \qquad \begin{bmatrix} \frac{1}{2} \\ \frac{1}{2} \end{bmatrix},$$

respectively. This means that half the time R should show heads and half the time he should show tails; likewise for player C. The value of the game is zero, so the game is fair.

Example 9. Consider a matrix game with payoff matrix

$$\begin{bmatrix} 2 & -5 \\ 1 & 3 \end{bmatrix}.$$

Again substituting in (8), (9), and (13), we obtain

$$p_1 = \frac{3-1}{2+3-1+5} = \frac{2}{9}, \qquad p_2 = \frac{2+5}{2+3-1+5} = \frac{7}{9},$$

$$q_2 = \frac{3+5}{2+3-1+5} = \frac{8}{9}, \qquad q_2 = \frac{2-1}{2+3-1+5} = \frac{1}{9},$$

$$v = \frac{6+5}{2+3-1+5} = \frac{11}{9}.$$

Thus optimal strategies for R and C are

$$\begin{bmatrix} \frac{2}{9} & \frac{7}{9} \end{bmatrix} \quad \text{and} \quad \begin{bmatrix} \frac{8}{9} \\ \frac{1}{9} \end{bmatrix},$$

respectively; when both players use their optimal strategies, the value of the game (the expected payoff to R) is $\frac{11}{9}$. The game is not fair and in the long run favors player R.

We can now generalize our discussion to a game with an $m \times n$ payoff matrix $\mathbf{A} = [a_{ij}]$. First, let us observe that if we add a constant r to every entry of A, then the optimal strategies for R and C do not change, and the value of the new game is r plus the value of the old game (Exercise T.2). Thus we can assume that, by adding a suitable constant to every entry of the payoff matrix, every entry of $\mathbf{A}$ is positive.

Now R seeks to find $p_1, p_2, \ldots, p_m$, and v such that

$$v \text{ is a maximum}$$

subject to

$$\begin{aligned} a_{11}p_1 + a_{21}p_2 + \cdots + a_{m1}p_1 - v &\geqslant 0 \\ a_{12}p_1 + a_{22}p_2 + \cdots + a_{m2}p_2 - v &\geqslant 0 \\ &\vdots \\ a_{1n}p_1 + a_{2n}p_2 + \cdots + a_{mn}p_m - v &\geqslant 0 \\ p_1 + p_2 + \cdots + p_m &= 1 \\ p_1 \geqslant 0, \quad p_2 \geqslant 0, \ldots, p_m \geqslant 0, v &\geqslant 0. \end{aligned} \tag{15}$$

Since every entry of $\mathbf{A}$ is positive, we may assume that $v > 0$. Now divide each of the constraints in (15) by v, and let

$$y_i = \frac{p_i}{v}.$$

Observe that

$$y_1 + y_2 + \cdots + y_m = \frac{p_1}{v} + \frac{p_2}{v} + \cdots + \frac{p_m}{v}$$

$$= \frac{1}{v}(p_1 + p_2 + \cdots + p_m) = \frac{1}{v}.$$

Thus v is a maximum if and only if $y_1 + y_2 + \cdots + y_m$ is a minimum. We can now restate problem (15), R's problem, as follows:

$$\text{Minimize } y_1 + y_2 + \cdots + y_m$$

subject to

$$\begin{aligned} a_{11}y_1 + a_{21}y_1 + \cdots + a_{m1}y_1 &\geq 1 \\ a_{12}y_1 + a_{22}y_2 + \cdots + a_{m2}y_2 &\geq 1 \\ &\vdots \\ a_{1n}y_1 + a_{2n}y_2 + \cdots + a_{mn}y_m &\geq 1 \\ y_1 \geq 0, \quad y_2 \geq 0, \ldots, y_m &\geq 0. \end{aligned} \tag{16}$$

Observe that (16) has one fewer constraint and one fewer variable than (15).

Turning next to C's problem, we note that he seeks to find $q_1, q_2, \ldots, q_n$ and v' such that

$$v' \text{ is a minimum}$$

subject to

$$\begin{aligned} a_{11}q_1 + a_{12}q_2 + \cdots + a_{1n}q_n - v' &\leq 0 \\ a_{21}q_1 + a_{22}q_2 + \cdots + a_{2n}q_n - v' &\leq 0 \\ &\vdots \\ a_{m1}q_1 + a_{m2}q_2 + \cdots + a_{mn}q_n - v' &\leq 0 \\ q_1 + q_2 + \cdots + q_n &= 1 \\ q_1 \geq 0, \quad q_2 \geq 0, \ldots, q_n \geq 0, v' &\geq 0. \end{aligned} \tag{17}$$

The fundamental theorem of matrix games, which we now state, says that every matrix game has a solution.

THEOREM 7.7 (Fundamental Theorem of Matrix Games). *Every matrix game has a solution. That is, there are optimal strategies for R and C. Moreover, $v = v'$.*

Since $v = v'$, we can divide each of the constraints in (17) by $v' = v$ and let

$$x_i = \frac{q_i}{v}.$$

Now

$$x_1 + x_2 + \cdots + x_n = \frac{1}{v},$$

so v is a minimum if and only if $x_1 + x_2 + \cdots + x_n$ is a maximum. We can now restate problem (17), C's problem, as follows:

$$\text{Maximize } x_1 + x_2 + \cdots + x_n$$

subject to

$$\begin{aligned} a_{11}x_1 + a_{12}x_2 + \cdots + a_{1n}x_n &\leqslant 1 \\ a_{21}x_1 + a_{22}x_2 + \cdots + a_{2n}x_n &\leqslant 1 \\ &\vdots \\ a_{m1}x_1 + a_{m2}x_2 + \cdots + a_{mn}x_n &\leqslant 1 \\ x_1 \geqslant 0, x_2 \geqslant 0, \ldots, x_n &\geqslant 0. \end{aligned} \tag{18}$$

It can be shown that when (17) is solved by the simplex method, the final tableau will contain the optimal strategies for R in the objective row under the columns of the slack variables. That is, y_1 is found in the objective row under the first slack variable, y_2 is found in the objective row under the second slack variable, and so on.

Example 10. Consider a game with payoff matrix

$$\begin{bmatrix} 2 & -3 & 0 \\ 3 & 1 & -2 \end{bmatrix}.$$

Adding 4 to each element of the matrix, we obtain a matrix $\mathbf{A}$ with positive entries.

$$\mathbf{A} = \begin{bmatrix} 6 & 1 & 4 \\ 7 & 5 & 2 \end{bmatrix}.$$

We now find optimal strategies for the game with payoff matrix **A**. Problem (18), C's problem, becomes:

$$\text{Maximize } x_1 + x_2 + x_3$$

subject to

$$\begin{aligned} 6x_1 + x_2 + 4x_3 &\leqslant 1 \\ 7x_1 + 5x_2 + 2x_3 &\leqslant 1 \end{aligned}$$

$$x_1 \geqslant 0, \quad x_2 \geqslant 0, \quad x_3 \geqslant 0.$$

If we introduce the slack variables x_4 and x_5, our problem becomes:

$$\text{Maximize } x_1 + x_2 + x_3$$

subject to

$$\begin{aligned} 6x_1 + x_2 + 4x_3 + x_4 &= 1 \\ 7x_1 + 5x_2 + 2x_3 + x_5 &= 1 \end{aligned}$$

$$x_1 \geqslant 0, \quad x_2 \geqslant 0, \quad x_3 \geqslant 0, \quad x_4 \geqslant 0, \quad x_5 \geqslant 0.$$

Using the simplex method, we obtain

		x_1	x_2	x_3 ↓	x_4	x_5	z	
←	x_4	6	1	(4)	1	0	0	1
	x_5	7	5	2	0	1	0	1
		−1	−1	−1	0	0	1	0

		x_1	x_2 ↓	x_3	x_4	x_5	z	
	x_3	$\frac{3}{2}$	$\frac{1}{4}$	1	$\frac{1}{4}$	0	0	$\frac{1}{4}$
←	x_5	4	($\frac{9}{2}$)	0	$-\frac{1}{2}$	1	0	$\frac{1}{2}$
		$\frac{1}{2}$	$-\frac{3}{4}$	0	$\frac{1}{4}$	0	1	$\frac{1}{4}$

	x_1	x_2	x_3	x_4	x_5	z	
x_3	$\frac{23}{18}$	0	1	$\frac{5}{18}$	$-\frac{1}{18}$	0	$\frac{2}{9}$
x_2	$\frac{8}{9}$	1	0	$-\frac{1}{9}$	$\frac{2}{9}$	0	$\frac{1}{9}$
	$\frac{7}{6}$	0	0	$\frac{1}{6}$	$\frac{1}{6}$	1	$\frac{1}{3}$

Thus we have

$$x_1 = 0, \qquad x_2 = \tfrac{1}{9}, \qquad \text{and} \qquad x_3 = \tfrac{2}{9}.$$

The maximum value of $x_1 + x_2 + x_3$ is $\frac{1}{3}$, so the minimum value of v is 3. Hence

$$q_1 = x_1 v = 0, \qquad q_2 = x_2 v = (\tfrac{1}{9})(3) = \tfrac{1}{3},$$

and

$$q_3 = x_3 v = (\tfrac{2}{9})(3) = \tfrac{2}{3}.$$

Thus an optimal strategy for C is

$$\mathbf{Q} = \begin{bmatrix} 0 \\ \frac{1}{3} \\ \frac{2}{3} \end{bmatrix}.$$

An optimal solution to (16), R's problem, is found in the objective row under the columns of the slack variables. Thus, under the slack variables x_4 and x_5, we find

$$y_1 = \tfrac{1}{6} \qquad \text{and} \qquad y_2 = \tfrac{1}{6}.$$

Since $v = 3$,

$$p_1 = y_1 v = (\tfrac{1}{6})(3) = \tfrac{1}{2} \qquad \text{and} \qquad p_2 = y_2 v = (\tfrac{1}{6})(3) = \tfrac{1}{2}.$$

Thus an optimal strategy for R is

$$\mathbf{P} = [\tfrac{1}{2} \quad \tfrac{1}{2}].$$

The game is not fair, since the value is 3, with player R having the advantage.

Sometimes, it is possible to solve a matrix game by reducing the size of the payoff matrix **A**. If each element of the rth *row* of **A** is *less than or equal* to the corresponding element of the sth row of **A**, then the rth row is called **recessive** and the sth row is said to **dominate** the rth row. If each element of the rth *column* of **A** is *greater than or equal* to the corresponding element in the sth column of **A**, then the rth column is called **recessive** and the sth column is said to **dominate** the rth column.

Example 11. In the payoff matrix

$$\begin{bmatrix} 2 & -1 & 3 \\ 0 & 1 & 4 \\ 3 & 2 & 4 \end{bmatrix},$$

the first row is recessive; the third row dominates the first row.

In the payoff matrix

$$\begin{bmatrix} -2 & 4 & 3 \\ 3 & -3 & -3 \\ 5 & 2 & 1 \end{bmatrix},$$

the second column is recessive; the third column dominates the second column.

Consider a matrix game in which the rth row is recessive and the sth row dominates the rth row. It is then obvious that player R will always tend to choose the sth row rather than the rth row, since he will be guaranteed a gain equal to or greater than the gain realized by choosing the rth row. Thus, since the rth row will never be chosen, it can be dropped from further consideration. Suppose now that the rth column is recessive and that the sth column dominates the rth column. Since player C wishes to keep his losses to a minimum, by choosing the rth column he will be guaranteed a loss equal to or smaller than the loss incurred by choosing the sth column. Since the rth column will never be chosen, it can be dropped from further consideration. These techniques, when applicable result in a smaller payoff matrix.

Example 12. Consider the matrix game with payoff matrix

$$\mathbf{A} = \begin{bmatrix} 2 & -1 & 3 \\ -2 & 2 & 4 \\ 3 & 0 & 4 \end{bmatrix}.$$

Since the third row of $\mathbf{A}$ dominates its first row, the latter can be dropped, obtaining

$$\mathbf{A}_1 = \begin{bmatrix} -2 & 2 & 4 \\ 3 & 0 & 4 \end{bmatrix}.$$

Since the second column of $\mathbf{A}_1$ dominates its third column, the latter can be dropped, obtaining

$$\mathbf{A}_2 = \begin{bmatrix} -2 & 2 \\ 3 & 0 \end{bmatrix},$$

which has no saddle point. The solution to the matrix game with payoff matrix $\mathbf{A}_2$ can be obtained from Equations (8), (9), and (13). We have

$$p_1 = \frac{0 - 3}{-2 + 0 - 2 - 3} = \frac{-3}{-7} = \frac{3}{7},$$

$$p_2 = \frac{-2 - 2}{-2 + 0 - 2 - 3} = \frac{-4}{-7} = \frac{4}{7},$$

$$q_1 = \frac{0 - 2}{-2 + 0 - 2 - 3} = \frac{-3}{-7} = \frac{2}{7},$$

$$q_2 = \frac{-2 - 3}{-2 + 0 - 2 - 3} = \frac{-5}{-7} = \frac{5}{7},$$

and

$$v = \frac{0 - 6}{-2 + 0 - 2 - 3} = \frac{-6}{-7} = \frac{6}{7}.$$

Since in $\mathbf{A}$, the original payoff matrix, the first row and third column were dropped, we obtain

$$\mathbf{P} = \begin{bmatrix} 0 & \frac{3}{7} & \frac{4}{7} \end{bmatrix}$$

as an optimal strategy for player R. Similarly,

$$\mathbf{Q} = \begin{bmatrix} \frac{2}{7} \\ \frac{5}{7} \\ 0 \end{bmatrix}$$

is an optimal strategy for player C.

FURTHER READING

McKinsey, J. C. C. *Introduction to the Theory of Games*. New York: McGraw-Hill Book Company, 1952.
Owen, G. *Game Theory*. Philadelphia: W. B. Saunders Company, 1968.

7.4 Exercises

In Exercises 1 through 4 write the payoff matrix for the given game.

1. Each of two players shows two or three fingers. If the sum of the fingers shown is even, then R pays C an amount equal to the sum of the numbers shown; if the sum is odd, then C pays R an amount equal to the sum of the numbers shown.

2. **Stone, scissors, paper.** Each of two players selects one of the words *stone*, *scissors*, *paper*. Stone beats scissors, scissors beats paper, and paper beats stone. In case of a tie, there is no payoff. In case of a win, the winner collects $1.
3. Firms A and B, both handling specialized sporting equipment, are planning to locate in either Abington or Wyncote. If they both locate in the same town, each will capture 50 per cent of the trade. If A locates in Abington and B locates in Wyncote, then A will capture 60 per cent of the business (and B will keep 40 per cent); if A locates in Wyncote and B locates in Abington, then A will hold on to 25 per cent of the business (and B to 75 per cent).
4. Player R has a nickel and a dime with him. He chooses one of the coins and player C must guess R's choice. If C guesses correctly, he keeps the coin; if he guesses incorrectly, he must give R an amount equal to the coin shown.
5. Find all saddle points for the following matrix games.

(a) $\begin{bmatrix} 5 & 4 \\ 3 & -2 \end{bmatrix}$. (b) $\begin{bmatrix} 2 & 1 & 0 \\ 3 & 1 & -2 \\ 4 & 2 & -4 \end{bmatrix}$.

(c) $\begin{bmatrix} 3 & 4 & 5 \\ -2 & 5 & 1 \\ -1 & 0 & 1 \end{bmatrix}$. (d) $\begin{bmatrix} 5 & 2 & 4 & 2 \\ 0 & -1 & 2 & 0 \\ 3 & 2 & 3 & 2 \\ 1 & 0 & -1 & -1 \end{bmatrix}$.

6. Find optimal strategies for the following strictly determined games. Give the payoff for R.

(a) $\begin{bmatrix} -3 & 4 \\ 3 & 5 \end{bmatrix}$. (b) $\begin{bmatrix} -1 & -3 & -2 \\ 3 & -1 & 4 \\ -1 & -2 & 5 \end{bmatrix}$.

(c) $\begin{bmatrix} -2 & 3 & -2 & 4 \\ -1 & 2 & -2 & 4 \\ -2 & 3 & -3 & 5 \\ -1 & 2 & -3 & 1 \end{bmatrix}$.

7. Find optimal strategies for the following strictly determined games. Give the payoff for R.

(a) $\begin{bmatrix} 2 & 1 & 3 \\ -2 & 0 & 2 \end{bmatrix}$. (b) $\begin{bmatrix} -2 & -2 & 4 & 5 \\ -2 & -2 & 1 & 0 \\ 0 & 1 & 1 & 2 \end{bmatrix}$.

(c) $\begin{bmatrix} 6 & 4 \\ 7 & 4 \end{bmatrix}$.

8. Consider a matrix game with payoff matrix

$$\begin{bmatrix} 2 & -3 & -2 \\ -4 & 5 & 6 \end{bmatrix}.$$

Find $E(\mathbf{P}, \mathbf{Q})$, the expected payoff to R, if

(a) $\mathbf{P} = [\frac{1}{4} \quad \frac{3}{4}]$ and $\mathbf{Q} = \begin{bmatrix} \frac{1}{3} \\ \frac{1}{6} \\ \frac{1}{2} \end{bmatrix}$.

(b) $p_1 = \frac{2}{3}, p_2 = \frac{1}{3}; q_1 = \frac{1}{2}, q_2 = \frac{1}{4}$, and $q_3 = \frac{1}{4}$.

9. Consider a matrix game with payoff matrix

$$\begin{bmatrix} 3 & -3 \\ 2 & 5 \\ 1 & 0 \end{bmatrix}.$$

Find $E(\mathbf{P}, \mathbf{Q})$, the expected payoff to R, if

(a) $\mathbf{P} = [\frac{1}{2} \quad \frac{1}{3} \quad \frac{1}{6}]$ and $\mathbf{Q} = \begin{bmatrix} \frac{1}{6} \\ \frac{5}{6} \end{bmatrix}$.

(b) $p_1 = 0, p_2 = 0, p_3 = 1; q_1 = \frac{1}{7}$ and $q_2 = \frac{6}{7}$.

In Exercises 10 and 11 solve the given matrix game using (8) and (13). Find the value of the game using (9).

10. $\begin{bmatrix} 4 & 8 \\ 6 & -2 \end{bmatrix}$.

11. $\begin{bmatrix} -3 & 2 \\ 4 & -5 \end{bmatrix}$.

In Exercises 12 and 13 solve the given matrix game by linear programming.

12. $\begin{bmatrix} -2 & 3 \\ 4 & 5 \\ 5 & 2 \end{bmatrix}$.

13. $\begin{bmatrix} 2 & -3 & 4 \\ 4 & 0 & 1 \\ 3 & 2 & -2 \end{bmatrix}$.

In Exercises 14 and 15 solve the given matrix game by using the method of Example 12.

14. $\begin{bmatrix} -3 & 1 & 3 \\ 1 & -2 & 2 \\ 2 & -1 & 3 \end{bmatrix}$.

15. $\begin{bmatrix} 0 & -4 & 3 & 0 \\ 2 & -3 & 4 & 1 \\ -1 & 2 & 2 & 2 \\ 1 & -4 & 3 & 0 \end{bmatrix}$.

16. Solve Exercise 1.

17. Solve Exercise 2.

18. Solve Exercise 3.

19. Solve Exercise 4.

20. In a labor–management dispute, labor can make one of three different moves L_1, L_2, and L_3 while management can make one of two moves, M_1

and M_2. Suppose that the following payoff matrix is obtained (the entries represent millions of dollars). Determine the best courses of action for both labor and management.

$$L \begin{array}{c} \\ \\ L_1 \\ L_2 \\ L_3 \end{array} \overset{\displaystyle M}{\overset{\begin{array}{cc} M_1 & M_2 \end{array}}{\begin{bmatrix} 2 & 4 \\ 3 & 2 \\ 5 & 1 \end{bmatrix}}}.$$

Theoretical Exercises

T.1. Consider a matrix game with $m \times n$ payoff matrix $\mathbf{A}$. Verify that if player R uses strategy $\mathbf{P}$ and player C uses strategy $\mathbf{Q}$, then the expected payoff to R is $\mathbf{PAQ}$.

T.2 Consider a matrix game with payoff matrix $\mathbf{A}$. Show that if a constant r is added to each entry of $\mathbf{A}$, then we get a new game whose optimal strategies are the same as those for the original game, and the value of the new game is r plus the value of the old game.

7.5 Least Squares

The problem of gathering and analyzing data is one that is present in almost every facet of human activity. Frequently, we measure a value of y for a given value of x and then plot the points (x, y) on graph paper. From the resulting graph we try to develop a relationship between the variables x and y that can then be used to predict new values of y for given values of x.

Example 1. In the manufacture of product XXX, the amount of the compound beta present in the product is controlled by the amount of the ingredient alpha used in the process. In manufacturing a gallon of XXX the amount of alpha used and the amount of beta present are recorded. The following data were obtained:

Alpha used (ounces/gallon)	3	4	5	6	7	8	9	10	11	12
Beta present (ounces/gallon)	4.5	5.5	5.7	6.6	7.0	7.7	8.5	8.7	9.5	9.7

The points in this table are plotted in Figure 7.18.

Suppose it is known that the relationship between the amount of alpha used and the amount of beta present is given by a linear equation so that the graph is a straight line. It would thus not be reasonable to graph the data by connecting the plotted points by drawing a curve that passes through every point. Moreover, the data are of a *probabilistic* nature. That is, it is not *deterministic* in the sense that if we repeated the experiment, we would expect slightly different values of beta for the same values of alpha. The technique for drawing the straight line that "best fits" the given data is called the method of **least squares**, which we now consider.

Suppose that we are given n points $(x_1, y_1), (x_2, y_2), \ldots, (x_n, y_n)$ with $x_1 < x_2 < \cdots < x_n$. We are interested in finding the line

$$y = b_0 + b_1 x, \tag{1}$$

which "best fits the data." This is the line which has the property that the

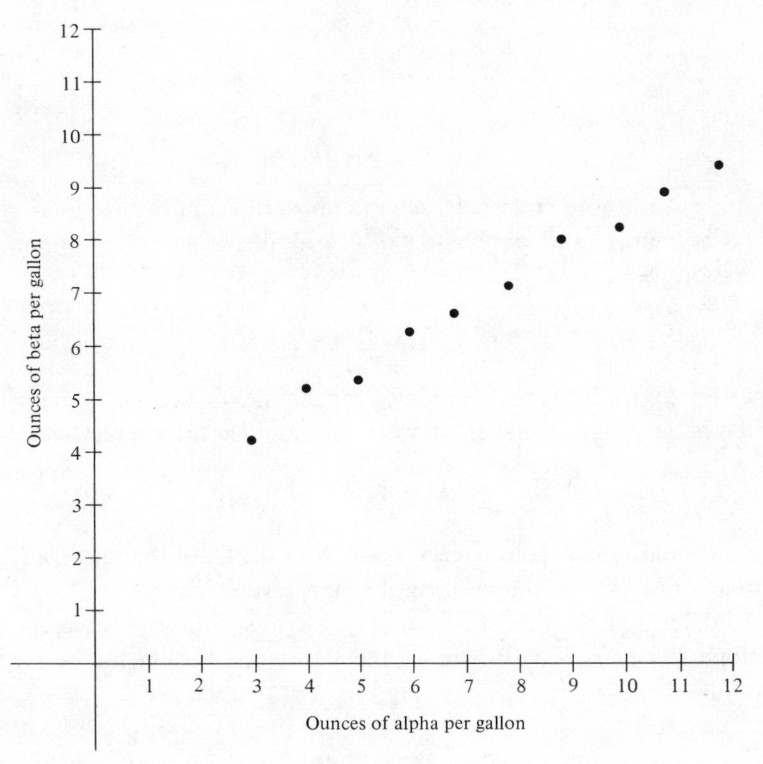

Figure 7.18

sum of the squares of the vertical distances from the points (x_i, y_i) to the line will be as small as possible. This line is called the **line of best fit**.

In Figure 7.19 we show the five data points (x_1, y_1), (x_2, y_2), (x_3, y_3), (x_4, y_4), and (x_5, y_5). Let the vertical distance from the point (x_i, y_i) to the (desired) line of best fit be d_i. Then the line of best fit is chosen so that

$$d_1^2 + d_2^2 + d_3^2 + d_4^2 + d_5^2 \tag{2}$$

is a minimum.

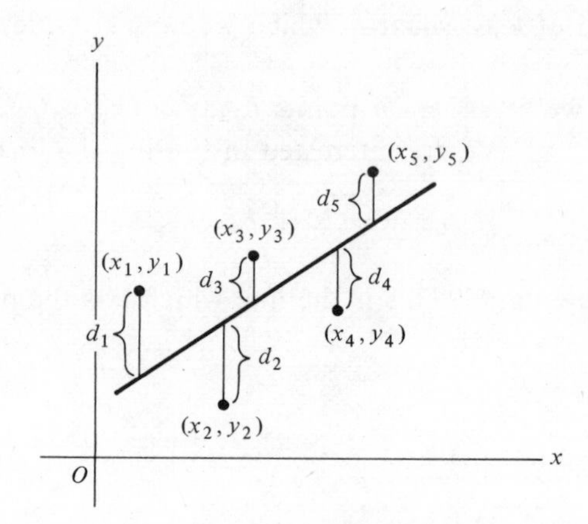

Figure 7.19

It is natural to wonder why we minimize the sum of the squares of the d_i and not some other expression. For example, if we try to minimize the expression

$$d_1 + d_2 + d_3 + d_4 + d_5,$$

then we may find several candidates for the line of best fit, some of which will be better than others. If we were to minimize the expression

$$|d_1| + |d_2| + |d_3| + |d_4| + |d_5|,$$

then the analysis becomes very complicated. Thus the most successful approach results from minimizing the sum of the squares.

To determine the line of best fit, (1), we must find its y-intercept b_0 and its slope b_1. The y-coordinate of the point on the line of best fit with x-coordinate x_i is

$$b_0 + b_1 x_i,$$

by Equation (1). Hence the vertical distance from the data point (x_i, y_i) to the point on the line of best fit with x-coordinate x_i is

$$y_i - b_0 - b_1 x_i.$$

We are trying to determine b_0 and b_1 so that the sum of the squares of these vertical distances will be minimized, that is, so that

$$\sum_{i=1}^{n} (y_i - b_0 - b_1 x_i)^2 \tag{3}$$

is minimized.

The first approach for finding b_0 and b_1 is for readers who have not studied calculus. The second approach presented uses calculus and is simpler and somewhat more direct.

Finding b_0 and b_1 Without Calculus

Suppose that

$$y = a_0 + a_1 x_i \tag{4}$$

is another line fitting the points. Then the sum of the squares of the vertical distances from the points (x_i, y_i) to the points on line (4) is

$$\sum_{i=1}^{n} (y_i - a_0 - a_1 x_i)^2. \tag{5}$$

Observe that (3) must be less than or equal to (5) for any choice of a_0 and a_1. That is,

$$\sum_{i=1}^{n} (y_i - a_0 - a_1 x_i)^2 \geqslant \sum_{i=1}^{n} (y_i - b_0 - b_1 x_i)^2. \tag{6}$$

Rewrite the left side of (6) as

$$\sum_{i=1}^{n} (y_i - a_0 - a_1 x_i)^2 = \sum_{i=1}^{n} [y_i - b_0 - b_1 x_i + (b_0 - a_0) + (b_1 - a_1) x_i]^2. \tag{7}$$

Letting

$$\epsilon_0 = b_0 - a_0 \qquad \text{and} \qquad \epsilon_1 = (b_1 - a_1),$$

we can write (7) as

$$\sum_{i=1}^{n} (y_i - a_0 - a_1x_i)^2 = \sum_{i=1}^{n} (y_i - b_0 - b_1x_i)^2 + 2\sum_{i=1}^{n} (\epsilon_0 + \epsilon_1x_i)(y_i - b_0 - b_1x_i) + \sum_{i=1}^{n} (\epsilon_0 + \epsilon_1x_i)^2.$$

It then follows that (6) holds if and only if

$$2\sum_{i=1}^{n} (\epsilon_0 + \epsilon_1x_i)(y_i - b_0 - b_1x_i) = 0$$

for every ϵ_0 and ϵ_1.

Letting $\epsilon_0 = 0$ and $\epsilon_1 = 1$, we obtain

$$2\sum_{i=1}^{n} x_i(y_i - b_0 - b_1x_i) = 0, \tag{8}$$

and letting $\epsilon_0 = 1$ and $\epsilon_1 = 0$, we obtain

$$2\sum_{i=1}^{n} (y_i - b_0 - b_1x_i) = 0. \tag{9}$$

Simplifying and expanding (8) and (9), we are led to having to solve the linear system

$$\begin{aligned} \left(\sum_{i=1}^{n} x_i\right)b_0 + \left(\sum_{i=1}^{n} x_i^2\right)b_1 &= \sum_{i=1}^{n} x_iy_i \\ nb_0 + \left(\sum_{i=1}^{n} x_i\right)b_1 &= \sum_{i=1}^{n} y_i, \end{aligned} \tag{10}$$

which has the solution (verify)

$$b_0 = \frac{\left(\sum_{i=1}^{n} y_i\right)\left(\sum_{i=1}^{n} x_i^2\right) - \left(\sum_{i=1}^{n} x_i\right)\left(\sum_{i=1}^{n} x_iy_i\right)}{n\left(\sum_{i=1}^{n} x_i^2\right) - \left(\sum_{i=1}^{n} x_i\right)^2}. \tag{A}$$

$$b_1 = \frac{n\left(\sum_{i=1}^{n} x_iy_i\right) - \left(\sum_{i=1}^{n} y_i\right)\left(\sum_{i=1}^{n} x_i\right)}{n\left(\sum_{i=1}^{n} x_i^2\right) - \left(\sum_{i=1}^{n} x_i\right)^2}. \tag{B}$$

Finding b_0 and b_1 with the Aid of Calculus

Since we are to minimize (3):

$$E = \sum_{i=1}^{n} (y_i - b_0 - b_1 x_i)^2,$$

we compute

$$\frac{\partial E}{\partial b_0} = 0 \qquad \text{and} \qquad \frac{\partial E}{\partial b_1} = 0,$$

yielding

$$-2 \sum_{i=1}^{n} (y_i - b_0 - b_1 x_i) = 0$$

and

$$-2x_i \sum_{i=1}^{n} (y_i - b_0 - b_1 x_i) = 0.$$

Simplifying and expanding these equations, we are led to having to solve the linear system

$$\begin{aligned} \left(\sum_{i=1}^{n} x_i\right) b_0 + \left(\sum_{i=1}^{n} x_i^2\right) b_1 &= \sum_{i=1}^{n} x_i y_i \\ n b_0 + \left(\sum_{i=1}^{n} x_i\right) b_1 &= \sum_{i=1}^{n} y_i, \end{aligned} \tag{10}$$

which has the solution

$$b_0 = \frac{\left(\sum_{i=1}^{n} y_i\right)\left(\sum_{i=1}^{n} x_i^2\right) - \left(\sum_{i=1}^{n} x_i\right)\left(\sum_{i=1}^{n} x_i y_i\right)}{n\left(\sum_{i=1}^{n} x_i^2\right) - \left(\sum_{i=1}^{n} x_i\right)^2}, \tag{A}$$

$$b_1 = \frac{n\left(\sum_{i=1}^{n} x_i y_i\right) - \left(\sum_{i=1}^{n} y_i\right)\left(\sum_{i=1}^{n} x_i\right)}{n\left(\sum_{i=1}^{n} x_i^2\right) - \left(\sum_{i=1}^{n} x_i\right)^2}. \tag{B}$$

Example 2. Find the line of best fit for the data (1, 2), (2, 5), (3, 3), and (4, 6).

solution. The given data points are $(x_1, y_1) = (1, 2)$; $(x_2, y_2) = (2, 5)$; $(x_3, y_3) = (3, 3)$; and $(x_4, y_4) = (4, 6)$. Then since $n = 4$, we have

$$\sum_{i=1}^{4} x_i = x_1 + x_2 + x_3 + x_4 = 1 + 2 + 3 + 4 = 10$$

$$\sum_{i=1}^{4} x_i^2 = x_1^2 + x_2^2 + x_3^2 + x_4^2 = 1^2 + 2^2 + 3^2 + 4^2 = 30$$

$$\sum_{i=1}^{4} y_i = y_1 + y_2 + y_3 + y_4 = 2 + 5 + 3 + 6 = 16$$

$$\sum_{i=1}^{4} x_i y_i = x_1 y_1 + x_2 y_2 + x_3 y_3 + x_4 y_4 = 2 + 10 + 9 + 24 = 45.$$

Substituting these values in Equations (A) and (B), we obtain

$$b_0 = \tfrac{3}{2} \quad \text{and} \quad b_1 = 1.$$

Thus the line of best fit is

$$y = x + \tfrac{3}{2},$$

which is shown in Figure 7.20 along with the given four data points.

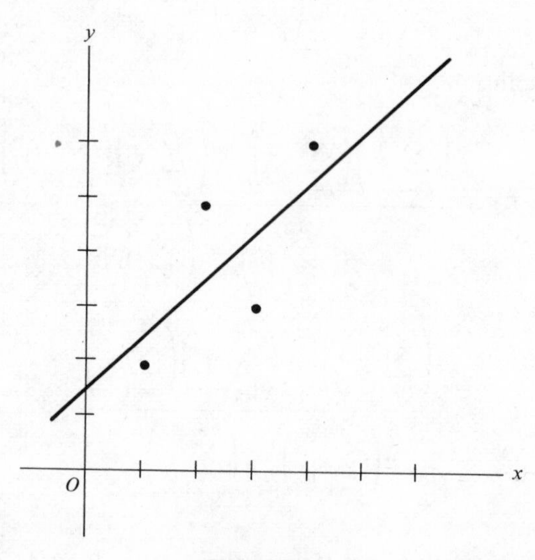

Figure 7.20

Example 3

(a) Find the equation of the line of best fit for the data in Example 1.

(b) Use the equation obtained in part (a) to predict the number of ounces of beta present in a gallon if 30 ounces of alpha are used per gallon.

solution

(a) We can arrange our computation as follows:

x_i (alpha)	x_i^2	y_i (beta)	x_iy_i
3	9	4.5	13.5
4	16	5.5	22.0
5	25	5.7	28.5
6	36	6.6	39.6
7	49	7.0	49.0
8	64	7.7	61.6
9	81	8.5	76.5
10	100	8.7	87.0
11	121	9.5	104.5
12	144	9.7	116.4
$\sum_{i=1}^{10} x_i = 75$	$\sum_{i=1}^{10} x_i^2 = 645$	$\sum_{i=1}^{10} y_i = 73.4$	$\sum_{i=1}^{10} x_iy_i = 598.6$

Substituting in Equations (A) and (B), we obtain

$$b_0 = 2.967 \qquad \text{and} \qquad b_1 = 0.583.$$

Thus the line of best fit, shown in Figure 7.21, is

$$y = 0.583x + 2.967, \tag{11}$$

where y is the amount of beta present and x is the amount of alpha used.

(b) If $x = 30$, then substituting in (11), we obtain

$$y = 20.457.$$

Thus there would be 20.457 ounces of *beta* present in a gallon of the compound.

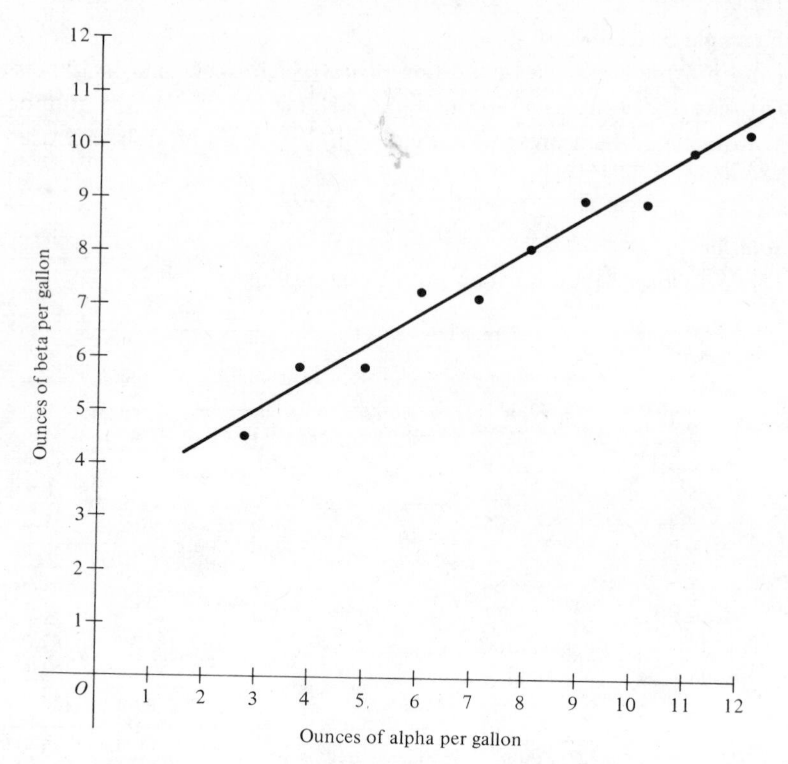

Figure 7.21

7.5 Exercises

In Exercises 1 through 4 find the line of best fit for the given data points.

1. (2, 1), (3, 2), (4, 3), (5, 2).

2. (3, 2), (4, 3), (5, 2), (6, 4), (7, 3).

3. (2, 3), (3, 4), (4, 3), (5, 4), (6, 3), (7, 4).

4. (3, 3), (4, 5), (5, 4), (6, 5), (7, 5), (8, 6), (9, 5), (10, 6).

5. In an experiment designed to determine the extent of a person's natural orientation, a subject is put in a special room and kept there for a certain length of time. He is then asked to find his way out of a maze and a record is made of the time it takes him to accomplish this task. The following data are obtained:

Time in room (hours)	1	2	3	4	5	6
Time to find way out of maze (minutes)	0.8	2.1	2.6	2.0	3.1	3.3

Let x denote the number of hours in the room and let y denote the numbers of minutes that it takes the subject to find his way out.
(a) Find the line of best fit relating x and y.
(b) Use the equation obtained in (a) to estimate the time it will take the subject to find his way out of the maze after 16 hours in the room.

6. A steel producer gathers the following data:

Year	1965	1966	1967	1968	1969	1970
Annual sales (millions of dollars)	1.2	2.3	3.2	3.6	3.8	5.1

Represent the years 1965, . . . , 1970 as 0, 1, 2, 3, 4, 5, respectively, and let x denote the year. Let y denote the annual sales (in millions of dollars).
(a) Find the line of best fit relating x and y.
(b) Use the equation obtained in (a) to estimate the annual sales for the year 1984.

7. A sales organization obtains the following data relating the number of salesmen to annual sales.

Number of salesmen	5	6	7	8	9	10
Annual sales (millions of dollars)	2.3	3.2	4.1	5.0	6.1	7.2

Let x denote the number of salesmen and let y denote the annual sales (in millions of dollars).
(a) Find the line of best fit relating x and y.
(b) Use the equation obtained in (a) to estimate annual sales when there are 20 salesmen.

Theoretical Exercise

T.1. Verify that Equations (A) and (B) give the solution to (10).

7.6 Linear Economic Models

As society has grown more complex, increasing attention has been paid to the analysis of economic behavior. Problems of economic behavior are more difficult to handle than problems in the physical sciences for many reasons. We may not know all the factors or variables that must be considered, we may not know what data need to be gathered, we may not have enough data, or the resulting mathematical problem may be too difficult to solve.

In the 1930s Wassily Leontief, a Harvard University economics professor, who was awarded the 1973 Nobel Prize in economics, developed a pioneering approach to the mathematical analysis of economic behavior. In this section we give a very brief introduction to the application of linear algebra to economics.

Our approach leans heavily on the presentations in the following books, which may be consulted by the reader for a more extensive treatment of this material: David Gale, *The Theory of Linear Economic Models* (New York: McGraw-Hill Book Company, 1960); and B. Johnston, G. Price, and F. S. van Vleck, *Linear Equations and Matrices* (Reading, Mass.: Addison-Wesley Publishing Co., Inc., 1966).

The Leontief Closed Model

Example 1.* Consider a simple society consisting of a farmer who produces all the food, a tailor who makes all the clothes, and a builder who builds all the homes in the community. Suppose that during the year the portion of each commodity consumed by each of the individuals is given in Table 7.2. Suppose also that one unit of each commodity is produced during the year. Let p_1, p_2, and p_3 be the annual incomes received by the farmer, builder, and tailor, respectively. We assume that every one pays the same price for a commodity. Thus the farmer pays the same price for his food as the tailor and builder, although he grew the food himself. We are interested in achieving a state of equilibrium which is defined as: *No one makes money or loses money*.

TABLE 7.2

		Goods produced by		
		Farmer	Carpenter	Tailor
Goods consumed by	Farmer	$\frac{7}{16}$	$\frac{1}{2}$	$\frac{3}{16}$
	Carpenter	$\frac{5}{16}$	$\frac{1}{6}$	$\frac{5}{16}$
	Tailor	$\frac{1}{4}$	$\frac{1}{3}$	$\frac{1}{2}$

*This example is due to Johnston, Price, and van Vleck. The general model for this example is also presented by Gale.

The farmer's expenditures are

$$\tfrac{7}{16}\,p_1 + \tfrac{1}{2}\,p_2 + \tfrac{3}{16}\,p_3,$$

and since it must equal his income, we have

$$\tfrac{7}{16}\,p_1 + \tfrac{1}{2}\,p_2 + \tfrac{3}{16}\,p_3 = p_1. \tag{1}$$

Similarly, for the carpenter we obtain

$$\tfrac{5}{16}\,p_1 + \tfrac{1}{6}\,p_2 + \tfrac{5}{16}\,p_3 = p_2, \tag{2}$$

and for the tailor we have

$$\tfrac{1}{4}\,p_1 + \tfrac{1}{3}\,p_2 + \tfrac{1}{2}\,p_3 = p_3. \tag{3}$$

Equations (1), (2), and (3) can be rewritten as

$$\mathbf{AP} = \mathbf{P}, \tag{4}$$

where

$$\mathbf{A} = \begin{bmatrix} \frac{7}{16} & \frac{1}{2} & \frac{3}{16} \\ \frac{5}{16} & \frac{1}{6} & \frac{5}{16} \\ \frac{1}{4} & \frac{1}{3} & \frac{1}{2} \end{bmatrix}, \qquad \mathbf{P} = \begin{bmatrix} p_1 \\ p_2 \\ p_3 \end{bmatrix}.$$

Equation (4) can be rewritten as

$$(\mathbf{I}_n - \mathbf{A})\mathbf{P} = \mathbf{0}, \tag{5}$$

which is a homogeneous system.

We are interested in a solution $\mathbf{P}$ of (5) whose components p_i will be nonnegative with at least one positive p_i, since $\mathbf{P} = \mathbf{0}$ means that all the prices are zero, which makes no sense.

Solving (4), we obtain

$$\mathbf{P} = r\begin{bmatrix} 4 \\ 3 \\ 4 \end{bmatrix},$$

where r is any real number. Letting r be a positive number, we obtain our desired solution.

Example 2 (The Exchange Model). Consider now the general problem where we have n manufacturers $M_1, M_2, \ldots, M_n$, and n goods $G_1, G_2, \ldots, G_n$, with M_i making only G_i. Consider a fixed interval of time, say one year, and suppose that M_i only makes one unit of G_i during the year.

In producing the goods G_i, manufacturer M_i may consume amounts of goods $G_1, G_2, \ldots, G_i, \ldots, G_n$. Thus iron, along with many other ingredients, is used in the manufacture of iron. Let a_{ij} be the amount of good G_j consumed by manufacturer M_i. Then

$$0 \leqslant a_{ij} \leqslant 1.$$

Suppose that the model is **closed**; that is, no goods leave or enter the system. This means that the total consumption of each good must equal its total production. Since the total production of G_j is 1, we obtain

$$a_{1j} + a_{2j} + \cdots + a_{nj} = 1 \qquad j = 1, 2, \ldots, n.$$

If the price per unit of G_k is p_k, then manufacturer M_i pays

$$a_{i1}p_1 + a_{i2}p_2 + \cdots + a_{in}p_n \tag{6}$$

for the goods he uses.

Now assume that no manufacturer makes money or loses money. That is, each manufacturer's income equals his expenses. Since M_i only manufactures one unit, his income is p_i. Thus from (6) we have

$$\begin{aligned} a_{11}p_1 + a_{12}p_2 + \cdots + a_{1n}p_n &= p_1 \\ a_{21}p_1 + a_{22}p_2 + \cdots + a_{2n}p_n &= p_2 \\ &\vdots \\ a_{n1}p_1 + a_{n2}p_2 + \cdots + a_{nn}p_n &= p_n, \end{aligned}$$

which can be written in matrix form as

$$\mathbf{AP} = \mathbf{P}, \tag{7}$$

where

$$\mathbf{A} = \left[a_{ij} \right] \qquad \text{and} \qquad \mathbf{P} = \begin{bmatrix} p_1 \\ p_2 \\ \vdots \\ p_n \end{bmatrix}.$$

Equation (7) can be rewritten as

$$(\mathbf{I}_n - \mathbf{A})\mathbf{P} = \mathbf{0}. \tag{8}$$

Thus our problem is that of finding a vector

$$\mathbf{P} \geqslant \mathbf{0},$$

with at least one component that is positive and satisfies Equation (8).

DEFINITION An $n \times n$ matrix $\mathbf{A} = [a_{ij}]$ is called an **exchange matrix** if it satisfies the following two properties:

(a) $a_{ij} \geqslant 0$.
(b) $a_{1j} + a_{2j} + \cdots + a_{nj} = 1$, for $j = 1, 2, \ldots, n$.

Example 3. Matrix **A** of Example 1 is an exchange matrix.

Our general problem can now be stated as: Given an exchange matrix **A**, find a vector $\mathbf{P} \geqslant 0$ with at least one positive component, satisfying Equation (8). It can be shown that this problem always has a solution.

In our general problem we required that each manufacturer's income equal his expenses. Instead, we could have required that each manufacturer's expenses not exceed his income. This would have led to

$$\mathbf{AP} \leqslant \mathbf{P} \tag{9}$$

instead of $\mathbf{AP} = \mathbf{P}$. However, it can be shown (Exercise T.1) that if Equation (9) holds, then Equation (7) will hold. Thus, if no manufacturer spends more than he earns, then everyone's income equals his expenses. An economic interpretation of this statement is that if some manufacturer is making a profit, then at least one manufacturer is taking a loss.

An International Trade Model

Example 4. Suppose that n countries $C_1, C_2, \ldots, C_n$ are engaged in trading with each other and that a common currency is in use. We assume that prices are fixed throughout the discussion and the C_j's income y_j comes entirely from selling its goods either internally or to other countries. We also assume that the fraction of C_j's income that is spent on imports from C_i is a fixed number a_{ij} which does not depend upon C_j's income y_j.

Since the a_{ij}'s are fractions of y_j, we have

$$a_{ij} \geqslant 0$$

$$a_{1j} + a_{2j} + \cdots + a_{nj} = 1,$$

so that $\mathbf{A} = [a_{ij}]$ is an exchange matrix. We now wish to determine the total income y_i for each country C_i. Since the value of C_i's exports to C_j is $a_{ij}y_j$, the total income of C_i is

$$a_{i1}y_1 + a_{i2}y_2 + \cdots + a_{in}y_n.$$

Hence y_i must satisfy

$$a_{i1}y_1 + a_{i2}y_2 + \cdots + a_{in}y_n = y_i.$$

In matrix notation we must find $\mathbf{P} = \begin{bmatrix} y_1 \\ y_2 \\ \vdots \\ y_n \end{bmatrix} \geqslant 0$, with at least one $y_i > 0$, so that

$$\mathbf{AP} = \mathbf{P},$$

which is our earlier problem.

A Linear Production Model

Suppose that we have n goods $G_1, G_2, \ldots, G_n$ and n activities $M_1, M_2, \ldots, M_n$. Assume that each activity M_i produces only one good G_i and that G_i is only produced by M_i. Let c_{ij} be the amount of G_j that has to be consumed to produce one unit of G_i. The matrix $\mathbf{C} = [c_{ij}]$ is called the **consumption matrix**. The ith row of C

$$\begin{bmatrix} c_{i1} & c_{i2} \cdots c_{in} \end{bmatrix} = \mathbf{C}_i$$

gives the inputs of the goods needed to produce one unit of G_i. Observe that c_{ii} may be positive, which means that we may require some amount of G_i to make one unit of G_i.

Let x_i be the number of units of G_i produced in a fixed period of time, say, one year. The vector

$$\mathbf{X} = \begin{bmatrix} x_1 & x_2 \cdots x_n \end{bmatrix}$$

is called the **production vector**. Since C_i gives the amounts necessary to produce one unit of G_i, the amounts necessary to produce x_i units of G_i are given by $x_i\mathbf{C}_i$. Thus the total amount consumed by the model is

$$x_1\mathbf{C}_1 + x_2\mathbf{C}_2 + \cdots + x_n\mathbf{C}_n = \mathbf{XC}.$$

The **net production** is the total produced minus the total consumed; that is,

$$\mathbf{X} - \mathbf{XC} = \mathbf{X}(\mathbf{I}_n - \mathbf{C}).$$

Now assume that there is an outside demand d_i for goods G_i and let

$$\mathbf{D} = [d_1 \quad d_2 \cdots d_n] \qquad d_i \geqslant 0$$

be the **demand vector**.

Our problem can now be stated: Given a demand vector **D**, can we find a production vector **X** such that the outside demand **D** is met without any surplus? That is, can we find a row vector $\mathbf{X} \geqslant \mathbf{0}$, so that we obtain the following equation?

$$\mathbf{X}(\mathbf{I}_n - \mathbf{C}) = \mathbf{D} \tag{10}$$

Example 5. Let

$$\mathbf{C} = \begin{bmatrix} -1 & 3 \\ 1 & -1 \end{bmatrix}$$

be a consumption matrix. Then

$$\mathbf{I}_2 - \mathbf{C} = \begin{bmatrix} 1 & 0 \\ 0 & 1 \end{bmatrix} - \begin{bmatrix} -1 & 3 \\ 1 & -1 \end{bmatrix} = \begin{bmatrix} 2 & -3 \\ -1 & 2 \end{bmatrix}.$$

Equation (10) becomes

$$[x_1 \quad x_2]\begin{bmatrix} 2 & -3 \\ -1 & 2 \end{bmatrix} = [d_1 \quad d_2]$$

and

$$\begin{aligned}[x_1 \quad x_2] &= [d_1 \quad d_2]\begin{bmatrix} 2 & -3 \\ -1 & 2 \end{bmatrix}^{-1} \\ &= [d_1 \quad d_2]\begin{bmatrix} 2 & 3 \\ 1 & 2 \end{bmatrix} \geqslant 0,\end{aligned}$$

since $d_1 \geqslant 0$ and $d_2 \geqslant 0$. Thus we can obtain a production vector for any given demand vector.

In general, if $(\mathbf{I}_n - \mathbf{C}) \geqslant 0$, then $\mathbf{X} = \mathbf{D}(\mathbf{I}_n - \mathbf{C})^{-1} \geqslant 0$ is a production vector for any given demand vector. However, for a given consumption matrix, there may be no solution.

Example 6. Consider the consumption matrix

$$\mathbf{C} = \begin{bmatrix} 0 & 2 \\ 2 & 0 \end{bmatrix}.$$

Then

$$\mathbf{I}_2 - \mathbf{C} = \begin{bmatrix} 1 & -2 \\ -2 & 1 \end{bmatrix}$$

and

$$(\mathbf{I}_2 - \mathbf{C})^{-1} = \begin{bmatrix} -\frac{1}{3} & -\frac{2}{3} \\ -\frac{2}{3} & -\frac{1}{3} \end{bmatrix},$$

so that

$$\mathbf{X} = \mathbf{D}(\mathbf{I}_2 - \mathbf{C})$$

is not a production vector for $\mathbf{D} \neq \mathbf{0}$, since all its components are not nonnegative. Thus the problem has no solution. If $\mathbf{D} = \mathbf{0}$, we do have a solution, namely, $\mathbf{X} = \mathbf{0}$, which means that if there is no outside demand, nothing is produced.

DEFINITION An $n \times n$ matrix $\mathbf{C}$ is called **productive** if $(\mathbf{I}_n - \mathbf{C})^{-1} \geqslant \mathbf{0}$. That is, $\mathbf{C}$ is productive if every entry of $(\mathbf{I}_n - \mathbf{C})^{-1}$ is nonnegative. The model is also called **productive**.

The basic result is contained in the following theorem.

THEOREM 7.8. *If* $\mathbf{C}$ *is productive, then for any* $\mathbf{D} \geqslant 0$, *the equation*

$$\mathbf{X}(\mathbf{I}_n - \mathbf{C}) = \mathbf{D}$$

has a unique solution $\mathbf{X} \geqslant \mathbf{0}$.

7.6 Exercises

1. Which of the following matrices are exchange matrices?

(a) $\begin{bmatrix} \frac{1}{3} & 0 & 1 \\ \frac{2}{3} & 1 & 0 \\ \frac{1}{2} & -\frac{1}{2} & 0 \end{bmatrix}$. (b) $\begin{bmatrix} \frac{1}{2} & \frac{1}{3} & \frac{3}{4} \\ \frac{1}{2} & \frac{1}{3} & \frac{1}{4} \\ 0 & \frac{1}{3} & 0 \end{bmatrix}$.

(c) $\begin{bmatrix} \frac{1}{3} & -\frac{2}{3} & \frac{1}{2} \\ \frac{2}{3} & \frac{2}{3} & \frac{1}{2} \\ 0 & 1 & 0 \end{bmatrix}$. (d) $\begin{bmatrix} 1 & \frac{1}{4} & \frac{5}{6} \\ 0 & \frac{1}{4} & \frac{1}{6} \\ 0 & \frac{1}{2} & 0 \end{bmatrix}$.

In Exercises 2 through 4 find a vector $\mathbf{P} \geqslant 0$, with at least one positive component, satisfying Equation (8) for the given exchange matrix.

2. $\begin{bmatrix} \frac{1}{3} & \frac{2}{3} & 0 \\ \frac{1}{3} & 0 & \frac{1}{4} \\ \frac{1}{3} & \frac{1}{3} & \frac{3}{4} \end{bmatrix}$. **3.** $\begin{bmatrix} \frac{1}{2} & 1 & \frac{2}{3} \\ 0 & 0 & 0 \\ \frac{1}{2} & 0 & \frac{1}{3} \end{bmatrix}$. **4.** $\begin{bmatrix} 0 & \frac{1}{3} & 1 \\ \frac{1}{6} & \frac{1}{6} & 0 \\ \frac{5}{6} & \frac{1}{2} & 0 \end{bmatrix}$.

5. Consider the simple economy of Example 1. Suppose that the farmer consumes $\frac{2}{5}$ of the food, $\frac{2}{5}$ of the shelter, and $\frac{1}{5}$ of the clothes; that the builder consumes $\frac{1}{3}$ of the food, $\frac{1}{3}$ of the shelter, and $\frac{1}{3}$ of the clothes; that the tailor consumes $\frac{1}{2}$ of the food, $\frac{1}{2}$ of the shelter, and none of the clothes. Find the exchange matrix $\mathbf{A}$ for this problem and a vector $\mathbf{P} \geqslant \mathbf{0}$, with at least one positive component satisfying Equation (8).

6. Consider the international trade model consisting of three countries, C_1, C_2, and C_3. Suppose that the fraction of C_1's income spent on imports from C_1 is $\frac{1}{4}$, from C_2 is $\frac{1}{2}$, and from C_3 is $\frac{1}{4}$; that the fraction of C_2's income spent on imports from C_1 is $\frac{2}{5}$, from C_2 is $\frac{1}{5}$, and from C_3 is $\frac{2}{5}$; that the fraction of C_3's income spent on imports from C_1 is $\frac{1}{2}$, from C_2 is $\frac{1}{2}$, and from C_3 is 0. Find the income of each country.

In Exercises 7 through 10 determine which matrices are productive.

7. $\begin{bmatrix} \frac{1}{2} & \frac{1}{3} & 0 \\ 0 & \frac{2}{3} & 0 \\ 1 & 0 & 2 \end{bmatrix}$. **8.** $\begin{bmatrix} 0 & \frac{2}{3} & 0 \\ 1 & 0 & 0 \\ 0 & -\frac{1}{3} & 0 \end{bmatrix}$. **9.** $\begin{bmatrix} 0 & \frac{1}{3} & \frac{1}{2} \\ \frac{1}{2} & 0 & \frac{1}{2} \\ \frac{1}{4} & \frac{1}{3} & 0 \end{bmatrix}$.

10. $\begin{bmatrix} 0 & \frac{1}{3} & \frac{1}{3} \\ \frac{1}{4} & 0 & \frac{1}{6} \\ \frac{1}{3} & \frac{2}{3} & 0 \end{bmatrix}.$

11. Suppose that the consumption matrix for the linear production model is

$$\begin{bmatrix} \frac{1}{2} & \frac{1}{2} \\ \frac{1}{2} & \frac{1}{4} \end{bmatrix}.$$

(a) Find the production vector for the demand vector $[1 \quad 3]$.
(b) Find the production vector for the demand vector $[2 \quad 0]$.

Theoretical Exercises

T.1. In the exchange model (Example 3), show that $\mathbf{AP} \leqslant \mathbf{P}$ implies that $\mathbf{AP} = \mathbf{P}$.

8 Numerical Linear Algebra

Almost all of the computational work done on applied problems is carried out on computers. These efforts have given added impetus to the relatively new area of *numerical linear algebra*, which seeks to evaluate various methods, improve existing methods, and find new methods. In this chapter we give a very brief sketch of some widely used numerical methods for solving linear systems and for finding eigenvalues and eigenvectors. Computer programs that implement these methods are widely available.

8.1 Error Analysis

Before we can discuss the evaluation of different numerical methods, we look at the way in which numbers are handled in a computer. Most of the computational work done by the computer is carried out in **floating-point**

arithmetic. In this system each number is written as

$$\pm .d_1 d_2 \cdots d_n \times 10^e, \tag{1}$$

where e is an integer that satisfies

$$-M \leqslant e \leqslant M,$$

and the digits d_i satisfy

$$1 \leqslant d_1 \leqslant 9 \quad \text{and} \quad 0 \leqslant d_i \leqslant 9 \qquad i = 2, \ldots, n.$$

The number M depends upon the computer being used. For the IBM 360 and 370 systems, $M = 75$, while for the Univac 1108 computer, $M = 38$. If $M = 75$, the smallest number that can be stored in the computer is 10^{-75}, and the largest is 10^{75}. The number of digits in (1) is called the **number of significant digits**. In the IBM 360 and 370 computers, seven significant digits are used; in the UNIVAC 1108 computer, eight digits are used. The fraction $.d_1 d_2 \cdots d_n$ in (1) is called the **mantissa** and e is called the **exponent**. A number of the form in (1) said to be a **floating-point number**.

Examples of floating-point numbers are

$$+0.231 \times 10^2, \quad -0.6844 \times 10^4, \quad -0.751 \times 10^{-2},$$
$$+0.4 \times 10^2, \quad +0.63 \times 10^0.$$

Suppose that we now want to represent a number N in the computer. If there are infinitely many digits in N (for example, $N = \frac{7}{3} = 2.33 \ldots$), then we must approximate the number as it is entered into the computer, for we only have n significant digits at our disposal. First, we write the number as

$$\pm .d_1 d_2 \cdots d_n d_{n+1} \cdots \times 10^e, \tag{2}$$

$1 \leqslant d_1 \leqslant 9$, and $0 \leqslant d_i \leqslant 9$. Now we proceed in one of two ways.

Truncation or chopping. Keep the first n digits in (2). Thus $\frac{7}{3} = 2.333 \ldots = 0.233 \ldots \times 10^1$, which when truncated to four significant figures becomes 0.2333×10^1. Similarly, $\frac{8}{3} = 2.66 \ldots = 0.2666 \ldots \times 10^0$, which when truncated to four significant figures becomes 0.2666×10^1.

Rounding. If $d_{n+1} \geqslant 5$, add 1 to d_n and truncate. Otherwise, merely truncate. Another way of stating this procedure is: Add 5 to d_{n+1} and then truncate. Thus $\frac{7}{3}$ rounded to four significant figures is 0.2333×10^1 and $\frac{8}{3}$

rounded to four significant figures is 0.2667×10^1. Some computers deal with the problem of representing numbers by rounding, others by truncating. Also, many computers make available **double-precision** arithmetic (through either hardware or software), thereby providing twice as many digits as in single-precision floating-point arithmetic.

Numerical errors that occur due to either rounding or truncating are called **roundoff errors**. There are two types of errors that can be studied. Let N be a real number and let $\hat{N}$ be an approximation to it. The **absolute error** ε_a is defined as

$$\varepsilon_a = \hat{N} - N,$$

and the **relative error** ε_r is

$$\varepsilon_r = \frac{\hat{N} - N}{N} \qquad \text{if } N \neq 0.$$

The relative error is not defined if $N = 0$. The relative error is the more important quantity, since a small error in a large number may not be too harmful, while the same error in a small number can be disastrous. Thus an error of 0.1234 in the number 20,642.3217 is rather trivial, but the same error in 0.8266 poses some serious problems. The three major causes of roundoff error are: (1) adding a large number to a relatively small number, (2) subtracting two numbers that are almost equal, and (3) dividing a number by a very small number.

Much effort in numerical linear algebra goes into the development of methods that reduce roundoff error and methods that estimate the roundoff error incurred in a numerical procedure.

8.1 Exercises

In Exercises 1 through 4 write the given number as a floating-point number.

1. 34.7213 **2.** -0.0002135 **3.** -284 **4.** 24.00

In Exercises 5 through 8 write the given number as a floating-point number with four significant digits, first by rounding and then by truncating.

5. 1.23 **6.** -4.25678 **7.** $\frac{17}{3}$ **8.** $\frac{13}{6}$

In Exercises 9 through 12 find the absolute error and the relative error.

9. $N = 12.341$, $\hat{N} = 12.362$. **10.** $N = -0.4821$, $\hat{N} = -0.4215$.
11. $N = 6482.0$, $\hat{N} = 6483.1$. **12.** $N = 0.00724$, $\hat{N} = 0.00742$.

8.2 Linear Systems

In Section 1.3 we discussed the Gauss–Jordan reduction method for solving a linear system of m equations in n unknowns by reducing the augmented matrix to reduced row echelon form. We first present a similar method, which is more efficient from a computational point of view.

DEFINITION An $m \times n$ matrix $\mathbf{A}$ is said to be in **row echelon form** if:

(a) Each of the first k rows $(1 \leqslant k \leqslant m)$ has at least one nonzero element, and if $k < m$, then rows $k + 1, k + 2, \ldots, m$ consist entirely of zeros.

(b) The first, counting from left to right, element in each of the first k rows is a 1.

(c) If the 1 in row i, $1 \leqslant i \leqslant k$, occurs in column j_i, then $j_1 < j_2 < \cdots < j_k$.

(d′) All the elements in column j_i, rows $i + 1, i + 2, \ldots, m$ are zero.

If $k = m$, there are no rows all of whose elements are zero. This definition differs from that for reduced row echelon form, in that property (d′) replaced property (d) in the definition of Section 1.3.

Example 1. The matrices

$$\begin{bmatrix} 1 & -3 & 2 & 4 \\ 0 & 1 & 4 & 5 \\ 0 & 0 & 1 & 2 \end{bmatrix}, \quad \begin{bmatrix} 1 & 2 & 3 & 1 & 2 \\ 0 & 1 & -2 & 4 & 3 \\ 0 & 0 & 1 & -2 & 6 \\ 0 & 0 & 0 & 0 & 0 \\ 0 & 0 & 0 & 0 & 0 \end{bmatrix}$$

are in row echelon form.

It is not difficult to see that using the first six steps of Theorem 1.5 of Section 1.3, we can show that every $m \times n$ matrix is row equivalent to a matrix in row echelon form, which need not be unique. We reduce a matrix to row echelon form by using the first six steps of Theorem 1.5 in the following example.

Example 2. Let $\mathbf{A}$ be the matrix of Example 5 in Section 1.3.

$$\mathbf{A} = \begin{bmatrix} 0 & 2 & 3 & -4 & 1 \\ 0 & 0 & 2 & 3 & 4 \\ 2 & 2 & -5 & 2 & 4 \\ 2 & 0 & -6 & 9 & 7 \end{bmatrix}.$$

Proceeding as far as step 6, we have

$$\begin{array}{c} \begin{matrix} 1 & 1 & -\frac{5}{2} & 1 & 2 \end{matrix} \\ \mathbf{B}_3 = \begin{bmatrix} 0 & 1 & \frac{3}{2} & -2 & \frac{1}{2} \\ 0 & 0 & 2 & 3 & 4 \\ 0 & 0 & 2 & 3 & 4 \end{bmatrix}. \end{array}$$

We now identify **C** by deleting but not erasing the first row of $\mathbf{B}_3$.

$$\begin{array}{c} \begin{matrix} 1 & 1 & -\frac{5}{2} & 1 & 2 \\ 0 & 1 & \frac{3}{2} & -2 & \frac{1}{2} \end{matrix} \\ \mathbf{C} = \begin{bmatrix} 0 & 0 & \textcircled{2} & 3 & 4 \\ 0 & 0 & 2 & 3 & 4 \end{bmatrix} \end{array}$$

$$\begin{array}{c} \begin{matrix} 1 & 1 & -\frac{5}{2} & 1 & 2 \\ 0 & 1 & \frac{3}{2} & -2 & \frac{1}{2} \end{matrix} \\ \mathbf{C}_1 = \mathbf{C}_2 = \begin{bmatrix} 0 & 0 & 1 & \frac{3}{2} & 2 \\ 0 & 0 & 2 & 3 & 4 \end{bmatrix} \end{array}$$

No row of **C** had to be interchanged. The first row of **C** was divided by 2.

$$\begin{array}{c} \begin{matrix} 1 & 1 & -\frac{5}{2} & 1 & 2 \\ 0 & 1 & \frac{3}{2} & -2 & \frac{1}{2} \end{matrix} \\ \mathbf{C}_3 = \begin{bmatrix} 0 & 0 & 1 & \frac{3}{2} & 2 \\ 0 & 0 & 0 & 0 & 0 \end{bmatrix} \end{array}$$

-2 times the first row of $\mathbf{C}_2$ was added to its second row.

The final matrix

$$\begin{bmatrix} 1 & 1 & -\frac{5}{2} & 1 & 2 \\ 0 & 1 & \frac{3}{2} & -2 & \frac{1}{2} \\ 0 & 0 & 1 & \frac{3}{2} & 2 \\ 0 & 0 & 0 & 0 & 0 \end{bmatrix}$$

is in row echelon form.

If we reduce the augmented matrix $[\mathbf{A} \mid \mathbf{B}]$ of the linear system $\mathbf{AX} = \mathbf{B}$ to row echelon form, we get another method for solving linear systems which is called **Gaussian elimination**.

Example 3. Consider Example 6 of Section 1.3:

$$\begin{aligned} x + 2y + 3z &= 9 \\ 2x - y + z &= 8 \\ 3x \phantom{{}+2y} - z &= 3. \end{aligned}$$

The augmented matrix is row equivalent to the matrix (verify),

$$\left[\begin{array}{ccc|c} 1 & 2 & 3 & 9 \\ 0 & 1 & 1 & 2 \\ 0 & 0 & 1 & 3 \end{array}\right], \tag{1}$$

which is in row echelon form. The solution can be obtained by using **back substitution** as follows:

From the last row of (1) we have

$$0x + 0y + 1z = 3,$$

so that $z = 3$. From the second row of (1),

$$0x + 1y + 1z = 2,$$

so

$$y = 2 - z = 2 - 3 = -1.$$

Now from the first row of (1),

$$x + 2y + 3z = 9,$$

so

$$x = 9 - 2y - 3z = 9 - 2(-1) - 3(3) = 2.$$

Thus the solution is $x = 2$, $y = -1$, $z = 3$.

Throughout the rest of this section we limit our discussion to linear systems in n equations with n unknowns.

One of the troubles with Gaussian elimination is that we have to divide by the pivot; if the latter is a very small number, then the roundoff error can cast considerable doubt upon the final answer. A partial remedy to this problem consists in using the technique called **partial pivoting**. This method calls for choosing the largest nonzero entry in the pivotal column as the pivot element.

Example 4. To solve the linear system

$$\begin{aligned} 14x_1 + 2x_2 + 4x_3 &= -10 \\ 16x_1 + 40x_2 - 4x_3 &= 55 \\ -2x_1 + 4x_2 - 16x_3 &= -38 \end{aligned} \tag{2}$$

by Gaussian elimination with partial pivoting, we proceed as follows. Form the augmented matrix

$$\left[\begin{array}{rrr|r} 14 & 2 & 4 & -10 \\ 16 & 40 & -4 & 55 \\ -2 & 4 & -16 & -38 \end{array}\right].$$

Our computations will be carried out to three decimal places and we round after each multiplication and division.

$$\left[\begin{array}{rrr|r} 16 & 40 & -4 & 55 \\ 14 & 2 & 4 & -10 \\ -2 & 4 & -16 & -38 \end{array}\right]$$

The first and second rows were interchanged.

$$\left[\begin{array}{rrr|r} 1 & 2.5 & -0.25 & 3.438 \\ 14 & 2 & 4 & -10 \\ -2 & 4 & -16 & -38 \end{array}\right]$$

The first row was divided by the pivot 16.

$$\left[\begin{array}{rrr|r} 1 & 2.5 & -0.25 & 3.438 \\ 0 & -33 & 7.5 & -58.132 \\ 0 & 9 & -16.5 & -31.124 \end{array}\right]$$

-14 times the first row was added to the second row; 2 times the first row was added to its third row.

$$\left[\begin{array}{rrr|r} 1 & 2.5 & -0.25 & 3.438 \\ 0 & 1 & -0.227 & 1.762 \\ 0 & 9 & -16.5 & -31.124 \end{array}\right]$$

The second row was divided by -33; that is, -33 is the pivot for the 2×4 submatrix obtained by removing the first row.

$$\left[\begin{array}{rrr|r} 1 & 2.5 & -0.25 & 3.438 \\ 0 & 1 & -0.227 & 1.762 \\ 0 & 0 & -14.457 & -46.982 \end{array}\right]$$

-9 times the second row was added to the third row.

$$\left[\begin{array}{ccc|c} 1 & 2.5 & -0.25 & 3.438 \\ 0 & 1 & -0.227 & 1.762 \\ 0 & 0 & 1 & 3.254 \end{array}\right]$$

The third row was divided by -14.457; that is, -14.457 is the pivot for the 1×4 submatrix obtained by removing the first two rows.

Then

$$z = 3.25$$

$$y = 1.762 + (0.227)(3.25) = 2.500$$

$$x = 3.438 + (0.25)(3.25) - (2.5)(2.500) = -1.999.$$

Thus the solution obtained is

$$x = -1.999, \qquad y = 2.500, \qquad \text{and} \qquad z = 3.250,$$

which agrees rather well with the exact solution:

$$x = -2, \qquad y = 2.5, \qquad \text{and} \qquad z = 3.25.$$

Full pivoting consists in choosing the largest nonzero entry in the entire augmented matrix as the pivot. Although this variant sometimes gives better results than partial pivoting, it requires a relabeling of the variables. The additional programming and bookkeeping required make Gaussian elimination with partial pivoting a more popular method than Gaussian elimination with full pivoting.

Ill-Conditioned Systems

There are linear systems, $\mathbf{AX} = \mathbf{B}$, where $\mathbf{A}$ is $n \times n$, in which relatively small changes in some of the elements of the augmented matrix $[\mathbf{A} \mid \mathbf{B}]$ may lead to relatively large changes in the solution. Such systems are called **ill-conditioned systems**. Gaussian elimination with partial pivoting or full pivoting may reduce the roundoff error for some of these systems. Sometimes **double-precision arithmetic** (requiring twice as much storage as single-precision and longer running times) may provide some help. However, for some ill-conditioned systems there may be no possible remedy.

Example 5. Consider the linear system

$$\begin{aligned} x - \quad\;\; y &= 1 \\ x - 1.01y &= 0 \end{aligned} \tag{3}$$

whose exact solution is

$$x = 101, \qquad y = 100.$$

Suppose that the coefficient of y in the second equation of (1) now changes from 1.01 to 0.99. The resulting system

$$\begin{aligned} x - \quad\;\; y &= 1 \\ x - 0.99y &= 0 \end{aligned} \tag{4}$$

has the exact solution

$$x = -99, \qquad y = -100.$$

Thus the systems (3) and (4), which are nearly identical, have vastly different solutions. System (3) is an ill-conditioned system. If we sketch the lines in (3) (Figure 8.1), we see that they are nearly parallel. Thus a minor shift of one of the lines results in a major shift of the point of intersection.

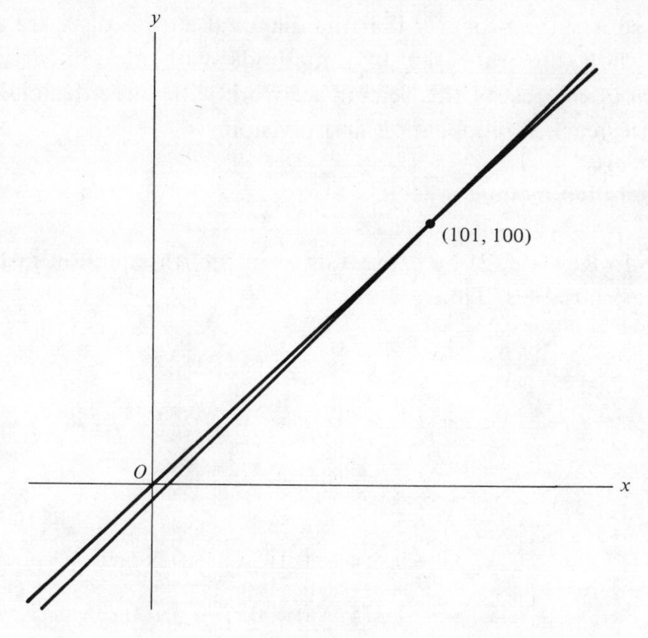

Figure 8.1

Iterative Methods

Gaussian elimination with partial pivoting is a **direct** method for solving $\mathbf{AX} = \mathbf{B}$, where $\mathbf{A}$ is $n \times n$. That is, the solution will be obtained after a finite number of steps. Moreover, the number of steps required to obtain the solution can be estimated in advance.

We shall now outline two methods, one due to Jacobi, the other to Gauss and Seidel, for solving such linear systems. These methods are **iterative** in nature: that is, we start out with an initial approximation to the solution, which we successively try to improve. If the successive approximations tend to approach the solution, we say that the method **converges**. Otherwise, we say that the method **diverges**.

The Jacobi and Gauss–Seidel methods do not always converge, and we give below a sufficient condition for their convergence. These methods are more efficient, when applicable, than Gaussian elimination with partial pivoting for very large matrices which are **sparse** matrices (very large matrices having a large number of zero elements).

To discuss these methods we start with the linear system $\mathbf{AX} = \mathbf{B}$, where $\mathbf{A}$ is $n \times n$. Assume that the determinant of $\mathbf{A}$ is nonzero. This means that we can rearrange the equations so that the diagonal entries of the coefficient matrix are all nonzero. Thus assume that our equations already satisfy the property that the diagonal entries of $\mathbf{A}$ are all nonzero.

We shall illustrate the two methods with the linear system (2). Throughout the rest of this section we work with three decimal places and *round* after each multiplication and division.

Jacobi iteration method

STEP 1. Rewrite (2) by expressing x_i in the ith equation in terms of the remaining variables. Thus

$$\begin{aligned} x_1 &= -\tfrac{10}{14} - \tfrac{2}{14}x_2 - \tfrac{4}{14}x_3 \\ x_2 &= \tfrac{55}{40} - \tfrac{16}{40}x_1 + \tfrac{4}{40}x_3 \\ x_3 &= \tfrac{38}{16} - \tfrac{2}{16}x_1 + \tfrac{4}{16}x_2 \end{aligned}$$

or

$$\begin{aligned} x_1 &= -0.714 - 0.143x_2 - 0.286x_3 \\ x_2 &= 1.375 - 0.400x_1 + 0.100x_3 \\ x_3 &= 2.375 - 0.125x_1 + 0.250x_2. \end{aligned} \tag{5}$$

STEP 2. Choose an initial approximation $x_1^{(0)}, x_2^{(0)}, \ldots, x_n^{(0)}$ to the solution. In the absence of other information, let $x_1^{(0)} = x_2^{(0)} = \cdots = x_n^{(0)} = 0$. In our example, $x_1^{(0)} = x_2^{(0)} = x_3^{(0)} = 0$.

STEP 3. Substitute the values of the variables calculated in the previous iteration [$(k-1)$th iteration] into the right side of (5) to obtain a new approximation, $x_1^{(k)}, x_2^{(k)}, \ldots, x_n^{(k)}$. For our example,

$$x_1^{(1)} = -0.714, \qquad x_2^{(1)} = 1.375, \qquad x_3^{(1)} = 2.375.$$

Using step 3 again, we now obtain our second approximation,

$$x_1^{(2)} = -1.590, \qquad x_2^{(2)} = 1.899, \qquad x_3^{(2)} = 2.808.$$

The first nine approximations ($k = 8$) to the solutions are shown in Table 8.1. Thus our approximation to the solution is

$$x \cong -1.998, \qquad y \cong 2.497, \qquad \text{and} \qquad z \cong 3.247$$

and was obtained in nine iterations. The exact solution is $x = -2$, $y = 2.5$, and $z = 3.25$. The symbol $\cong$ means "approximately."

TABLE 8.1

Iteration	$x_1^{(k)}$	$x_2^{(k)}$	$x_3^{(k)}$
0	0	0	0
1	−0.714	1.375	2.375
2	−1.590	1.899	2.808
3	−1.789	2.292	3.049
4	−1.914	2.396	3.172
5	−1.964	2.458	3.213
6	−1.984	2.482	3.236
7	−1.994	2.493	3.244
8	−1.998	2.497	3.247

Gauss–Seidel iteration method. In step 3 of the Jacobi iteration method, we substitute the values of the variables calculated in the last iteration into the right side of (5) to obtain a new approximation. However, in the next iteration, once we have a new value for x_1, we could use it to calculate x_2 in Equation (5); once we have new values for x_1 and x_2, we could use them to calculate x_3; and so on. Thus we can replace step 3 of the Jacobi iteration method by the following procedure.

STEP 3′. Substitute the most recently calculated values of the variables, in the current iteration, into the right side of (5) to obtain a new approximation $x_1^{(k)}, x_2^{(k)}, \ldots, x_n^{(k)}$. The resulting method is called the **Gauss–Seidel iteration method.**

To solve (2) by the Gauss–Seidel iteration method, we proceed as follows. Start with the initial approximation

$$x_1^{(0)} = x_2^{(0)} = x_3^{(0)} = 0,$$

substitute these values into the right side of the first equation in (5), and obtain

$$x_1^{(1)} = -0.714.$$

We next substitute

$$x_1^{(1)} = -0.714 \quad \text{and} \quad x_3^{(0)} = 0$$

into the right side of the second equation of (5) to obtain

$$x_2^{(1)} = 1.661.$$

Now substitute

$$x_1^{(1)} = -0.714 \quad \text{and} \quad x_2^{(1)} = 1.661$$

into the right side of the third equation of (5) to obtain

$$x_3^{(1)} = 2.879.$$

We next use the values of $x_2^{(1)} = 1.661$ and $x_3^{(1)} = 2.879$ in the first equation of (5) to obtain $x_1^{(2)} = -1.775$. Now use $x_1^{(2)} = -1.775$ and $x_2^{(1)} = 1.661$ in the second equation of (5) to obtain $x_2^{(2)} = 2.373$. The first six approximations ($k = 5$) to the solutions are shown in Table 8.2. Thus our approximation to the solution is

$$x \cong -2.000, \quad y \cong 2.500, \quad \text{and} \quad z \cong 3.250,$$

and was obtained in six iterations.

Although Gauss–Seidel iteration took fewer iterations than the Jacobi iteration method for our illustrative problem, there are problems where the opposite is true. Moreover, one cannot predict, in advance, which method is best for a given problem.

TABLE 8.2

Iteration	$x_1^{(k)}$	$x_2^{(k)}$	$x_3^{(k)}$
0	0	0	0
1	− 0.714	1.661	2.879
2	− 1.775	2.373	3.190
3	− 1.965	2.480	3.241
4	− 1.996	2.497	3.249
5	− 2.000	2.500	3.250

Convergence

Consider the linear system $\mathbf{AX} = \mathbf{B}$, where $\mathbf{A}$ is $n \times n$. A sufficient condition for the convergence of the Jacobi and Gauss–Seidel iteration methods is

$$|a_{ii}| > \sum_{\substack{j=1 \\ j \neq i}}^{n} |a_{ij}| \qquad (i = 1, 2, \ldots, n). \tag{6}$$

Equation (6) means that every diagonal element is in absolute value larger than the sum of the absolute values of the other entries in its row. A matrix $\mathbf{A}$ satisfying (6) is called **diagonally dominant**.

If $\mathbf{A}$ is diagonally dominant, then the Jacobi and Gauss–Seidel methods converge. Of course, the methods may converge even if this condition fails to hold (see Exercise 15).

Example 6. The matrix

$$\begin{bmatrix} 3 & -2 & 0 \\ 2 & -6 & 3 \\ 5 & 2 & 8 \end{bmatrix}$$

is diagonally dominant, whereas

$$\begin{bmatrix} 4 & 2 & -2 \\ 2 & 4 & 1 \\ 5 & 2 & 8 \end{bmatrix}$$

is not, since $|4| = |2| + |-2|$.

Sometimes it is necessary to interchange the equations so that the coefficient matrix will be diagonally dominant.

Example 7. Consider the linear system

$$\begin{aligned} 12x_1 - 14x_2 + x_3 &= 4 \\ -6x_1 + 4x_2 + 11x_3 &= -6 \\ 6x_1 + 3x_2 + 2x_3 &= 8. \end{aligned}$$

The coefficient matrix is obviously not diagonally dominant. However, if we interchange the first and second rows and then the first and third rows of the resulting system (verify), we obtain the coefficient matrix

$$\begin{bmatrix} 6 & 3 & 2 \\ 12 & -14 & 1 \\ -6 & 4 & 11 \end{bmatrix},$$

which is diagonally dominant: $|6| > |3| + |2|$; $|-14| > |12| + |1|$; $|11| > |-6| + |4|$.

8.2 Exercises

In Exercises 1 and 2 transform the given matrix to row echelon form.

1. $\begin{bmatrix} 1 & -2 & 0 \\ 2 & -3 & -1 \\ 1 & 3 & 2 \end{bmatrix}.$

2. $\begin{bmatrix} 0 & -1 & 2 & 3 \\ 2 & 3 & 4 & 5 \\ 1 & 3 & -1 & 2 \\ 3 & 2 & 4 & 1 \end{bmatrix}.$

In Exercises 3 and 4 solve the given linear systems by Gaussian elimination.

3. $$\begin{aligned} x_1 + 2x_2 + x_3 &= 0 \\ -3x_1 + 3x_2 + 2x_3 &= -7 \\ 4x_1 - 2x_2 - 3x_3 &= 2. \end{aligned}$$

4. $$\begin{aligned} 3x_1 - 3x_2 + 2x_3 &= 12 \\ 4x_1 + 6x_2 - 4x_3 &= -2 \\ 5x_1 - 9x_2 - 6x_3 &= 6. \end{aligned}$$

In Exercises 5 through 8 solve the given linear system by Gaussian elimination with partial pivoting. Carry out your computations to three decimal places and round after each multiplication and division.

5.
$$\begin{aligned} 3x_1 - 2x_2 + 3x_3 &= -8 \\ 6x_1 - 4x_2 + 5x_3 &= -14 \\ -12x_1 + 6x_2 + 7x_3 &= -8. \end{aligned}$$

6.
$$\begin{aligned} -2x_1 + 3x_2 + 4x_3 &= 8 \\ 3x_1 + 6x_2 - x_3 &= -8 \\ x_1 + 2x_2 + x_3 &= 4. \end{aligned}$$

7.
$$\begin{aligned} 2.5x_1 + 3.5x_2 - 4.25x_3 &= 37.3 \\ 3.4x_1 + 2.5x_2 - 2.01x_3 &= 26.8 \\ 5.3x_1 - 2.4x_2 + 6.21x_3 &= -20.68. \end{aligned}$$

8.
$$\begin{aligned} 2.2x_1 - 3.5x_2 + 1.2x_3 &= -18.3 \\ 4.4x_1 + 4.6x_2 - 3.8x_3 &= -25.6 \\ -6.3x_1 + 3.8x_2 + 2.4x_3 &= 57.1. \end{aligned}$$

9. Verify that the linear system

$$\begin{aligned} 2.121x + 3.421y &= 13.205 \\ 2.12x + 3.42y &= 13.200 \end{aligned}$$

is ill-conditioned by solving the linear system

$$\begin{aligned} 2.121x + 3.421y &= 13.205 \\ 2.12x + 3.42y &= 13.203. \end{aligned}$$

10. Which of the following matrices are diagonally dominant?

(a) $\begin{bmatrix} 2 & 1 \\ 4 & -5 \end{bmatrix}$ (b) $\begin{bmatrix} 3 & 2 \\ -4 & 3 \end{bmatrix}$ (c) $\begin{bmatrix} 3 & 1 & 2 \\ 0 & 4 & -2 \\ 1 & -2 & 4 \end{bmatrix}$.

(d) $\begin{bmatrix} -4 & 3 & 0 \\ 4 & 6 & 1 \\ 2 & -2 & 8 \end{bmatrix}$. (e) $\begin{bmatrix} 3 & -2 & -1 \\ 4 & 5 & 1 \\ -2 & -4 & 6 \end{bmatrix}$.

In Exercises 11 through 14 solve the given linear system by (1) Jacobi's iteration method and (2) Gauss–Seidel's iteration method. Carry out the computations to three decimal places for six iterations ($k = 5$) and round after each multiplication and division.

11.
$$\begin{aligned} 16x + 5y &= -7 \\ 4x + 15y &= 67. \end{aligned}$$

12.
$$\begin{aligned} 6x_1 - x_2 - 2x_3 &= 12 \\ 3x_1 + 5x_2 + x_3 &= 3 \\ 2x_1 - x_2 + 7x_3 &= 36. \end{aligned}$$

13. $$\begin{aligned} 9x_1 - 2x_2 + 6x_3 &= 9 \\ 3x_1 + 6x_2 - 2x_3 &= 15 \\ 6x_1 - x_2 + 8x_3 &= 4.5. \end{aligned}$$

14. $$\begin{aligned} 15x_1 - 4x_2 - 2x_3 &= -2 \\ 5x_1 - 8x_2 + x_3 &= -9 \\ 10x_1 - 2x_2 + 15x_3 &= -50. \end{aligned}$$

15. (a) Consider the linear system

$$\begin{aligned} 2x + 4y &= 8 \\ x - y &= -5. \end{aligned}$$

The coefficient matrix is not diagonally dominant. Show that the Gauss–Seidel iteration method diverges. Use seven iterations ($k = 6$).

(b) Consider the linear system

$$\begin{aligned} 2x - y &= -8 \\ x + y &= -1. \end{aligned}$$

The coefficient matrix is not diagonally dominant. Show that the Gauss–Seidel iteration method converges. Use seven iterations ($k = 6$).

8.3 Eigenvalues and Eigenvectors

In this section we present two methods for finding eigenvalues and eigenvectors of matrices. The first method finds the eigenvalue of largest absolute value and an associated eigenvector. The second method finds all the eigenvalues and associated eigenvectors of a symmetric matrix.

The Power Method

Suppose that **A** is an $n \times n$ matrix which is diagonalizable and that the eigenvalues $\lambda_1, \lambda_2, \ldots, \lambda_n$ are such that

$$|\lambda_1| > |\lambda_2| \geqslant |\lambda_3| \geqslant \cdots \geqslant |\lambda_n|,$$

so that λ_1 is the eigenvalue of largest magnitude. The power method, an easily implemented technique, computes λ_1 and an associated eigenvector.

The key idea in the power method is the following result, whose proof we omit.

If $\mathbf{U}_0$ is an arbitrary vector, then the vectors $\mathbf{A}^k\mathbf{U}_0$ (k a nonnegative integer) form a sequence of approximations to $\mathbf{X}_1$, an eigenvector of $\mathbf{A}$ associated with the dominant eigenvalue λ_1.

If $\mathbf{X}_1$ were exactly known, then using the inner product in R^n, we form the ratio

$$\frac{(\mathbf{A}\mathbf{X}_1, \mathbf{X}_1)}{(\mathbf{X}_1, \mathbf{X}_1)} = \frac{(\lambda_1\mathbf{X}_1, \mathbf{X}_1)}{(\mathbf{X}_1, \mathbf{X}_1)} = \frac{\lambda_1(\mathbf{X}_1, \mathbf{X}_1)}{(\mathbf{X}_1, \mathbf{X}_1)} = \lambda_1,$$

and thus obtain λ_1.

Suppose now that for k sufficiently large, we have a good approximation $\hat{\mathbf{X}}_1$ to $\mathbf{X}_1$. Then

$$\frac{(\mathbf{A}\hat{\mathbf{X}}_1, \hat{\mathbf{X}}_1)}{(\hat{\mathbf{X}}_1, \hat{\mathbf{X}}_1)}$$

is an approximation to λ_1.

The components of $\mathbf{A}^k\mathbf{U}_0$ become increasingly large numbers as k increases which leads to a large roundoff error. This problem can be avoided by multiplying $\mathbf{A}^k\mathbf{U}_0$ by a suitable scalar for each k. The power method thus consists of the following steps.

STEP 1. Choose an arbitrary vector $\mathbf{U}_0$ as an initial approximation to an eigenvector of $\mathbf{A}$ associated with λ_1. For example, $\mathbf{U}_0$ can be taken as the vector all of whose components are 1.

STEP 2. Compute $\mathbf{A}\mathbf{U}_0 = \mathbf{U}_1$. Let $k = 1$.

STEP 3. Let u_{kr} be the component of $\mathbf{U}_k$ with largest absolute value. Define

$$\mathbf{V}_k = \frac{1}{|u_{kr}|}\mathbf{U}_k \qquad k \geqslant 1.$$

Thus $\mathbf{V}_k$ is an approximation to an eigenvector for $\mathbf{A}$ associated with λ_1.

STEP 4. Form an approximation to λ_1:

$$\frac{(\mathbf{A}\mathbf{V}_k, \mathbf{V}_k)}{(\mathbf{V}_k, \mathbf{V}_k)} \cong \lambda_1 \qquad k \geqslant 1.$$

STEP 5. Increase k by 1, let $\mathbf{U}_{k+1} = \mathbf{A}\mathbf{V}_k$, and return to step 3.

Example 1. Let

$$\mathbf{A} = \begin{bmatrix} 4 & 1 \\ 2 & 5 \end{bmatrix}.$$

If we carry three decimal places and round after each multiplication and division, our results for nine iterations ($k = 8$) are shown in Table 8.3.

TABLE 8.3

Iteration	$\mathbf{V}_k$	Approximate value of λ_1
0	$[1 \quad 1]^T$	—
1	$[0.714 \quad 1]^T$	6.080
2	$[0.600 \quad 1]^T$	6.059
3	$[0.548 \quad 1]^T$	6.035
4	$[0.524 \quad 1]^T$	6.016
5	$[0.512 \quad 1]^T$	6.010
6	$[0.506 \quad 1]^T$	6.005
7	$[0.503 \quad 1]^T$	6.002
8	$[0.501 \quad 1]^T$	6.001

We conclude that the dominant eigenvalue is approximately 6.001 and that an associated eigenvector is approximately

$$\hat{\mathbf{X}}_1 = \begin{bmatrix} 0.501 \\ 1 \end{bmatrix}.$$

The exact answers are

$$\lambda_1 = 6 \quad \text{and} \quad \mathbf{X}_1 = \begin{bmatrix} \frac{1}{2} \\ 1 \end{bmatrix}.$$

The rate of convergence of the power method depends upon the ratio $|\lambda_2/\lambda_1|$ and upon the choice of the vector $\mathbf{U}_0$. If the ratio $|\lambda_2/\lambda_1|$ is close to 1, then convergence is very slow. There is no way of knowing how to choose a good $\mathbf{U}_0$. There are many ways of extending the power method to handle cases excluded in our discussion. These and methods for improving the rate of convergence are discussed in the book by Fadeev and Fadeeva listed in Further Reading at the end of the chapter.

If $\mathbf{A}$ is an $n \times n$ symmetric matrix with eigenvalues $\lambda_1, \lambda_2, \ldots, \lambda_n$, where

$$|\lambda_1| > |\lambda_2| > |\lambda_3| \geqslant \cdots \geqslant |\lambda_n|,$$

it is possible to find λ_2 and an associated eigenvector $\mathbf{X}_2$ by a method called **deflation**, which applies the power method to the matrix $\mathbf{A}_1 = \mathbf{A} - \lambda_1\mathbf{X}_1\mathbf{X}_1^T$. This procedure works very well if we have a *very good* approximation to λ_1 and $\mathbf{X}_1$. Otherwise, the results can be rather poor. A much better procedure for finding all the eigenvalues and associated eigenvectors of a *symmetric* matrix is Jacobi's method, which is widely used.

Jacobi's Method

Let $\mathbf{A}$ be a symmetric $n \times n$ matrix and let $\mathbf{E}_{pq}$ be the $n \times n$ matrix obtained from the identity matrix $\mathbf{I}_n$ by replacing the following elements of $\mathbf{I}_n$ as indicated: p, pth element replaced by $\cos\theta$; p, qth element replaced by $\sin\theta$; q, pth element replaced by $-\sin\theta$; q, qth element replaced by $\cos\theta$. Thus, if we start with $\mathbf{I}_4$, then

$$\mathbf{E}_{24} = \begin{bmatrix} 1 & 0 & 0 & 0 \\ 0 & \cos\theta & 0 & \sin\theta \\ 0 & 0 & 1 & 0 \\ 0 & -\sin\theta & 0 & \cos\theta \end{bmatrix}.$$

The $\mathbf{E}_{pq}$ matrices are called **rotation matrices**, because for $n = 2$ they represent a rotation of axes through the angle θ.

The matrices $\mathbf{E}_{pq}$ are orthogonal; that is, $\mathbf{E}_{pq}^T\mathbf{E}_{pq} = \mathbf{I}_n$ (Exercise T.1). If we now form the matrix

$$\mathbf{B}_1 = \mathbf{E}_{pq}^T\mathbf{A}\mathbf{E}_{pq}, \tag{1}$$

then the entries of $\mathbf{B}_1$ differ from those of $\mathbf{A}$ only in the pth and qth rows and in the pth and qth columns. More specifically, the p, qth element (element in pth row, qth column) of $\mathbf{B}_1$ is

$$(a_{pp} - a_{qq})\cos\theta\sin\theta + a_{pq}(\cos^2\theta - \sin^2\theta). \tag{2}$$

Since

$$\cos\theta\sin\theta = \tfrac{1}{2}\sin 2\theta \qquad \text{and} \qquad \cos^2\theta - \sin^2\theta = \cos 2\theta,$$

we can write (2) as

$$\tfrac{1}{2}(a_{pp} - a_{qq})\sin 2\theta + a_{pq}\cos 2\theta. \tag{3}$$

Let a_{pq} be the largest nonzero element in absolute value of $\mathbf{A}$ that is not on the main diagonal, and choose θ so that the p, qth element of $\mathbf{B}_1$ is zero.

From (3) we see that θ is chosen so that

$$\tan 2\theta = -\frac{2a_{pq}}{a_{pp} - a_{qq}} \qquad \text{for } a_{pp} - a_{qq} \neq 0. \tag{4}$$

If $a_{pp} - a_{qq} = 0$, then from (3) the p, qth element of $\mathbf{B}_1$ is

$$a_{pq}(\cos^2\theta - \sin^2\theta) = a_{pq}\cos 2\theta,$$

which can be made zero by taking $\theta = \pi/4$. We now repeat the same procedure with $\mathbf{B}_1$ taking the place of $\mathbf{A}$. That is, if b_{rs} is the largest off-diagonal element of $\mathbf{B}_1$, then we form the matrix

$$\mathbf{B}_2 = \mathbf{E}_{rs}^T\mathbf{B}_1\mathbf{E}_{rs},$$

where $\mathbf{E}_{rs}$ is the rotation matrix chosen so that the r, sth element of $\mathbf{B}_2$ will be zero; that is, θ is chosen so that

$$\tan 2\theta = -\frac{2a_{rs}}{a_{rr} - a_{ss}} \qquad \text{for } a_{rr} - a_{ss} \neq 0.$$

If $a_{rr} - a_{ss} = 0$, let $\theta = \pi/4$.

It can then be proved that as r increases, we get a sequence of matrices $\mathbf{B}_1, \mathbf{B}_2, \ldots, \mathbf{B}_t, \ldots$, which converge to a diagonal matrix $\mathbf{D}$, whose diagonal elements are the eigenvalues of $\mathbf{A}$.

Thus, when we stop our computations, we have

$$\mathbf{E}_{k+1}^T \cdots \mathbf{E}_2^T\mathbf{E}_1^T\mathbf{A}\mathbf{E}_1\mathbf{E}_2 \cdots \mathbf{E}_{k+1} = \mathbf{B}_{k+1}, \tag{5}$$

where $\mathbf{E}_1, \mathbf{E}_2, \ldots, \mathbf{E}_{k+1}$ are rotation matrices. If we let

$$\mathbf{E}_1\mathbf{E}_2 \cdots \mathbf{E}_{k+1} = \mathbf{P}, \tag{6}$$

then we can write (5) as

$$\mathbf{P}^T\mathbf{A}\mathbf{P} = \mathbf{B}_{k+1}. \tag{7}$$

Now $\mathbf{P}$ is an orthogonal matrix and we may recall from Section 5.2 that the columns of $\mathbf{P}$ are eigenvectors of $\mathbf{A}$. Thus, to find approximations to the eigenvectors of $\mathbf{A}$ associated with $\lambda_1, \lambda_2, \ldots, \lambda_n$, we have to form the products of the rotation matrices as in (6).

From (7) it follows that

$$\mathbf{A} = \mathbf{P}\mathbf{B}_{k+1}\mathbf{P}^T.$$

As an additional check on the accuracy of our results, we may compute $\mathbf{P}\mathbf{B}_{k+1}\mathbf{P}^T$ and see how close it is to the given matrix **A**.

In using the Jacobi method, it is not necessary to evaluate trigonometric functions to compute $\sin\theta$ and $\cos\theta$ from (4); instead we proceed as follows. Let

$$f = -a_{pq} \qquad \text{and} \qquad g = \tfrac{1}{2}(a_{pp} - a_{qq}).$$

Compute

$$h = \operatorname{sign}(g)\frac{f}{\sqrt{f^2 + g^2}}. \tag{8}$$

Then

$$\sin\theta = \frac{h}{\sqrt{2(1 + \sqrt{1 - h^2})}} \qquad \cos\theta = \sqrt{1 - \sin^2\theta}. \tag{9}$$

Observe that if $a_{pp} - a_{qq} = 0$, then $h = 1$ in (8), and from (9) $\sin\theta = 1/\sqrt{2}$, so that $\theta = \pi/4$, as has already been mentioned earlier. Thus we can use (8) and (9) in all cases, without having to check whether $a_{pp} - a_{qq} \neq 0$.

For a detailed discussion of a computer implementation of Jacobi's method, the reader is referred to the article by Greenstadt listed in Further Reading at the end of the chapter.

Example 2. Let

$$\mathbf{A} = \begin{bmatrix} -1 & 1 & -2 \\ 1 & -1 & -2 \\ -2 & -2 & -2 \end{bmatrix}.$$

We now find the eigenvalues and associated eigenvectors of **A** by Jacobi's method. We carry three decimal places and round after each multiplication and division. Moreover, the matrix product **ABC** will be calculated by first computing **BC** and then **A**(**BC**). Proceed as follows.

We can reduce the 1, 3th, 2, 3th, 3, 1th, or 3, 2th element of **A** to zero. We choose to reduce the 1, 3th element to zero, obtaining

$$\mathbf{E}_{13} = \begin{bmatrix} 0.789 & 0 & 0.615 \\ 0 & 1 & 0 \\ -0.615 & 0 & 0.789 \end{bmatrix} \qquad \begin{aligned} f &= 2.000 \\ g &= 0.500 \\ h &= 0.970; \end{aligned}$$

$$\mathbf{B}_1 = \begin{bmatrix} 0.562 & 2.019 & -0.003 \\ 2.019 & -1.000 & -0.963 \\ -0.004 & -0.963 & -3.565 \end{bmatrix}.$$

We now reduce the 2, 1th element of $\mathbf{B}_1$ to zero, obtaining

$$\mathbf{E}_{21} = \begin{bmatrix} 0.825 & -0.565 & 0 \\ 0.565 & 0.825 & 0 \\ 0 & 0 & 1 \end{bmatrix} \qquad \begin{aligned} f &= -2.019 \\ g &= -0.781 \\ h &= 0.933; \end{aligned}$$

$$\mathbf{B}_2 = \begin{bmatrix} 1.946 & 0.001 & -0.546 \\ 0.001 & -2.384 & -0.792 \\ -0.547 & -0.792 & -3.565 \end{bmatrix}.$$

We next reduce the 2, 3th element of $\mathbf{B}_2$ to zero, obtaining

$$\mathbf{E}_{23} = \begin{bmatrix} 1 & 0 & 0 \\ 0 & 0.893 & 0.449 \\ 0 & -0.449 & 0.893 \end{bmatrix} \qquad \begin{aligned} f &= 0.792 \\ g &= 0.591 \\ h &= 0.802; \end{aligned}$$

$$\mathbf{B}_3 = \begin{bmatrix} 1.946 & 0.246 & -0.488 \\ 0.247 & -1.984 & 0.002 \\ -0.488 & 0.002 & -3.959 \end{bmatrix}.$$

We now reduce the 1, 3th element of $\mathbf{B}_3$ to zero, obtaining

$$\mathbf{E}_{13} = \begin{bmatrix} 0.996 & 0 & 0.082 \\ 0 & 1 & 0 \\ -0.082 & 0 & 0.996 \end{bmatrix} \qquad \begin{aligned} f &= 0.488 \\ g &= 2.953 \\ h &= 0.163; \end{aligned}$$

$$\mathbf{B}_4 = \begin{bmatrix} 1.983 & 0.245 & 0.002 \\ 0.246 & -1.984 & 0.022 \\ 0.002 & 0.022 & -3.994 \end{bmatrix}.$$

We next reduce the 2, 1th element of $\mathbf{B}_4$ to zero, obtaining

$$\mathbf{E}_{21} = \begin{bmatrix} 0.998 & -0.062 & 0 \\ 0.062 & 0.998 & 0 \\ 0 & 0 & 1 \end{bmatrix} \qquad \begin{aligned} f &= -0.246 \\ g &= -1.984 \\ h &= 0.123; \end{aligned}$$

$$\mathbf{B}_5 = \begin{bmatrix} 1.998 & -0.002 & 0.003 \\ -0.001 & -1.999 & 0.022 \\ 0.003 & 0.022 & -3.994 \end{bmatrix}.$$

We now stop; the diagonal elements of $\mathbf{B}_5$ are approximations to the eigenvalues of $\mathbf{A}$. Thus we have $\lambda_1 \cong 1.998$, $\lambda_2 \cong -1.999$, and $\lambda_3 \cong -3.994$, which agree rather well with the exact values:

$$\lambda_1 = 2, \quad \lambda_2 = -2, \quad \lambda_3 = -4.$$

To find associated eigenvectors, we form $\mathbf{E}_{13}\mathbf{E}_{21}\mathbf{E}_{23}\mathbf{E}_{13}\mathbf{E}_{21} = \mathbf{P}$. We have

$$\mathbf{E}_{13}\mathbf{E}_{21}\mathbf{E}_{23}\mathbf{E}_{13}\mathbf{E}_{21} = \mathbf{P} = \begin{bmatrix} 0.576 & -0.707 & 0.409 \\ 0.578 & 0.708 & 0.407 \\ -0.578 & -0.001 & 0.816 \end{bmatrix}.$$

The columns of $\mathbf{P}$ give the approximate eigenvectors

$$\mathbf{X}_1 \cong \begin{bmatrix} 0.576 \\ 0.578 \\ -0.578 \end{bmatrix}, \qquad \mathbf{X}_2 \cong \begin{bmatrix} -0.707 \\ 0.708 \\ -0.001 \end{bmatrix}, \qquad \mathbf{X}_3 \cong \begin{bmatrix} 0.409 \\ 0.407 \\ 0.816 \end{bmatrix},$$

which are associated with λ_1, λ_2, and λ_3, respectively.

As a final check on our results, we compute $\mathbf{P}\mathbf{B}_5\mathbf{P}^T$, obtaining

$$\mathbf{P}\mathbf{B}_5\mathbf{P}^T = \begin{bmatrix} -1.015 & 1.002 & -2.010 \\ 1.001 & -0.982 & -1.981 \\ -2.010 & -1.981 & -1.995 \end{bmatrix},$$

which is in very close agreement with the given matrix $\mathbf{A}$.

FURTHER READING

Fadeev, D. K. and V. N. Faddeeva. *Computational Methods of Linear Algebra*. San Francisco: W. H. Freeman Co., Publishers, 1963.

Forsythe, G. and C. B. Moler. *Computer Solution of Linear Algebraic Systems*. Englewood Cliffs, N.J.: Prentice-Hall, Inc. 1967.

Fox, L. *An Introduction to Numerical Linear Algebra*. New York: Oxford University Press, Inc. 1965.

Franklin, J. N. *Matrix Theory*. Englewood Cliffs, N.J.: Prentice-Hall, Inc., 1968.

Greenstadt, John. "The Determination of the Characteristic Roots of a Matrix by the Jacobi Method." In *Mathematical Methods for Digital Computers*, ed. by A. Ralston and H. S. Wilf. New York: John Wiley & Sons, Inc., 1960.

Noble, Ben. *Applied Linear Algebra*. Englewood Cliffs, N.J.: Prentice-Hall, Inc., 1969.

Stewart, G. W. *Introduction to Matrix Computations*. New York: Academic Press, Inc., 1973.

Wilkinson, J. H. *The Algebraic Eigenvalue Problem*. New York: Oxford University Press, Inc., 1965.

8.3 Exercises

In Exercises 1 through 5 compute an approximation to the eigenvalue of largest magnitude and the associated eigenvector by the power method. Carry out the computations to three decimal places and round after each multiplication and division for six iterations ($k = 5$).

1. $\begin{bmatrix} -4 & 2 \\ 3 & 1 \end{bmatrix}$.

2. $\begin{bmatrix} 0 & -1 \\ 2 & 3 \end{bmatrix}$.

3. $\begin{bmatrix} 4 & 3 \\ 2 & 3 \end{bmatrix}$.

4. $\begin{bmatrix} 6 & 2 \\ -4 & -3 \end{bmatrix}$.

5. $\begin{bmatrix} 8 & 8 \\ 3 & -2 \end{bmatrix}$.

In Exercises 6 through 9 find the eigenvalues and associated eigenvectors by Jacobi's method. Carry out the computations to three decimal places and round after each multiplication and division. In each case calculate **ABC** by first calculating **BC**.

6. $\begin{bmatrix} 2 & 4 \\ 4 & 2 \end{bmatrix}$. Find $\mathbf{B}_3$.

7. $\begin{bmatrix} 2 & 2 \\ 2 & 2 \end{bmatrix}$. Find $\mathbf{B}_3$.

8. $\begin{bmatrix} -1 & -4 & -8 \\ -4 & -7 & 4 \\ -8 & 4 & -1 \end{bmatrix}$. Find $\mathbf{B}_5$.

9. $\begin{bmatrix} 8 & -2 & 0 \\ -2 & 9 & -2 \\ 0 & -2 & 10 \end{bmatrix}$. Find $\mathbf{B}_5$.

Theoretical Exercise

T.1. Prove that the matrices $\mathbf{E}_{pq}$ are orthogonal.

Appendix: The Computer in Linear Algebra

It would be difficult at this time to ignore the use of the computer in linear algebra, since most computations involving linear algebra in real problems are carried out on computers. This relationship is briefly explored in this Appendix. First, we provide a list of 53 *computer projects* grouped by sections corresponding to the text. These projects are of widely differing degrees of difficulty, and the student is expected to have adequate programming skills to handle these tasks. Next, we discuss the availability of *canned software* to implement linear algebra techniques. The use of many of these programs requires some programming skill, although some of them only require that the user know how to use a keypunch machine to punch up the data. Finally, we outline some of the features of APL, a relatively new programming language, which is especially suitable for matrix manipulations and is easy to learn, and we make several remarks about BASIC, a general purpose computer language, which is being widely used with time-sharing systems.

Computer Projects

Write programs to carry out the following projects.

Chapter 1

Section 1.2

1. Add two matrices.
2. Multiply a matrix by a scalar.
3. Subtract two matrices.
4. Multiply two matrices.
5. Form the transpose of a matrix.

Section 1.3

5. Reduce an $m \times n$ matrix to reduced row echelon form.
6. Solve a linear system by Gauss–Jordan reduction.

Section 1.4

7. Find the inverse of a matrix.

Chapter 2

Section 2.1

8. Given n, list all the permutations of $S = \{1, 2, \ldots, n\}$ and determine which are even and which are odd.

Section 2.2

9. Evaluate the determinant of a square matrix.

Section 2.3

10. Find the inverse of a matrix by cofactors (inefficient for $n > 4$).
11. Solve $\mathbf{AX} = \mathbf{B}$, where $\mathbf{A}$ is $n \times n$, by Cramer's rule (inefficient for $n > 4$).

Chapter 3

Section 3.2

12. Find the inner product of two vectors in R^n.
13. Find the cosine of the angle and the angle between two nonzero vectors in R^n.
14. Find the length of a given vector in R^n.

15. Determine whether a triangle with three given vertices in R^3 is a right triangle.
16. Determine whether a triangle with three given vertices in R^3 is isosceles.
17. Determine whether a quadrilateral with four given vertices in R^3 is a parallelogram.
18. Determine whether a quadrilateral with four given vertices in R^3 is a rectangle.
19. Determine whether a quadrilateral with four given vertices in R^3 is a square.

Section 3.3

20. Find the cross product of two vectors in R^3.

Section 3.5

21. Test a given set of vectors in R^n for linear independence.
22. Extend a given set of linearly independent vectors in R^n to a basis for R^n.
23. If a given set of r vectors in R^n spans a subspace V of R^n of dimension m, find a basis for V.
24. Determine whether a given vector $\mathbf{X}$ in R^n is a linear combination of a given set of vectors in R^n.
25. If a given vector $\mathbf{X}$ in R^n is a linear combination of a given set of vectors in R^n, find the coefficients of the linear combination.
26. Find a basis for the solution space of a given homogeneous system $\mathbf{AX} = \mathbf{0}$.

Section 3.6

27. Transform a given basis for a subspace V of R^n to an orthonormal basis for V by the Gram–Schmidt process.

Chapter 4

Section 4.2

28. Find a basis for the kernel of a linear transformation $\mathbf{L} : R^n \to R^m$ given by $\mathbf{L}(\mathbf{X}) = \mathbf{AX}$.
29. Find a basis for the range of a linear transformation $\mathbf{L} : R^n \to R^m$ given by $\mathbf{L}(\mathbf{X}) = \mathbf{AX}$.

Section 4.3

30. Find the coordinate vector of a given vector $\mathbf{X}$ with respect to a given basis S for R^n.

31. Let $\mathbf{L} : R^3 \to R^4$ be the linear transformation defined by

$$\mathbf{L}\left(\begin{bmatrix} a_1 \\ a_2 \\ a_3 \end{bmatrix}\right) = \begin{bmatrix} a_1 + 2a_2 - a_3 \\ 2a_1 + a_3 \\ 3a_1 - a_2 \\ 2a_1 + a_2 + a_3 \end{bmatrix}.$$

Let

$$S = \left\{\begin{bmatrix} 1 \\ 1 \\ 0 \end{bmatrix}, \begin{bmatrix} 0 \\ 1 \\ 2 \end{bmatrix}, \begin{bmatrix} 1 \\ 1 \\ 1 \end{bmatrix}\right\} \quad \text{and} \quad T = \left\{\begin{bmatrix} 1 \\ 2 \\ -1 \\ 0 \end{bmatrix}, \begin{bmatrix} 0 \\ 0 \\ 1 \\ 0 \end{bmatrix}, \begin{bmatrix} 1 \\ 1 \\ 0 \\ 0 \end{bmatrix}, \begin{bmatrix} 0 \\ 1 \\ 1 \\ 1 \end{bmatrix}\right\}$$

be bases for R^3 and R^4, respectively. Find the matrix representing $\mathbf{L}$ with respect to S and T.

Section 4.4

32. Find the rank of a given $m \times n$ matrix.

Chapter 5

Section 5.2

33. Determine if a given matrix is orthogonal.

Chapter 6

34. Solve a standard linear programming problem by the simplex method.

Chapter 7

Section 7.1

35. Find parametric equations for the line through two given points.
36. Find equations in symmetric form for a line through two given points.
37. Find an equation of the plane passing through three given points.

Section 7.2

38. Diagonalize a given quadratic form.

Section 7.3

39. Given the incidence matrix for a digraph with n vertices, find the number of ways in which a specified vertex P_i influences another specified vertex P_j in a specified number (r) of stages.
40. Given the incidence matrix for a digraph G, find all cliques in G.
41. Given the incidence matrix for a digraph G, determine whether G is connected or disconnected.
42. Given the incidence matrix for a digraph G with n vertices, determine if there is a path from a specified vertex P_i to another specified vertex P_j.
43. Given the incidence matrix for a digraph G with n vertices, determine the number of paths from a specified vertex P_i to another specified vertex P_j.

Section 7.4

44. Determine whether a given matrix game is strictly determined. If it is, find all saddle points.
45. Solve a matrix game with given 2×2 payoff matrix.

Section 7.5

46. Find the line of best fit for a given set of points.

Section 7.6

47. Given an exchange matrix $\mathbf{A}$, find a vector $\mathbf{P} \geqslant 0$ such that $(\mathbf{I}_n - \mathbf{A})\mathbf{P} = \mathbf{0}$, with at least one positive component.

Chapter 8

Section 8.2

48. Reduce an $m \times n$ matrix to row echelon form.
49. Implement Gaussian elimination with partial pivoting.
50. Implement the Jacobi iteration method.
51. Implement the Gauss–Seidel iteration method.

Section 8.3

52. Implement the power method for finding the eigenvalue of largest absolute value and a corresponding eigenvector of a given matrix.
53. Implement the Jacobi method for finding the eigenvalues and eigenvectors of a given symmetric matrix.

Canned Software

An enormous number of programs implementing linear algebra techniques have been developed by computer manufacturers and by computer users. Many of these are available through exchange organizations—SHARE (IBM users), USE (UNIVAC users), VIM (CDC users), BUG (Burroughs users). They range from very simple to very sophisticated programs. Many of them are available as subroutines, and to be used as a main program require the user to write his own input and output subroutines. Thus the use of canned programs often requires some programming skill. Of course, a canned program that has its own input and output routines generally only requires that the user know how to use a keypunch machine to punch up the data.

An example of canned software in linear algebra is provided by System/360 Scientific Subroutine Package (for IBM 360 Computers), written in FORTRAN IV language. Some of the many subroutines included are:

1. GMADD—add two matrices
2. GMSUB—subtract two matrices
3. GMPRD—multiply two matrices
4. GMTRA—form the transpose of a matrix
5. MINV—invert a matrix
6. GELG, DGELG—solve a linear system by Gaussian elimination (see Section 8.2)
7. EIGEN—find the eigenvalues and eigenvectors of a symmetric matrix
8. EISPACK—find the eigenvalues and eigenvectors of a matrix (symmetric, nonsymmetric, real, complex)

APL

APL (**A** **P**rogramming **L**anguage) is a relatively new programming language, developed in 1962 by Kenneth Iverson. APL is an **interactive** language in the sense that the user communicates with the computer by means of a **terminal**, a special typewriter linked to the computer by telephone lines. The terminal is used to enter the instructions to the computer as well as to receive the answers, which are usually received within seconds after the last instruction has been entered.

An easy language to learn, APL is especially suitable for matrix manipulations. We give several examples to illustrate the capability of the language in linear algebra.

Assign

The symbol ← means that we assign what follows it to what precedes it. This symbol is used to enter data to the computer: Thus

```
X←4
```

means that X is assigned the value 4.

Vectors

In APL a vector is denoted by a string of numbers. Thus if we type in

```
X←2 0 1 3
```

we enter the vector $\mathbf{X} = (2, 0, 1, 3)$ into the computer.

Vector Operations

To add vectors **X** and **Y**, we write `X+Y`.

To form the product of a vector **X** by the scalar K, we write `K×X` or `X×K`. Thus, if $\mathbf{X} = (2, 1, 0, 3)$ and $\mathbf{Y} = (3, 2, 1, 4)$, and we type in

```
X←2 1 0 3
Y←3 2 1 4
X + Y,
```

we have formed $\mathbf{X} + \mathbf{Y}$. If we type in

```
X←2 1 0 3
K←3
K×X
```

we form **KX**.

Matrices

To enter the 3×2 matrix

$$\mathbf{A} = \begin{bmatrix} 2 & 1 \\ 3 & 2 \\ 0 & 1 \end{bmatrix}$$

into the computer we type in

```
A← 3 2ρ 2 1 3 2 0 1.
```

This instruction tells the computer to assign to A the 3×2 matrix with entries listed row by row after the special symbol ρ.

Matrix Operations

To form the sum of the matrices **A** and **B**, we write A + B.
To form the product of a matrix **A** by a scalar K, we write K A.
To form the product of the matrices **A** and **B**, we write A + · ×B.
To form the inverse of a square matrix **A**, we write ⌹ A.
Thus let

$$\mathbf{A} = \begin{bmatrix} 2 & 1 & 3 \\ 4 & 6 & 4 \end{bmatrix}, \qquad \mathbf{B} = \begin{bmatrix} 2 & 6 & 8 \\ 3 & 5 & 2 \end{bmatrix},$$

$$\mathbf{C} = \begin{bmatrix} 2 & 3 \\ 4 & 1 \\ 2 & 5 \end{bmatrix}, \qquad \mathbf{D} = \begin{bmatrix} 1 & 2 \\ 3 & 4 \end{bmatrix},$$

and $K = 2$.

To form **A** + **B**, we type in

```
A ← 2 3ρ 2 1 3 4 6 4
B← 2 3ρ 2 6 8 3 5 2
A + B
```

To form **KA**, we type in

```
A←2 3ρ 2 1 3 4 6 4
K ← 2
K × A
```

To form **AC**, we type in

```
A←2 3ρ 2 1 3 4 6 4
C←3 2ρ 2 3 4 1 2 5
A + ·× C
```

To form $\mathbf{D}^{-1}$, we type in

```
D←2 2ρ1 2 3 4
⌹ D
```

In APL it is quite simple to solve a linear system of n equations in n unknowns. Thus if

$$\mathbf{E} = \begin{bmatrix} 1 & 2 \\ 3 & 4 \end{bmatrix} \quad \text{and} \quad \mathbf{F} = \begin{bmatrix} 8 \\ 9 \end{bmatrix},$$

then to solve $\mathbf{EX} = \mathbf{F}$ we type in

```
E←2 2ρ1 2 3 4
F←8 9
F ⌹ E
```

There are many other instructions in APL that enable the user to readily manipulate vectors and matrices.

FURTHER READING

Gilman, Leonard and Allan J. Rose. *APL, An Interactive Approach*, 2nd ed. New York: John Wiley & Sons, Inc., 1974.

Basic

BASIC is a general purpose computer language that is being widely used with time-sharing systems. Typical matrix operations are

`MAT C = A + B`	Form the sum **C** of two given matrices **A** and **B**.
`MAT C = (K)*A`	Form the scalar multiple **C** of a given matrix **A** by a given scalar K.
`MAT C = A*B`	Form the product **C** of two given matrices **A** and **B**.
`MAT C = TRN(A)`	Form the transpose **C** of a given matrix **A**.
`MAT C = INV(A)`	Form the inverse **C** of a given matrix **A**.

FURTHER READING

Kemeny, John G. and Thomas E. Kurtz. *BASIC Programming*. New York: John Wiley & Sons, Inc., 1967.

Answers to Odd-Numbered Exercises

Chapter 1

Section 1.1, p. 7

1. $x = 4, y = 2$.

3. $x = -4, y = 2, z = 10$.

5. $x = 2, y = -1, z = -2$.

7. $x = -20$, $y =$ any real number, $z = 4y - 32$.

9. No solution.

11. $x = 5, y = 1$.

13. No solution.

Section 1.2, p. 24

1. (a) $-3, -5, 4$. (b) $4, 5$. (c) $2, 6, -1$.

3. (a) $\begin{bmatrix} 58 & 4 \\ 66 & 4 \end{bmatrix}$. (b) $\begin{bmatrix} 28 & 8 & 38 \\ 34 & 4 & 41 \end{bmatrix}$. (c) $\begin{bmatrix} 5 & 10 & 15 \\ 10 & 5 & 20 \end{bmatrix}$.

(d) $\begin{bmatrix} -2 & 17 \\ 16 & 31 \end{bmatrix}$. (e) Impossible. (f) $\begin{bmatrix} -36 & 12 & -36 \\ -48 & -12 & -60 \\ -24 & -12 & -36 \end{bmatrix}$.

5. (a) $\begin{bmatrix} 1 & 2 \\ 2 & 1 \\ 3 & 4 \end{bmatrix}, \begin{bmatrix} 1 & 2 & 3 \\ 2 & 1 & 4 \end{bmatrix}$. (b) $\begin{bmatrix} 5 & 4 & 5 \\ -5 & 2 & 3 \\ 8 & 9 & 4 \end{bmatrix}$. (c) $\begin{bmatrix} 14 & 16 \\ 8 & 9 \end{bmatrix}$.

(d) $\begin{bmatrix} -6 & 10 \\ 11 & 17 \end{bmatrix}$. (e) Impossible. (f) $\begin{bmatrix} 18 & 6 & 25 \\ 10 & 4 & 15 \end{bmatrix}$.

7. $\mathbf{A} + \mathbf{B} = \begin{bmatrix} 3 & 2 & -1 \\ 6 & 2 & 10 \end{bmatrix}$, $\mathbf{A} + \mathbf{B} + \mathbf{C} = \begin{bmatrix} -1 & -4 & 0 \\ 8 & 5 & 10 \end{bmatrix}$.

9. $\mathbf{A}(\mathbf{B} + \mathbf{C}) = \begin{bmatrix} -10 & -8 & 16 \\ 10 & 14 & -28 \end{bmatrix}$.

11. $\mathbf{A}(r\mathbf{B}) = \begin{bmatrix} -6 & 18 & -42 \\ 9 & -27 & 0 \end{bmatrix}$.

13. $(\mathbf{AB})^T = \begin{bmatrix} 11 & 5 \\ 15 & -4 \end{bmatrix}$.

15. $a = 0, b = 2, c = 1, d = 2$.

21.
$$\begin{aligned} -2x - y + 4w &= 5 \\ -3x + 2y + 7z + 8w &= 3 \\ x + 2w &= 4 \\ 3x + z + 3w &= 6. \end{aligned}$$

23. (a) $\begin{bmatrix} 3 & -1 & 2 \\ 2 & 1 & 0 \\ 0 & 1 & 3 \\ 4 & 0 & -1 \end{bmatrix}$. (b) $\begin{bmatrix} 3 & -1 & 2 \\ 2 & 1 & 0 \\ 0 & 1 & 3 \\ 4 & 0 & -1 \end{bmatrix}\begin{bmatrix} x \\ y \\ z \end{bmatrix} = \begin{bmatrix} 4 \\ 2 \\ 7 \\ 4 \end{bmatrix}$.

(c) $\left[\begin{array}{ccc|c} 3 & -1 & 2 & 4 \\ 2 & 1 & 0 & 2 \\ 0 & 1 & 3 & 7 \\ 4 & 0 & -1 & 4 \end{array}\right]$.

Section 1.3, p. 42

1. **A**, **E**, and **G**.

3. (a) $\begin{bmatrix} 2 & 0 & 4 & 2 \\ -1 & 3 & 1 & 1 \\ 3 & -2 & 5 & 6 \end{bmatrix}$. (b) $\begin{bmatrix} 2 & 0 & 4 & 2 \\ -12 & 8 & -20 & -24 \\ -1 & 3 & 1 & 1 \end{bmatrix}$.

(c) $\begin{bmatrix} 0 & 6 & 6 & 4 \\ 3 & -2 & 5 & 6 \\ -1 & 3 & 1 & 1 \end{bmatrix}$.

5. Possible answers:

(a) $\begin{bmatrix} 4 & 3 & 7 & 5 \\ 2 & 0 & 1 & 4 \\ -2 & 4 & -2 & 6 \end{bmatrix}$. (b) $\begin{bmatrix} 3 & 5 & 6 & 8 \\ -4 & 8 & -4 & 12 \\ 2 & 0 & 1 & 4 \end{bmatrix}$.

(c) $\begin{bmatrix} 4 & 3 & 7 & 5 \\ -1 & 2 & -1 & 3 \\ 0 & 4 & -1 & 10 \end{bmatrix}$.

7. $\begin{bmatrix} 1 & 0 & 0 & 0 \\ 0 & 1 & 0 & 0 \\ 0 & 0 & 1 & 0 \\ 0 & 0 & 0 & 1 \end{bmatrix}$.

9. (a) $x = -2 + r, y = -1, z = 8 - 2r, r =$ any real number.
(b) $x = 1, y = \frac{2}{3}, z = -\frac{2}{3}$.
(c) No solution.

11. (1) $a = -2$, (2) $a \neq \pm 2$, (3) $a = 2$.

13. (1) $a = \pm\sqrt{6}$, (2) $a \neq \pm\sqrt{6}$, (3) none.

15. (a) $x = -1, y = 4, z = -3$. (b) $x = 0, y = 0, z = 0$.

17. (a) $x = 1 - r, y = 2, z = 1, r =$ any real number. (b) No solution.

Section 1.4, p. 57

1. $\mathbf{A}^{-1} = \begin{bmatrix} \frac{3}{8} & -\frac{1}{8} \\ \frac{1}{4} & \frac{1}{4} \end{bmatrix}$.

3. Nonsingular; $\mathbf{A}^{-1} = \begin{bmatrix} 4 & -1 \\ -3 & 1 \end{bmatrix}$.

5. (a) $\begin{bmatrix} \frac{1}{2} & -\frac{1}{4} \\ \frac{1}{6} & \frac{1}{12} \end{bmatrix}$. (b) $\begin{bmatrix} 0 & 1 & -1 \\ 2 & -2 & -1 \\ -1 & 1 & 1 \end{bmatrix}$.

(c) $\begin{bmatrix} \frac{7}{3} & -\frac{1}{3} & -\frac{1}{3} & -\frac{2}{3} \\ \frac{4}{9} & -\frac{1}{9} & -\frac{4}{9} & \frac{1}{9} \\ -\frac{1}{9} & -\frac{2}{9} & \frac{1}{9} & \frac{2}{9} \\ -\frac{5}{3} & \frac{2}{3} & \frac{2}{3} & \frac{1}{3} \end{bmatrix}$.

7. (a) $\begin{bmatrix} -2 & \frac{3}{2} \\ 1 & -\frac{1}{2} \end{bmatrix}$. (b) Singular. (c) $\begin{bmatrix} \frac{3}{2} & -1 & \frac{1}{2} \\ \frac{1}{2} & 0 & -\frac{1}{2} \\ -\frac{3}{2} & 1 & \frac{1}{2} \end{bmatrix}$.

9. (a) Singular. (b) $\begin{bmatrix} 1 & -1 & 0 \\ 1 & -2 & 1 \\ -\frac{3}{2} & \frac{5}{2} & -\frac{1}{2} \end{bmatrix}$. (c) $\begin{bmatrix} -1 & \frac{3}{2} & \frac{1}{2} \\ 1 & -\frac{3}{2} & \frac{1}{2} \\ 0 & \frac{1}{2} & -\frac{1}{2} \end{bmatrix}$.

11. (a) and (b).

13. $\begin{bmatrix} \frac{4}{5} & -\frac{3}{5} \\ -\frac{1}{5} & \frac{2}{5} \end{bmatrix}$.

Chapter 2

Section 2.1, p. 72

1. (a) 5. (b) 7. (c) 4. (d) 4. (e) 7. (f) 0.

3. (a) −. (b) +. (c) −. (d) −. (e) +. (f) +.

5. (a) 7. (b) −24. (c) 4.

9. $|\mathbf{B}| = 3$, $|\mathbf{C}| = 9$, $|\mathbf{D}| = -3$.

11. (a) 72. (b) 0. (c) −24.

13. (a) -30. (b) 0. (c) 6.

Section 2.2, p. 87

1. $A_{11} = -11, \quad A_{12} = 29, \quad A_{13} = 1$
$A_{21} = -4, \quad A_{22} = 7, \quad A_{23} = -2$
$A_{31} = 2, \quad A_{32} = -10, \quad A_{33} = 1.$

3. (a) -43. (b) 75. (c) 0.

5. (a) 0. (b) -6. (c) -36.

9. (a) $\begin{bmatrix} 24 & -42 & -30 \\ 19 & -2 & -30 \\ -4 & 32 & 30 \end{bmatrix}$. (b) 150.

11. (a) Singular. (b) $\begin{bmatrix} \frac{2}{7} & -\frac{3}{7} \\ \frac{1}{7} & \frac{2}{7} \end{bmatrix}$. (c) $\begin{bmatrix} \frac{1}{4} & -\frac{1}{20} & \frac{3}{20} \\ 0 & \frac{1}{5} & \frac{2}{5} \\ 0 & \frac{1}{10} & -\frac{3}{10} \end{bmatrix}$.

13. (a) $\begin{bmatrix} 0 & \frac{1}{2} \\ 1 & \frac{3}{2} \end{bmatrix}$. (b) $\begin{bmatrix} \frac{1}{4} & 0 & 0 \\ 0 & -\frac{1}{3} & 0 \\ 0 & 0 & \frac{1}{2} \end{bmatrix}$.

(c) $\begin{bmatrix} \frac{15}{14} & \frac{5}{28} & -\frac{9}{28} & -\frac{23}{14} \\ \frac{8}{7} & -\frac{1}{7} & -\frac{1}{7} & -\frac{9}{7} \\ \frac{3}{7} & \frac{1}{14} & \frac{1}{14} & -\frac{6}{7} \\ -\frac{4}{7} & \frac{1}{14} & \frac{1}{14} & \frac{8}{7} \end{bmatrix}$.

15. (a) and (d) are singular.

17. (a) Trivial. (b) Nontrivial.

19. $x = 1, \quad y = -1, \quad z = 0, \quad w = 2.$

21. No solution.

Chapter 3

Section 3.1, p. 108

1.

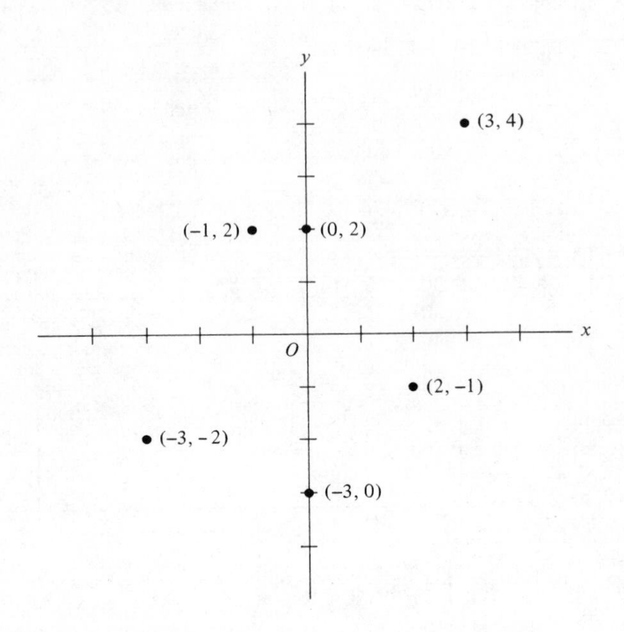

3. (a) $\begin{bmatrix} 1, & 7 \end{bmatrix}$. (b) $\begin{bmatrix} 3 \\ 2 \end{bmatrix}$. (c) $\begin{bmatrix} -2 \\ -4 \end{bmatrix}$. (d) $\begin{bmatrix} -7 \\ 3 \end{bmatrix}$.

5. (a) $\mathbf{X} + \mathbf{Y} = (0, 8)$; $\mathbf{X} - \mathbf{Y} = (4, -2)$; $2\mathbf{X} = (4, 6)$; $3\mathbf{X} - 2\mathbf{Y} = (10, -1)$.
(b) $\mathbf{X} + \mathbf{Y} = (3, 5)$; $\mathbf{X} - \mathbf{Y} = (-3, 1)$; $2\mathbf{X} = (0, 6)$; $3\mathbf{X} - 2\mathbf{Y} = (-6, 5)$.
(c) $\mathbf{X} + \mathbf{Y} = (5, 8)$; $\mathbf{X} - \mathbf{Y} = (-1, 4)$; $2\mathbf{X} = (4, 12)$; $3\mathbf{X} - 2\mathbf{Y} = (0, 14)$.

7. (a) $x = 2$. (b) $y = \frac{8}{3}$. (c) $x = 3, y = -2$.

9. (a) $\sqrt{5}$. (b) 5. (c) 2. (d) 5.

11. (a) $\sqrt{8}$. (b) 5. (c) $\sqrt{10}$. (d) $\sqrt{13}$.

13. (a) $(\frac{3}{5}, \frac{4}{5})$. (b) $(-2/\sqrt{13}, -3/\sqrt{13})$. (c) $(1, 0)$.

15. (a) -4. (b) 0. (c) 0. (d) -5.

17. (a) $\dfrac{-4}{\sqrt{5} \cdot \sqrt{13}}$. (b) 0. (c) 0. (d) -1.

21. (a) $\mathbf{i} + 3\mathbf{j}$. (b) $-2\mathbf{i} - 3\mathbf{j}$. (c) $-2\mathbf{i}$. (d) $3\mathbf{j}$.

Section 3.2, p. 125

1. (a) $\mathbf{X} + \mathbf{Y} = (1, 3, -5)$, $\mathbf{X} - \mathbf{Y} = (1, 1, -1)$, $2\mathbf{X} = (2, 4, -6)$, $3\mathbf{X} - 2\mathbf{Y} = (3, 4, -5)$.
(b) $\mathbf{X} + \mathbf{Y} = (3, 0, 6, -1)$, $\mathbf{X} - \mathbf{Y} = (5, -4, -4, 7)$, $2\mathbf{X} = (8, -4, 2, 6)$,
$3\mathbf{X} - 2\mathbf{Y} = (14, -10, -7, 17)$.

3. (1, 7).

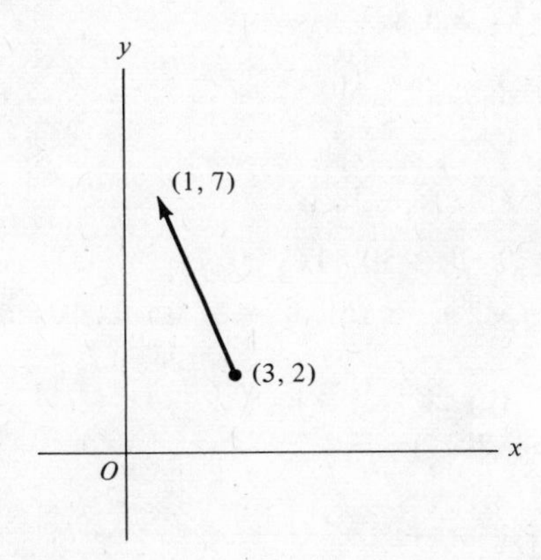

7.

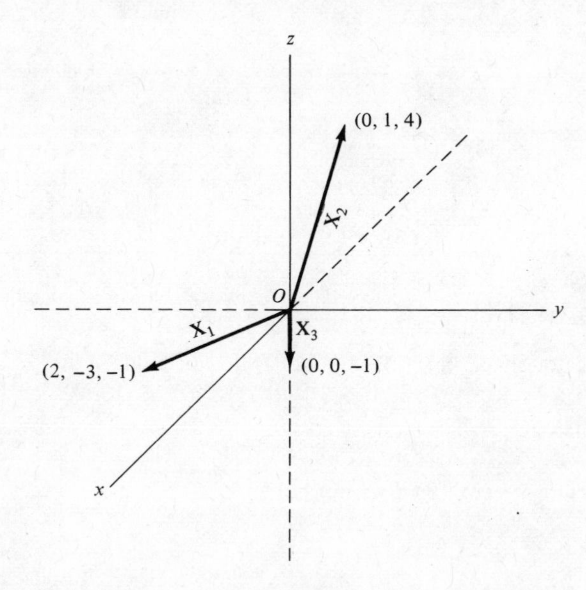

9. (4, 2, 2).

11. (a) $\sqrt{29}$. (b) $\sqrt{14}$. (c) $\sqrt{5}$. (d) $\sqrt{30}$.

13. (a) $\sqrt{18}$. (b) $\sqrt{6}$. (c) $\sqrt{50}$. (d) $\sqrt{10}$.

15. (a) 0. (b) -5. (c) 3. (d) 0.

19. (a) 0. (b) $-\dfrac{1}{\sqrt{3}}$. (c) $\dfrac{1}{\sqrt{5}}$. (d) 0.

21. (a) $\mathbf{X}_1$ and $\mathbf{X}_2$, $\mathbf{X}_1$ and $\mathbf{X}_6$, $\mathbf{X}_2$ and $\mathbf{X}_3$, $\mathbf{X}_3$ and $\mathbf{X}_6$, $\mathbf{X}_4$ and $\mathbf{X}_6$.
(b) $\mathbf{X}_1$ and $\mathbf{X}_3$.
(c) None.

23. (a) $(2/\sqrt{14}, -1/\sqrt{14}, 3/\sqrt{14})$.
(b) $(1/\sqrt{30}, 2/\sqrt{30}, 3/\sqrt{30}, 4/\sqrt{30})$.
(c) $(0, 1/\sqrt{2}, -1/\sqrt{2})$. (d) $(0, -1/\sqrt{6}, 2/\sqrt{6}, -1/\sqrt{6})$.

25. (a) $\mathbf{i} + 2\mathbf{j} - 3\mathbf{k}$. (b) $2\mathbf{i} + 3\mathbf{j} - \mathbf{k}$.
(c) $\mathbf{j} + 2\mathbf{k}$. (d) $-2\mathbf{k}$.

Section 3.3, p. 134

1. (a) $-15\mathbf{i} - 2\mathbf{j} + 9\mathbf{k}$. (b) $-3\mathbf{i} + 3\mathbf{j} + 3\mathbf{k}$. (c) $7\mathbf{i} + 5\mathbf{j} - \mathbf{k}$.
(d) $0\mathbf{i} + 0\mathbf{j} + 0\mathbf{k}$.

9. $\frac{1}{2}\sqrt{478}$.

11. $\sqrt{150}$.

13. 39.

Section 3.4, p. 141

5. Not a vector space; (a), (c), (d), and (f) do not hold.

7. Vector space.

9. Not a vector space; (β) does not hold.

11. Vector space.

13. (b).

15. (b) and (c).

19. (a) and (c).

Section 3.5, p. 156

1. (b) and (c).

3. (a), (c), and (d).

5. (a) and (d).

7. (a) $\mathbf{X}_3 = 3\mathbf{X}_1 + \mathbf{X}_2 + \mathbf{X}_4$.
(d) $\mathbf{X}_2 = \mathbf{X}_1 - \mathbf{X}_3$.

9. (a) and (d).

11. (a) and (d).

13. (c).

15. (a); $\mathbf{X} = \frac{3}{2}\mathbf{X}_1 + \frac{1}{2}\mathbf{X}_2 - \frac{3}{2}\mathbf{X}_3$.

17. Possible answer $\{\mathbf{X}_1, \mathbf{X}_2\}$; $\dim W = 2$.

19. (a) 2. (b) 1. (c) 2. (d) 2.

21. (a) 4. (b) 3. (c) 3. (d) 4.

23. Possible answer $\{(1, 0, 1, 0), (0, 1, -1, 0), (1, 0, 0, 0), (0, 0, 0, 1)\}$.

25. Zero.

Section 3.6, p. 165

1. (b).

3. $\{(1/\sqrt{2}, -1/\sqrt{2}), (1/\sqrt{2}, 1/\sqrt{2})\}$.

5. (a) $\{(1, 2), (-4, 2)\}$. (b) $\{(1/\sqrt{5}, 2/\sqrt{5}), (-2/\sqrt{5}, 1/\sqrt{5})\}$.

7. $\{(\frac{2}{3}, -\frac{2}{3}, \frac{1}{3}), (\frac{2}{3}, \frac{1}{3}, -\frac{2}{3}), (\frac{1}{3}, \frac{2}{3}, \frac{2}{3})\}$.

9. Possible answer

$$\left\{(1/\sqrt{2}, 1/\sqrt{2}, 0, 0), \ (3/\sqrt{22}, -3/\sqrt{22}, 0, -2/\sqrt{22}), \ (1/\sqrt{11}, -1/\sqrt{11}, 0, 3/\sqrt{11})\right\}.$$

Chapter 4

Section 4.1, p. 172

1. (b).

3. (a).

7. (a) $\begin{bmatrix} 4 \\ -13 \end{bmatrix}$. (b) $\begin{bmatrix} 2a + b \\ -3a + 2b \end{bmatrix}$.

Section 4.2, p. 182

1. (a) $\left\{\begin{bmatrix} 0 \\ 0 \end{bmatrix}\right\}$. (b) Yes. (c) No.

3. (a) Possible answer $\left\{\begin{bmatrix} -2 \\ 0 \\ 1 \\ 1 \\ 0 \end{bmatrix}, \begin{bmatrix} 0 \\ 1 \\ 0 \\ 0 \\ 0 \end{bmatrix}\right\}$.

(b) Possible answer $\left\{\begin{bmatrix} 1 \\ 1 \\ 2 \\ 0 \end{bmatrix}, \begin{bmatrix} -1 \\ 0 \\ -1 \\ -1 \end{bmatrix}, \begin{bmatrix} -1 \\ -1 \\ -1 \\ 0 \end{bmatrix}\right\}$.

5. (a) Yes. (b) 1.

9. (a) 2. (b) 1.

Section 4.3, p. 191

1. (a) $\begin{bmatrix} 1 \\ -2 \end{bmatrix}$. (b) $\begin{bmatrix} -1 \\ 2 \end{bmatrix}$. (c) $\begin{bmatrix} 2 \\ 5 \end{bmatrix}$. (d) $\begin{bmatrix} 0 \\ 1 \end{bmatrix}$.

3. (a) $\begin{bmatrix} 1 \\ 0 \\ 1 \end{bmatrix}$. (b) $\begin{bmatrix} -1 \\ 2 \\ 1 \end{bmatrix}$. (c) $\begin{bmatrix} 0 \\ 1 \\ 0 \end{bmatrix}$. (d) $\begin{bmatrix} -2 \\ 1 \\ 3 \end{bmatrix}$.

5. (a) $\begin{bmatrix} 1 & -2 \\ 2 & 1 \\ 1 & 1 \end{bmatrix}$. (b) $\begin{bmatrix} \frac{7}{3} & -\frac{4}{3} \\ -\frac{2}{3} & \frac{5}{3} \\ \frac{2}{3} & -\frac{2}{3} \end{bmatrix}$. (c) $\begin{bmatrix} -3 \\ 4 \\ 3 \end{bmatrix}$.

7. (a) $[\mathbf{L}(\mathbf{X})]_T = \begin{bmatrix} 1 & 1 & 0 \\ 0 & 1 & -1 \end{bmatrix}[\mathbf{X}]_S$. (b) $[\mathbf{L}(\mathbf{X})]_{T'} = \begin{bmatrix} -1 & -\frac{1}{3} & 0 \\ 1 & \frac{2}{3} & 0 \end{bmatrix}[\mathbf{X}]_{S'}$.

(c) $\begin{bmatrix} 3 \\ -1 \end{bmatrix}$.

9. (a) $[\mathbf{L}(\mathbf{X}_1)]_S = \begin{bmatrix} 2 \\ -1 \end{bmatrix}$ $[\mathbf{L}(\mathbf{X}_2)]_S = \begin{bmatrix} -3 \\ 4 \end{bmatrix}$.

(b) $\mathbf{L}(\mathbf{X}_1) = \begin{bmatrix} 1 \\ 5 \end{bmatrix}$, $\mathbf{L}(\mathbf{X}_2) = \begin{bmatrix} -4 \\ -5 \end{bmatrix}$.

(c) $\begin{bmatrix} \frac{29}{3} \\ \frac{40}{3} \end{bmatrix}$.

11. (a) $\begin{bmatrix} 1 & 2 & 1 \\ 1 & 0 & 0 \\ 0 & 1 & 1 \end{bmatrix}$. (b) $\begin{bmatrix} 8 \\ 1 \\ 5 \end{bmatrix}$.

Section 4.4, p. 201

1. 3.

3. 2.

5. 3.

7. 2.

9. 3.

11. Singular.

13. Nonsingular.

15. Linearly dependent.

17. Nontrivial solution.

19. Has a solution.

21. Has no solution.

Chapter 5

Section 5.1, p. 221

1. $\lambda^3 - 4\lambda^2 + 7$.

3. $(\lambda - 4)(\lambda - 2)(\lambda - 3) = \lambda^3 - 9\lambda^2 + 26\lambda - 24$.

5. $f(\lambda) = (\lambda - 1)(\lambda - 3)(\lambda + 2)$.

7. $f(\lambda) = \lambda^2 - 5\lambda + 6$; $\lambda_1 = 2, \lambda_2 = 3$; $\mathbf{X}_1 = \begin{bmatrix} 1 \\ -1 \end{bmatrix}, \mathbf{X}_2 = \begin{bmatrix} 1 \\ -2 \end{bmatrix}$.

9. $f(\lambda) = \lambda^3 - 5\lambda^2 + 2\lambda + 8$; $\lambda_1 = -1, \lambda_2 = 2, \lambda_3 = 4$;

$$\mathbf{X}_1 = \begin{bmatrix} 1 \\ 0 \\ -1 \end{bmatrix}, \mathbf{X}_2 = \begin{bmatrix} -2 \\ -3 \\ 2 \end{bmatrix}, \mathbf{X}_3 = \begin{bmatrix} 8 \\ 5 \\ 2 \end{bmatrix}.$$

11. $f(\lambda) = (\lambda - 1)(\lambda + 1)(\lambda - 3)(\lambda - 2)$; $\lambda_1 = 1, \lambda_2 = -1, \lambda_3 = 3, \lambda_4 = -2$;

$$\mathbf{X}_1 = \begin{bmatrix} 1 \\ 0 \\ 0 \\ 0 \end{bmatrix}, \quad \mathbf{X}_2 = \begin{bmatrix} 1 \\ -1 \\ 0 \\ 0 \end{bmatrix}, \quad \mathbf{X}_3 = \begin{bmatrix} 9 \\ 3 \\ 4 \\ 0 \end{bmatrix}, \quad \mathbf{X}_4 = \begin{bmatrix} 29 \\ 7 \\ 9 \\ -3 \end{bmatrix}.$$

13. Not diagonalizable; $\lambda_1 = \lambda_2 = 1$.

15. Diagonalizable; $\lambda_1 = 1, \lambda_2 = -1, \lambda_3 = 2$.

17. Not diagonalizable; $\lambda_1 = 1, \lambda_2 = 1, \lambda_3 = 3$.

19. $\mathbf{P} = \begin{bmatrix} 1 & -3 & 1 \\ 0 & 0 & -6 \\ 1 & 2 & 4 \end{bmatrix}$; $\lambda_1 = 4, \lambda_2 = -1, \lambda_3 = 1$.

21. $\mathbf{P} = \begin{bmatrix} 1 & 2 & 1 \\ 0 & 1 & 0 \\ 0 & 0 & -3 \end{bmatrix}$; $\lambda_1 = 3, \lambda_2 = 2, \lambda_3 = 0.$

23. Basis for eigenspace associated with $\lambda_1 = \lambda_2 = 2$ is
$\left\{ \begin{bmatrix} 1 \\ 0 \\ 0 \\ 0 \end{bmatrix} \right\}.$
Basis for eigenspace associated with $\lambda_3 = \lambda_4 = 1$ is
$\left\{ \begin{bmatrix} 3 \\ -3 \\ 1 \\ 0 \end{bmatrix} \right\}.$

Section 5.2, p. 234

7. $\begin{bmatrix} 0 & 0 \\ 0 & 4 \end{bmatrix}$; $\mathbf{P} = \begin{bmatrix} 1/\sqrt{2} & 1/\sqrt{2} \\ -1/\sqrt{2} & 1/\sqrt{2} \end{bmatrix}.$

9. $\begin{bmatrix} 0 & 0 & 0 \\ 0 & 0 & 0 \\ 0 & 0 & 4 \end{bmatrix}$; $\mathbf{P} = \begin{bmatrix} 1 & 0 & 0 \\ 0 & -1/\sqrt{2} & 1/\sqrt{2} \\ 0 & 1/\sqrt{2} & 1/\sqrt{2} \end{bmatrix}.$

11. $\begin{bmatrix} -2 & 0 & 0 \\ 0 & 1 & 0 \\ 0 & 0 & 1 \end{bmatrix}$; $\mathbf{P} = \begin{bmatrix} 1/\sqrt{3} & -1/\sqrt{2} & -1/\sqrt{6} \\ 1/\sqrt{3} & 1/\sqrt{2} & -1/\sqrt{6} \\ 1/\sqrt{3} & 0 & 2/\sqrt{6} \end{bmatrix}.$

13. $\begin{bmatrix} 3 & 0 \\ 0 & 1 \end{bmatrix}.$

15. $\begin{bmatrix} 1 & 0 & 0 \\ 0 & 2 & 0 \\ 0 & 0 & 0 \end{bmatrix}.$

17. $\begin{bmatrix} 1 & 0 & 0 \\ 0 & 0 & 0 \\ 0 & 0 & 2 \end{bmatrix}.$

19. $\begin{bmatrix} 2 & 0 & 0 \\ 0 & -2 & 0 \\ 0 & 0 & 4 \end{bmatrix}.$

Chapter 6

Section 6.1, p. 257

1. Maximize $z = 120x + 100y$
subject to

$$
\begin{aligned}
2x + 2y &\leqslant 8 \\
5x + 3y &\leqslant 15 \\
x \geqslant 0, y &\geqslant 0.
\end{aligned}
$$

3. Maximize $z = 0.08x + 0.10y$
subject to

$$
\begin{aligned}
x + y &\leqslant 6000 \\
x &\geqslant 1500 \\
y &\leqslant 4000 \\
y &\leqslant \tfrac{1}{2}x \\
x \geqslant 0, y &\geqslant 0.
\end{aligned}
$$

5. Maximize $z = 40{,}000x + 45{,}000y$
subject to

$$
\begin{aligned}
x + y &\leqslant 30 \\
y &\geqslant 24 \\
x &\geqslant 2 \\
x &\leqslant 4 \\
x \geqslant 0, y &\geqslant 0.
\end{aligned}
$$

7. Maximize $z = 4x + 6y$
subject to

$$
\begin{aligned}
x + 2y &\leqslant 10 \\
x + y &\leqslant 7 \\
x \geqslant 0, y &\geqslant 0.
\end{aligned}
$$

9. Minimize $z = 10x + 12y$
subject to

$$
\begin{aligned}
2x + 3y &\geqslant 18 \\
x + 3y &\geqslant 12 \\
80x + 60y &\geqslant 480 \\
x \geqslant 0, y &\geqslant 0.
\end{aligned}
$$

11.

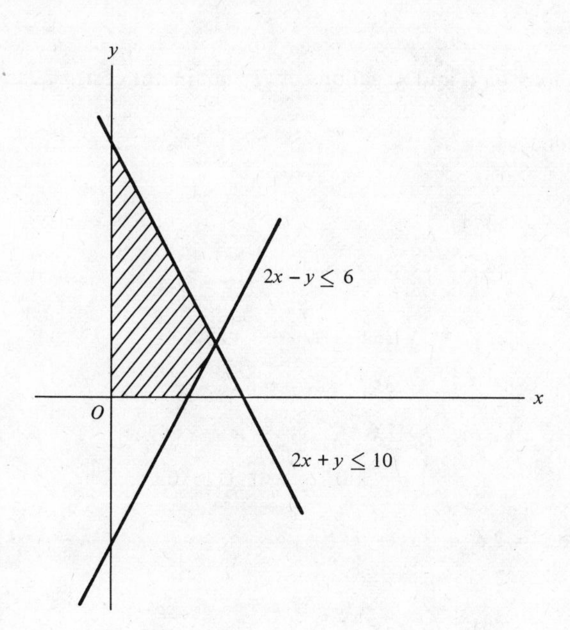

13.

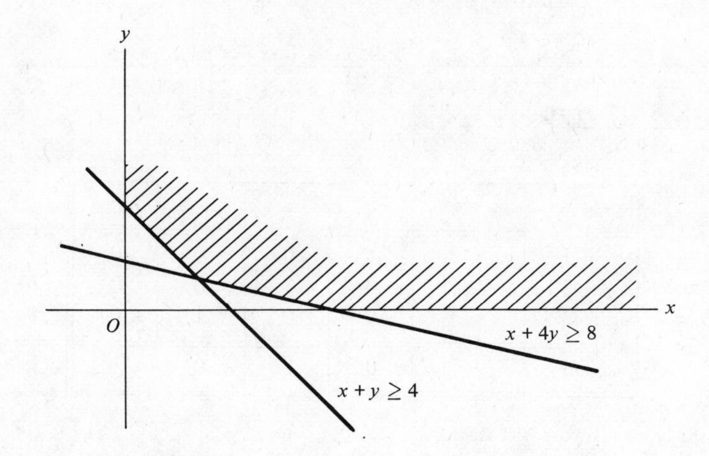

15. $x = \frac{8}{11}$, $y = \frac{45}{11}$, optimal value of $z = -\frac{21}{11}$.

17. Invest \$2000 in bond A and \$4000 in bond B; maximum return is \$560.

19. Carry no containers from the Smith Corporation and 1500 containers from the Johnson Corporation, or 120 containers from the Smith Corporation and

1440 containers from the Johnson Corporation. In either case the maximum revenue is \$900.

21. Use $\frac{5}{2}$ gallons of L and $\frac{3}{2}$ gallons of H; minimum cost is \$2.25.

23. No solution.

25. (a).

27. Maximize $z = 3x_1 - x_2 + 6x_3$
subject to

$$\begin{aligned} 2x_1 + 4x_2 + x_3 &\leq 4 \\ 3x_1 - 2x_2 + 3x_3 &\leq 4 \\ 2x_1 + x_2 - x_3 &\leq 8 \\ x_1 \geq 0, x_2 \geq 0, x_3 &\geq 0. \end{aligned}$$

29. Maximize $z = 2x_1 + 3x_2 + 7x_3$
subject to

$$\begin{aligned} 3x_1 + x_2 - 4x_3 + x_4 \qquad\qquad &= 3 \\ x_1 - 2x_2 + 6x_3 \qquad + x_5 \qquad &= 21 \\ x_1 - x_2 - x_3 \qquad\qquad + x_6 &= 9 \end{aligned}$$

$$x_1 \geq 0, x_2 \geq 0, x_3 \geq 0, x_4 \geq 0, x_5 \geq 0, x_6 \geq 0.$$

Section 6.2, p. 279

1.

	x	y	u	v	w	z	
u	3	−2	1	0	0	0	7
v	2	5	0	1	0	0	6
w	2	3	0	0	1	0	8
	−3	−7	0	0	0	1	0

3.

	x_1	x_2	x_3	x_4	x_5	x_6	x_7	z	
x_5	3	−2	1	1	1	0	0	0	6
x_6	1	1	1	1	0	1	0	0	8
x_7	2	−3	−1	2	0	0	1	0	10
	−2	−2	−3	−1	0	0	0	1	0

5. $x = 2, \quad y = 0, \quad$ optimal $z = 4$.

7. No finite optimal solution.

9. $x_1 = 0, x_2 = \frac{33}{20}, x_3 = \frac{27}{10}$, optimal $z = \frac{69}{10}$.

11. $x_1 = 0, \quad x_2 = 0, \quad x_3 = 49, \quad x_4 = 41, \quad$ optimal $z = 156$.

13. Carry 1500 containers from the Johnson Corporation and none from the Smith Corporation, or 120 containers from the Smith Corporation and 1440 containers from the Johnson Corporation. In either case the maximum revenue is $900.

15. Use 4 tons of gas, no coal and no oil; maximum power generated is 2000 kilowatts.

Chapter 7

Section 7.1, p. 289

1. (d).

3. (a) $x = 3 + 4t, y = 4 - 5t, z = -2 + 2t$.
(b) $x = 3 - 2t, y = 2 + 5t, z = 4 + t$.
(c) $x = t, y = t, z = t$.
(d) $x = -2 + 2t, y = -3 + 3t, z = 1 + 4t$.

5. (a) $\frac{x-2}{2} = \frac{y+3}{5} = \frac{z-1}{4}$.
(b) $\frac{x+3}{8} = \frac{y+2}{7} = \frac{z+2}{6}$.
(c) $\frac{x+2}{4} = \frac{y-3}{-6} = \frac{z-4}{1}$.
(d) $\frac{x}{4} = \frac{y}{5} = \frac{z}{2}$.

7. (a) $3x - 2y + 4z + 16 = 0$.
(b) $y - 3z + 3 = 0$.
(c) $-z + 4 = 0$.
(d) $-x - 2y + 4z - 3 = 0$.

9. (a) $x = \frac{8}{13} + 23t$, $y = -\frac{27}{13} + 2t$, $z = 13t.$ (b) $x = 7t$, $y = -8 + 22t$, $z = 4 + 13t.$

(c) $x = -\frac{13}{4} + 5t$, $y = -1 + 4t$, $z = -5 - 3t.$

11. Yes.

13. $(5, 1, 2).$

17. $4x - 4y + z + 16 = 0.$

19. $7x + 2y - 2z - 19 = 0.$

21. $x = -2 + 2t, \quad y = 5 - 3t, \quad z = -3 + 4t.$

Section 7.2, p. 302

1. (a) $[x \quad y]\begin{bmatrix} -3 & \frac{5}{2} \\ \frac{5}{2} & -2 \end{bmatrix}\begin{bmatrix} x \\ y \end{bmatrix}.$

(b) $[x_1 \quad x_2 \quad x_3]\begin{bmatrix} 2 & \frac{3}{2} & -\frac{5}{2} \\ \frac{3}{2} & 0 & \frac{7}{2} \\ -\frac{5}{2} & \frac{7}{2} & 0 \end{bmatrix}\begin{bmatrix} x_1 \\ x_2 \\ x_3 \end{bmatrix}.$

(c) $[x_1 \quad x_2 \quad x_3]\begin{bmatrix} 3 & \frac{1}{2} & -1 \\ \frac{1}{2} & 1 & -2 \\ -1 & -2 & -2 \end{bmatrix}\begin{bmatrix} x_1 \\ x_2 \\ x_3 \end{bmatrix}.$

3. (a) $\begin{bmatrix} -1 & 0 & 0 \\ 0 & 2 & 0 \\ 0 & 0 & 0 \end{bmatrix}.$ (b) $\begin{bmatrix} 3 & 0 & 0 \\ 0 & 0 & 0 \\ 0 & 0 & 0 \end{bmatrix}.$

5. $x_1'^2 + 2x_2'^2.$

7. $4x_3'^2.$

9. $y'^2.$

11. $y_1^2 + y_2^2 - y_3^2.$

13. $y_1^2 - y_2^2.$

15. $y_1^2 + y_2^2 + y_3^2 = 1$, an ellipsoid.

17. $\mathbf{Q}_1$, $\mathbf{Q}_2$, and $\mathbf{Q}_4$ are equivalent. The eigenvalues of the matrices associated with the quadratic forms are: for $\mathbf{Q}_1$: 0, 1, 2; for $\mathbf{Q}_2$: 0, 1, 3; for $\mathbf{Q}_3$: 0, 5, − 5; for $\mathbf{Q}_4$: 0, 1, 5. The rank r and signature s of $\mathbf{Q}_1$, $\mathbf{Q}_2$, and $\mathbf{Q}_4$ are $r = 2$ and $s = 2p - r = 2$.

Section 7.3, p. 318

1.

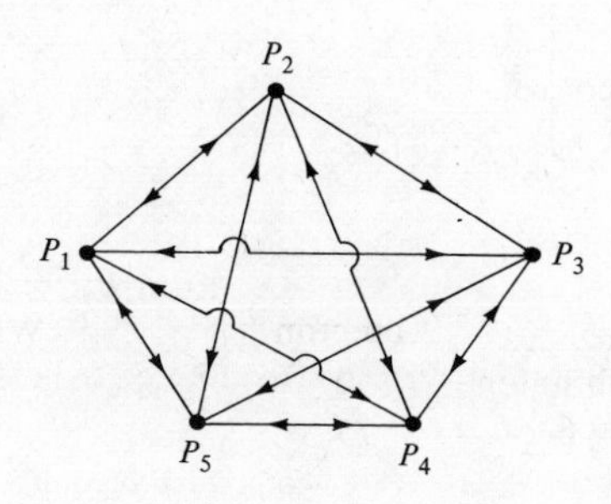

3. (a) and (c).

5. (a)

	P_1	P_2	P_3	P_4	P_5
P_1	0	1	0	0	0
P_2	1	0	1	0	1
P_3	1	0	0	1	0
P_4	0	1	0	0	0
P_5	0	0	0	1	0

(b)

	P_1	P_2	P_3	P_4	P_5	P_6
P_1	0	1	1	0	0	0
P_2	1	0	0	1	0	0
P_3	0	1	0	0	0	0
P_4	0	0	1	0	1	1
P_5	0	0	1	1	0	1
P_6	1	0	0	0	0	0

7. (b).

9. (a) One way: $P_2 \to P_5 \to P_1$.
(b) Two ways: $P_2 \to P_5 \to P_1 \to P_2$
$P_2 \to P_5 \to P_3 \to P_2$.

11. P_2, P_3, and P_4.

13. There is no clique.

15. (a) Connected. (b) Disconnected.

17. No.

Section 7.4, p. 339

1.

$$R \begin{array}{c} \\ \text{2 fingers shown} \\ \text{3 fingers shown} \end{array} \overset{C}{\begin{array}{cc} \text{2 fingers shown} & \text{3 fingers shown} \\ \left[\begin{array}{c} -4 \\ 5 \end{array}\right. & \left.\begin{array}{c} 5 \\ -6 \end{array}\right] \end{array}}.$$

3.

$$\text{Firm } A \begin{array}{c} \\ \text{Abington} \\ \text{Wyncote} \end{array} \overset{\text{Firm } B}{\begin{array}{cc} \text{Abington} & \text{Wyncote} \\ \left[\begin{array}{c} 50 \\ 25 \end{array}\right. & \left.\begin{array}{c} 60 \\ 50 \end{array}\right] \end{array}}.$$

5. (a) $\begin{bmatrix} 5 & \textcircled{4} \\ 3 & -2 \end{bmatrix}$. (b) $\begin{bmatrix} 2 & 1 & \textcircled{0} \\ 3 & 1 & -2 \\ 4 & 2 & -4 \end{bmatrix}$.

(c) $\begin{bmatrix} \textcircled{3} & 4 & 5 \\ -2 & 5 & 1 \\ -1 & 0 & 1 \end{bmatrix}$. (d) $\begin{bmatrix} 5 & 2 & 4 & 2 \\ 0 & -1 & 2 & 0 \\ 3 & \textcircled{2} & 3 & \textcircled{2} \\ 1 & 0 & -1 & -1 \end{bmatrix}$.

7. (a) $\mathbf{P} = [1 \quad 0]$, $\mathbf{Q} = \begin{bmatrix} 0 \\ 1 \\ 0 \end{bmatrix}$, $v = 1$.

(b) $\mathbf{P} = [0 \quad 0 \quad 1]$, $\mathbf{Q} = [1 \quad 0 \quad 0 \quad 0]$, $v = 0$.

(c) $\mathbf{P} = [0 \quad 1]$, $\mathbf{Q} = \begin{bmatrix} 0 \\ 1 \end{bmatrix}$, $v = 4$.

9. (a) $\frac{19}{36}$. (b) $\frac{1}{7}$.

11. (a) $p_1 = \frac{9}{14}$, $p_2 = \frac{5}{14}$, $q_1 = \frac{1}{2}$, $q_2 = \frac{1}{2}$, $v = \frac{1}{2}$.

13. $p_1 = \frac{4}{13}$, $p_2 = 0$, $p_3 = \frac{9}{13}$, $q_1 = 0$,
$q_2 = \frac{5}{13}$, $q_3 = \frac{8}{13}$, $v = \frac{13}{58}$.

15. $\mathbf{P} = [0 \quad \frac{3}{8} \quad \frac{5}{8} \quad 0]$, $\mathbf{Q} = \begin{bmatrix} \frac{5}{8} \\ \frac{3}{8} \\ 0 \\ 0 \end{bmatrix}$, $v = \frac{7}{8}$.

17. $\mathbf{P} = [\frac{1}{3} \quad \frac{1}{3} \quad \frac{1}{3}]$, $\mathbf{Q} = \begin{bmatrix} \frac{1}{3} \\ \frac{1}{3} \\ \frac{1}{3} \end{bmatrix}$, $v = 0$.

19. $\mathbf{P} = [\frac{2}{3} \quad \frac{1}{3}]$, $\mathbf{Q} = \begin{bmatrix} \frac{1}{2} \\ \frac{1}{2} \end{bmatrix}$, $v = 0$.

Section 7.5, p. 350

1. $y = 0.4x + 0.6$.

3. $y = 0.086x + 3.114$.

5. (a) $y = 0.426x + 0.827$. (b) 7.638.

7. (a) $y = 0.974x - 2.657$. (b) 16.829.

Section 7.6, p. 359

1. (b) and (d).

3. $\begin{bmatrix} 4 \\ 0 \\ 3 \end{bmatrix}$.

5.

$$\begin{array}{l} \\ \text{Farmer} \\ \text{Builder} \\ \text{Tailor} \end{array} \begin{array}{c} \begin{array}{ccc} \text{Farmer} & \text{Builder} & \text{Tailor} \end{array} \\ \begin{bmatrix} \frac{2}{5} & \frac{1}{3} & \frac{1}{2} \\ \frac{2}{5} & \frac{1}{3} & \frac{1}{2} \\ \frac{1}{5} & \frac{1}{3} & 0 \end{bmatrix} \end{array}; \quad \mathbf{P} = \begin{bmatrix} 75 \\ 75 \\ 40 \end{bmatrix}.$$

7. Not productive.

9. Productive.

11. (a) [18 6]. (b) [12 8].

Chapter 8

Section 8.1, p. 363

1. 0.347213×10^2.

3. $-.284 \times 10^3$.

5. 0.123×10, 0.123×10.

7. 0.5666×10, 0.5667×10.

9. $\varepsilon_a = 0.21 \times 10^{-1}$, $\varepsilon_r = 0.1702 \times 10^{-2}$.

11. $\varepsilon_a = 0.11 \times 10$, $\varepsilon_r = 0.1697 \times 10^{-3}$.

Section 8.2, p. 374

1. $\begin{bmatrix} 1 & -2 & 0 \\ 0 & 1 & -1 \\ 0 & 0 & 1 \end{bmatrix}$.

3. $x_1 = 2, x_2 = -3, x_3 = 4$.

5. $x_1 \cong 0.004, x_2 \cong 1.006, x_3 \cong -2.000$

7. $x_1 \cong 2.393, x_2 \cong 4.565, x_3 \cong -3.607$.

11.

	Jacobi	Gauss–Seidel	Exact
x	-1.993	-2.004	-2
y	4.998	5.002	5

13.

	Jacobi	Gauss–Seidel	Exact
x_1	1.559	1.658	$\frac{5}{3}$
x_2	1.685	1.513	$\frac{3}{2}$
x_3	-0.313	-0.492	$-\frac{1}{2}$

15. (a) Exact solution is $x = -2, y = 3$.
(b) Exact solution is $x = -3, y = 2$.

Section 8.3, p. 384

	Approximate		Exact	
	Eigenvalue	*Eigenvector*	*Eigenvalue*	*Eigenvector*
1.	-4.974	$[-1 \quad 0.527]^T$	-5	$[-1 \quad \frac{1}{2}]^T$
3.	6.000	$[1 \quad 0.667]^T$	6	$[1 \quad \frac{2}{3}]^T$
5.	9.976	$[1 \quad 0.245]^T$	10	$[1 \quad \frac{1}{4}]^T$
7.	3.998	$[0.707 \quad 0.707]^T$	4	$[1/\sqrt{2} \quad 1/\sqrt{2}\,]^T$
	0	$[-0.707 \quad 0.707]^T$	0	$[-1\sqrt{2} \quad 1/\sqrt{2}\,]^T$
9.	6.001	$[0.666 \quad 0.666 \quad 0.334]^T$	6	$[\frac{2}{3} \quad \frac{2}{3} \quad \frac{1}{3}]^T$
	11.989	$[-0.333 \quad 0.666 \quad -0.666]^T$	12	$[-\frac{1}{3} \quad \frac{2}{3} \quad -\frac{2}{3}]^T$
	8.998	$[-0.666 \quad 0.331 \quad 0.667]^T$	9	$[-\frac{2}{3} \quad \frac{1}{3} \quad \frac{2}{3}]^T$

Index

$A \begin{vmatrix} 1 & 2 \\ 1 & 1 \end{vmatrix}$ $|AB| \begin{vmatrix} 0 & 5 \\ 0 & 3 \end{vmatrix}$ -2

$B \begin{vmatrix} 0 & 1 \\ 0 & 2 \end{vmatrix}$ $BA \begin{vmatrix} 1 & 1 \\ 2 & 2 \end{vmatrix} = 0$

$\begin{vmatrix} 2 & 2 & 1 \\ 3 & 2 & 0 \\ 1 & 2 & 1 \end{vmatrix} \begin{matrix} 2 & 2 \\ 3 & 2 \\ 1 & 2 \end{matrix}$

$\begin{pmatrix} 1 & 2 & 3 \\ -1 & 0 & 7 \\ 1 & 2 & 3 \end{pmatrix} \begin{matrix} 1 & 2 \\ -1 & 0 \\ 1 & 2 \end{matrix}$

$8 - 8 = 0$

$4 + 6 - 6 - 2 = 2$

$\begin{vmatrix} 1 & 2 & 2 \\ 0 & 2 & 3 \\ 1 & 2 & 1 \end{vmatrix} \begin{matrix} 1 & 2 \\ 0 & 2 \\ 1 & 2 \end{matrix}$

$2 + 6 - 6 - 4 - 2$